GRE/GMAT Math

A Systematic Approach

Sid Thatte

Sid Thatte has an MBA in general management from the University of South Carolina, Columbia. He also has a bachelor's degree in mechanical engineering, and once designed nuclear power plants. He lives in Seattle, and teaches math for competitive exams.

GRE/GMAT Math: A Systematic Approach

ISBN-10: 1-4536-3398-7

ISBN-13: 978-1-4536-3398-4

GRE is a registered trademark of the Educational Testing Service, which was not involved in the production of, and does not endorse, this book.

GMAT is a registered trademark of the Graduate Management Admission Council, which was not involved in the production of, and does not endorse, this book.

CONTENTS

PREFACE

This book is dedicated to you, the user. During my learning and test-taking years, I often wished for a textbook that would fully cover all the math topics included on the GRE/GMAT. A few years out of graduate school, I decided to fulfill my long-time wish. This work is the result of my sincere attempt to formulate a one-stop manual for the math portion of these tests.

I have presented this book in a textbook/workbook hybrid format; solved examples are followed by practice examples provided with just enough space for calculations. Therefore, you may not need an additional notebook or scratch paper with this book; you will only need a pen or a pencil. On the computer-based GRE/GMAT, you will be given an electronic notepad to work your math. It is therefore important to get used to a limited amount of space to work your numbers with.

Since August 1, 2011, the new format of the computer-based GRE includes <u>two</u> math sections, each containing about 20 questions to be answered in 35 minutes. That's just around 1¾ minutes per question. Some questions are numeric entry type, which have no answer choices; you have to type your answer in the blank box provided. Also, there are the familiar multiple choice questions which are now of two types: the ones with only one correct answer, and the ones with several correct answers. An on-screen calculator is <u>provided</u> on the GRE. In contrast, the math section of the computer-based GMAT has 37 multiple-choice questions to be answered in 75 minutes. That's only about 2 minutes per question. Calculators are <u>not allowed</u> on the GMAT. But on either test, pacing is clearly important. To build up that speed, this book contains an exhaustive number of solved as well as practice examples.

Math questions on the GRE/GMAT test your ability to solve problems based on what you have learned in your grade school years. Several questions can be answered merely by tracing patterns without actually going the whole nine yards. However, to trace any pattern, you have to have your theory in place. Otherwise, no amount of shortcuts or tricks will prove useful. This book will help you develop a strong understanding of mathematics as a subject. I have tried to avoid being too technical or too boring. But at the same time, I have provided in-depth coverage of theory.

While studying from this book, you may notice that most questions do not have multiple-choice answers. It is my philosophy that solving a problem is more important than simply eliminating choices. In the real world, when we're faced with issues, we aren't always given five answer-choices, four of which are wrong! We have to originate the solutions ourselves. In my teaching experience, it has turned out to be the best way to learn.

The topics in this book progress in increasing order of difficulty. Moreover, the concepts presented build on top of one another. Therefore, <u>please study the chapters in the sequence in which they appear, without skipping any sections</u>. While discussing algebra, I have used more variables than might be considered normal! It is to eliminate your fear of variables. Techniques matter; variables don't. You may notice that some questions in this book are quite detailed, i.e. they have sub-questions (i), (ii), (iii) etc. I have done so only to group related concepts together. On the actual test, you may not have multiple questions within the same question.

Before starting with this book, please take a diagnostic test to gauge where you stand. There's no paper based test that compares to a computer-adaptive test. I have therefore not included any in this book. If you have taken the actual GRE/GMAT before, then you already have a score for reference. If not, then there are several websites where you can take a full-length math test (either free or paid), my personal choice being www.800score.com. My book is neither affiliated with nor endorsed by www.800score.com. I recommend this website because I believe their tests quite closely resemble the actual GRE/GMAT in their level of difficulty and their coverage of topics.

While using this book, you may notice that Chapter 7 (Quantitative Comparison) is specific to the GRE, whereas Chapter 8 (Data Sufficiency) is specific to the GMAT. Regardless, I advise studying them both; it will broaden your perspective on tackling mathematical reasoning type questions. Also, please try to memorize the multiplication tables and integer squares at the end of the book (Appendix B). It will help considerably speed up your calculations. Those of you planning on an MBA will need to be good at analytical math for life, whether you like it or not. As business executives, or analysts, or even entrepreneurs you will often base your decisions on numerical analyses. Those aiming for graduate degrees in most other fields will need math skills too, ranging from basic to expert levels. It might therefore be a good idea to retain this textbook for future reference after the GRE/GMAT.

I have done the entire typesetting as well as the proofreading of this book all by myself. I am responsible for any errors that may have escaped my attention. But it is my hope that I have met your expectations. I'd encourage you to write a review of this book on Amazon.com so I can stay informed of user experiences. Good luck with the preparation!

Sid Thatte, MBA

S Y M B O L S

MATHEMATICAL SYMBOLS USED IN THIS BOOK

$+$	plus
$-$	minus
$\pm$	plus or minus
$\times$	multiplied by
$\div$	divided by
$>$	is greater than
$\ngtr$	is not greater than
$<$	is less than
$\nless$	is not less than
$=$	is equal to
$\neq$	is not equal to
$\approx$	is approximately equal to
$\geq$	is greater than or equal to
$\leq$	is less than or equal to
$\Rightarrow$	implies (it follows that)
∞	infinity
$\lvert x \rvert$	modulus of x (absolute value of x)
$\sqrt{x}$	square root of x
Σ	sum of
$!$	factorial
$\parallel$	is parallel to
$\perp$	is perpendicular to
$\angle$	angle
Δ	triangle
$\sim$	is similar to
π	Greek letter pi, a constant $\left(\approx \frac{22}{7} \approx 3.14 \right)$
$^{\circ}$	degrees

OTHER SYMBOLS USED IN THIS BOOK

$\rightarrow$	converted to
$\rightleftarrows$	converted to and from

CHAPTER 1

ARITHMETIC

THE ORIGIN OF NUMBERS

Ever since the dawn of civilization, human beings have had the curiosity to document what they've observed. Sometimes they observed repetitive patterns, e.g. the same number of times the moon rising and setting before completely disappearing for a night, the same number of fingers in all humans, the same number of 'foot lengths' to cover a certain distance etc. It's these patterns that, in time, changed mankind's mere curiosity into the *need* to document occurrences. That's when the numbering system was born.

The very first numbers were devised to count what humans saw in their natural surroundings: the number of trees, the number of family members, the number of stars etc. Such numbers were termed **natural numbers**, or **counting numbers**. Clearly, those numbers did not include partial components, because there's no such thing as 13½ stars, 8¼ fingers, 2¾ sheep etc. Nor did they include negative numbers. You can imagine why. What about the number 'zero'? Mathematicians today debate whether zero should be a natural number. Those who say 'yes' believe so because zero is obtained by subtracting a natural number from itself. So, zero must be a natural number. Others disagree, saying that zero is not used to *count* anything; having *zero* sisters does not include counting anyone, having zero dollars in pocket does not include counting any currency etc. From the GRE/GMAT point of view, the definition of natural numbers or counting numbers is not important.

BASIC TYPES OF NUMBERS

The set of numbers that includes all counting numbers, their negatives, and the number zero comprises the set of **integers**. Sometimes, integers are also called **whole numbers**. But some mathematicians exclude negative numbers from whole numbers. Once again, from the GRE/GMAT point of view, the term 'whole numbers' is not as important, but the term 'integers' *very much is*.

Examples of **integers** are 5, 12, −7, 0, −1 etc. Clearly, integers do not include partial components. What if a number *includes* a partial component? Then it would be called a **fraction**. Examples would be 4½, $^{11}/_{19}$, −6¾, $^{71}/_{50}$, etc. What if a number has a *dot* and some numbers *after* it? Well, that 'dot' is called a 'decimal point,' and those numbers after it form the 'decimal fraction.' Together, such a number is termed a **decimal**. Some examples of decimals are 8.45, 0.6199, −3.007 etc.

A number can be expressed in the form of a **percent**, or a fraction of 100. "Per cent" comes from the Latin phrase "per centum" meaning "out of a hundred". The sign "%" is used to denote it. Examples would be 25%, 12.68%, −0.9% etc.

Finally, a number can also be expressed as a **radical**, or a root. For example, $\sqrt{5}$, $-\sqrt[2]{21}$, $\sqrt[-3]{19}$ etc. The $\sqrt{}$ sign is called the **radical sign**, and the number underneath it is called the **radicand**. The little number in the pocket of the radical sign is called the **index**. In the number $\sqrt[-3]{19}$, 19 is the radicand, and −3 is the index.

Each of the above types of numbers we've discussed falls under the category of **real numbers**. The definition of 'real numbers' varies depending upon whom you ask. From a practical point of view, I think real numbers are those that exist in the real world, and can be used to perform calculations. So if there are real numbers, then are there 'unreal' or 'imaginary' numbers too? Actually, yes. Square roots of negative numbers don't exist, at least not in the real world. Those fall under the category of **imaginary numbers**, as do all even roots of negative numbers. Imaginary numbers comprise a very interesting branch of mathematics, which is not included on the GRE/GMAT. Therefore, we'll not be discussing it in this book.

THE NUMBER LINE

Each real number—whether an integer, a fraction, a decimal or expressed as a percent or a root—can be shown to exist on the **number line**, sometimes also referred to as the **real number line**.

In its simplest form, a number line is a line on which all integers can be shown evenly spaced in increasing order from left (negative infinity "−∞") to right (positive infinity "∞"). All non-integers can also be shown in their approximate positions on it. The figure below illustrates a number line.

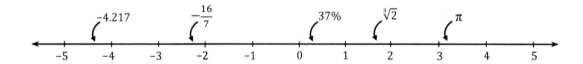

Now let's study each type of number in detail.

INTEGERS

As we just saw, integers can be positive or negative. When numbers contain positive or negative signs, + or −, they're called **signed numbers**. From a realistic point of view, a positive number does not need the positive sign written before it; it is assumed to be positive. Negative numbers, on the other hand, do need the negative sign written out.

Before we discuss addition or subtraction of signed numbers, let's first lay out the all-important rules for dealing with multiplication and division of signed numbers. The reason will be clear in just a moment.

MULTIPLICATION OF SIGNED NUMBERS

Positive Number × Positive Number = Positive Number   Symbolically, $(+) \times (+) = (+)$

Positive Number × Negative Number = Negative Number $(+) \times (-) = (-)$

Negative Number × Negative Number = Positive Number $(-) \times (-) = (+)$

DIVISION OF SIGNED NUMBERS

Positive Number ÷ Positive Number = Positive Number $(+) \div (+) = (+)$

Positive Number ÷ Negative Number = Negative Number $(+) \div (-) = (-)$

Negative Number ÷ Negative Number = Positive Number $(-) \div (-) = (+)$

You may notice that the rules for multiplication and division of signed numbers are identical.

ADDITION/SUBTRACTION OF SIGNED NUMBERS

When adding or subtracting signed numbers, it is always a good practice to *minimize the signs first*. If there are two adjacent signs, use the rules of multiplication/division from above to eliminate one of the signs. Then do the arithmetic. Consider the following examples.

1.001 $(+7) + (-8) + (-3)$

> There's a + sign adjacent to a – sign at two different places. We know that $(+) \times (-) = (-)$. So let's minimize those signs to only a – sign between 7 and 8, and one between 8 and 3. Also, 7 is not being added to anything before it. So let's discard the redundant + sign before 7.
>
> 7 – 8 – 3   Now, simply do the arithmetic.
>
> −4   This is the final answer.

1.002 $(-4) + (-7) + (-2)$

> Exactly the same logic as above, but this time we would need to keep the – sign before 4. Why? Because it is a negative number, therefore the sign is necessary.
>
> $-4 - 7 - 2$
>
> −13

1.003 $(-6) - (-3) + (+9)$

> Recall that $(-) \times (-) = (+)$. Also, the two + signs before 9 can be reduced to just one. The − sign before 6 remains.
>
> $-6 + 3 + 9$
>
> 6

1.004 $(+5) - (4) - (-12) + (-8)$

$5 - 4 + 12 - 8$

5

PRACTICE EXAMPLES

1.005 $(+5) + (-3) + (-6)$

1.006 $(-2) + (-11) - (1)$

1.007 $(-2) - (-3) + (4) + (-5)$

1.008 $(+13) - (-6) + (-3) - (+7)$

SPACE FOR CALCULATIONS

Let's try some examples involving multiplication/division, and then some involving all four arithmetic operations together.

1.009 $(-3)(-5)(-1)(-8)$

Keep in mind that we're multiplying four numbers here. First, decide whether the final sign will be + or −. Then do the arithmetic. Here, it will be +.

$3 \times 5 \times 1 \times 8$

120

1.010 $\dfrac{(-8)(+7)}{(-4)(-14)}$

The final sign will be −. The operation will now look like this:

$$-\frac{8 \times 7}{4 \times 14}$$

At this point you're probably wondering whether you should carry out the multiplications at the top and the bottom. NO. Notice that 4 goes twice into 8, and 7 also goes twice into 14. So let's cancel some numbers, and then deal with smaller numbers. Multiplying or dividing larger numbers only means there's a greater chance of making a calculation mistake!

$$-\frac{\cancel{8}^2 \times \cancel{7}^1}{\cancel{4}^1 \times \cancel{14}^2}$$ Now simply divide.

-1

1.011 $(4)(-7) - \dfrac{(-15)}{(+3)}$

The final sign for the left two numbers will be −, and that for the right two numbers will also be −.

$$-(4 \times 7) - \left(-\frac{15}{3}\right)$$

Now the two $-$ signs on the right can be combined into a $+$ sign, since $(-) \times (-) = (+)$.

$-28 + 5$

-23

1.012 $\dfrac{(-3)(+4)}{(-2)(-6)} - \dfrac{(-12)(-3)}{(+9)}$

The final sign for the left four numbers will be $-$, whereas that for the right three numbers will be $+$.

$$-\frac{3 \times 4}{2 \times 6} - \frac{12 \times 3}{9}$$

If you're wondering whether to cancel first or multiply first, I'd say it's your choice. The numbers aren't too big to multiply. Remember, you should cancel first if you think it could save you time. On the GRE/GMAT, saving time is as important as being accurate.

$-1 - 4$

-5

PRACTICE EXAMPLES

1.013 $(2)(-4)(+3)(-3)$

1.014 $\dfrac{(-3)(-8)}{(+2)(-12)}$

1.015 $(6)(-4) + \left(\dfrac{12}{-4}\right)$

1.016 $\dfrac{(-8)(-3)}{(-12)} + \dfrac{(6)(-4)}{(-1)(+12)}$

SPACE FOR CALCULATIONS

ODD AND EVEN NUMBERS

Another classification of integers includes odd and even numbers. An **even number** is a number that is divisible by 2. For example, 2, 4, 6, 8, 0, -2, -4, -6, -8 etc. are even numbers. An **odd number** is a number that is not divisible by 2. Examples would be 1, 3, 5, 7, -1, -3, -5, -7 etc.

Listed next are certain clear-cut rules we must remember when adding, subtracting, or multiplying odd and/or even integers. These rules do not apply to dividing odd or even integers, because dividing one integer by another may not always result in an integer; sometimes it may result in a fraction.

Odd Number + Odd Number = Even Number
Odd Number − Odd Number = Even Number

Even Number + Even Number = Even Number
Even Number − Even Number = Even Number

Odd Number + Even Number = Odd Number
Odd Number − Even Number = Odd Number

Odd Number × Odd Number = Odd Number
Even Number × Even Number = Even Number
Odd Number × Even Number = Even Number

In short,   Odd ± Odd = Even

Even ± Even = Even

Odd ± Even = Odd

Odd × Odd = Odd
Even × Even = Even
Odd × Even = Even

PRIME AND COMPOSITE NUMBERS

The final classification of integers includes prime numbers (also called 'primes') and composite numbers (or 'composites'). A **prime number** is a number that is divisible only by itself and by the number 1. It is not divisible by any other number. For example, 2, 3, 5, 7, 11, 13 etc. are prime. Clearly, a prime number has *two* distinct divisors: itself and 1. Both divisors *must be positive*. Prime numbers are *never* negative. A **composite number** is a number that has at least three distinct non-zero divisors. All even numbers except the numbers 0 and 2 are composite. Many odd numbers are also composite, for instance 15, 25, 39 etc.

So what about the number *one* itself? It is divisible only by 1, which is itself. So it has only one divisor, not two or more distinct divisors. Therefore, *one is neither prime nor composite*. What about *zero*? Zero has infinitely many divisors, i.e. any number times zero is zero. So it is not prime. For zero to be written as a product of two numbers, at least one of them will *have to be* zero itself. Therefore, zero can never be written as a product of two non-zero numbers. So it is not composite either. Hence, *zero is neither prime nor composite*.

Before we move on to the next section, let's understand three important terms from the perspective of mathematical reasoning: Non-zero numbers, non-negative numbers, and positive numbers. **Non-zero numbers** are all positive and negative numbers. Zero is excluded. **Non-negative numbers** are all positive numbers *and zero*. Negative numbers are excluded. Finally, positive numbers do not include zero. Zero is neither positive nor negative. We'll talk about zero a little later in this chapter.

PRIME FACTORIZATION OF INTEGERS

Every integer can be expressed as a product of prime numbers alone. Each of those prime numbers that can divide the original integer is called a **prime factor**. Consider the following example.

1.017   **Express 40 as a product of primes.**

4×10

$2 \times 2 \times 5 \times 2$ Because 4 itself is 2×2, and 10 itself is 5×2. Can we break it down any further? I don't think so. So, 40 is prime factorized as $2 \times 2 \times 5 \times 2$. Notice that each of these four numbers is a prime number.

Just for your information, if you initially broke down 40 as 20×2, you would still reach the same prime factors.

Generally speaking, a **factor** (also referred to as a **divisor**) is a number that can divide another number without leaving behind a remainder. Below is a quick review of some of the terms we learned back in middle school, which we may now have forgotten. These pertain to a long division.

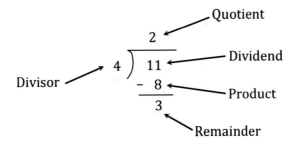

Also, some terms pertaining to the digits in an integer are the following.

➤ **Units digit**: First digit from the right in an integer, e.g. **6** in the number 71, 496
➤ **Tens digit**: Second digit from the right in an integer, e.g. **9** in the number 71, 496
➤ **Hundreds digit**: Third digit from the right in an integer, e.g. **4** in the number 71, 496
➤ **Thousands digit**: Fourth digit from the right in an integer, e.g. **1** in the number 71, 496

and so on...

Memorizing these terms will prove useful in some of the later chapters. Let's go to the next example.

1.018 **Prime factorize 84.**

2×42

$2 \times 2 \times 21$ Because 42 is 2×21

$2 \times 2 \times 7 \times 3$ Because 21 is 7×3

1.019 **Factorize 36 into primes.**

6×6

$3 \times 2 \times 3 \times 2$ Because 6 is 3×2

These examples seem quite easy. What if I ask you to prime factorize 1791? Where do you start? How to tell if there are any numbers that can divide it, or is it already a prime number? This is where certain tests of divisibility help in narrowing down choices to reach a starting point.

1791

TESTS OF DIVISIBILITY

1) A number is **divisible by 2** if it is an even number.
2) A number is **divisible by 3** if the sum of its digits is divisible by 3.
3) A number is **divisible by 4** if the units and tens digit are in the form of a 2-digit number that is divisible by 4.
4) A number is **divisible by 5** if its units digit is either 5 or 0.

Now consider the number 1791, and apply each of these tests to it. It's not divisible by 2 because it is not an even number. What is the sum of its digits? $1 + 7 + 9 + 1 = 18$. Is 18 divisible by 3? Yes, it is. Therefore, 1791 is divisible by 3. What is the 2-digit number formed by its units place and tens place? 91. Is 91 divisible by 4? No. Therefore, 1791 is not divisible by 4. It is clearly not divisible by 5.

Let's factorize 1791.

3×597

$3 \times 3 \times 199$ Because 597 is 3×199

199 is not divisible by 2, 3, 4 or 5. Nor is it divisible by any number I can think of. It is prime.

Please note that there's no test that will tell you whether a number is prime. You have to use your intuition and various tests of divisibility to figure it out. Only one fact helps: prime numbers (greater than 10) *always* end with **1, 3, 7** or **9**.

1.020 **What is the smallest prime factor of 390?**

390 is divisible by 10. Quite an easy starting point!

39×10

$13 \times 3 \times 5 \times 2$ Because 39 is 13×3, and 10 is 5×2

Therefore, the smallest prime factor of 390 is 2.

As a rule of thumb, *the smallest prime factor of any even number is 2.*

1.021 **What is the largest prime factor of 700?**

7×100

$7 \times 10 \times 10$ Because 100 is 10×10

$7 \times 5 \times 2 \times 5 \times 2$ Because 10 is 5×2

Therefore, the largest prime factor of 700 is 7.

PRACTICE EXAMPLES

1.022 Prime factorize 90.

1.023 Factorize 42 into primes.

1.024 Prime factorize 24.

1.025 Find the smallest prime factor of 147.

1.026 What is the greatest prime factor of 510?

HIGHEST COMMON FACTOR (HCF) AND LOWEST COMMON MULTIPLE (LCM)

Let's now build another concept. Consider the numbers 16 and 24. What are the different numbers that can divide each of them? In other words, what are their factors or divisors?

Factors of 16:   1, 2, 4, 8, 16

Factors of 24:   1, 2, 3, 4, 6, 8, 12, 24

Notice that there are several factors that appear in both lists, viz. 1, 2, 4, 8. These are called common factors. Which one of these common factors is the greatest (or the highest)? 8, of course. So, for 16 and 24, 8 is their **highest common factor** (also known as the **greatest common factor**, or **greatest common divisor**). Accordingly, it may be abbreviated as **HCF**, **GCF**, or **GCD**. As a general rule, when determining the HCF of two numbers, list their factors and pick the highest one that occurs in both the lists.

Starting over, let's list the multiples of 16 and 24 this time.

Multiples of 16:   16, 32, 48, 64, 80, 96,................

Multiples of 24:   24, 48, 72, 96, 120,..................

Notice that several of these multiples, viz. 48, 96,........ appear in both lists. Which one of these common multiples is the lowest (or the least)? 48. So, for 16 and 24, 48 is their **lowest common multiple** (also known as **least common multiple**). **LCM,** for short. Determining the LCM of two numbers can get tedious if the numbers are large, or unfamiliar, for example 135 and 35.

There's a formula to determine the LCM of two numbers. For any two integers A and B,

$$\text{LCM of } A \text{ and } B = \frac{A \times B}{\text{Their HCF}}$$

Let's verify this formula for 16 and 24.

$$\text{LCM of 16 and 24} = \frac{16 \times 24}{\text{Their HCF}} = \frac{\cancel{16}^{2} \times 24}{\cancel{8}^{1}} = 48$$

That is exactly what we determined using the regular method.

PRACTICE EXAMPLES

1.027 Find the HCF and LCM of 125 and 50.	**1.029** Find the GCF and LCM of 81 and 54.
1.028 Find the GCD and LCM of 36 and 24.	**1.030** Find the HCF and LCM of 49 and 98.

SPACE FOR CALCULATIONS

SOME FACTS ABOUT MULTIPLES

➤ If a number is divisible by two or more integers, then it is also divisible by *their LCM*. (But it's not always divisible by their product.) For example, 36 is divisible by 9 (4 times) and by 6 (6 times). But it is not divisible by their product 54 (9 × 6 = 54). However, it is divisible by their LCM 18.

➤ The product of multiples of two integers is a multiple of the product of those two integers. For example, 12 is a multiple of 3, and 8 is a multiple of 4. Hence, 12 × 8 is a multiple of 3 × 4, i.e. 96 is a multiple of 12. (Generally, if x is a multiple of a, and y is a multiple of b, then xy is a multiple of ab.)

FRACTIONS

A **fraction** is a quotient of two quantities. It means *a part* out of *the whole*. Fractions are generally of the form $\frac{N}{D}$, where N is called the **numerator** (a part), and can be any number. D is called the **denominator** (the whole), and is a non-zero number. Fractions can be positive, negative, or zero. When two or more fractions have the same denominator, they are called **like fractions**. Fractions with different denominators are called **unlike fractions**. Fractions can be of three types: **proper** fractions, **improper** fractions, and **mixed** fractions (also called mixed numbers).

Proper fractions	$N < D$	Examples: $\frac{11}{14}, \frac{3}{7}, \frac{1}{1000}, \frac{2}{3}$
Improper fractions	$N \geq D$	Examples: $\frac{10}{10}, \frac{5}{3}, \frac{2119}{517}, \frac{7}{7}$
Mixed fractions are of the form:	$A\frac{N}{D}$ (where $N \leq D$)	Examples: $2\frac{1}{2}, 5\frac{3}{4}, 6\frac{11}{29}, 3\frac{4}{7}$

MIXED FRACTIONS ⇌ IMPROPER FRACTIONS

There exists interconvertibility *only between improper fractions and mixed fractions*. If you were to convert a mixed fraction $5\frac{2}{3}$ into an improper fraction, you would multiply 3 by 5 (giving you 15), and then add it to 2 (giving you 17). That would be your new numerator. *The denominator would remain the same.* So, the equivalent improper fraction would be $\frac{17}{3}$. The process is depicted below.

$$5\frac{2}{3} \qquad \xrightarrow{\quad\dfrac{3 \times 5 + 2 = 17}{3}\quad} \qquad \frac{17}{3}$$

The value of the mixed fraction $5\frac{2}{3}$ is equal to $5 + \frac{2}{3}$, not $5 \times \frac{2}{3}$. Think about this: When I say "I worked there for 4½ years," it means 4 + ½ years, not 4 × ½ years. 4 × ½ years is just 2 years! Clearly, that's not the same as 4½ years.

If you were to convert the improper fraction $\frac{17}{3}$ back into a mixed fraction, you would simply carry out the actual division, i.e. 17 ÷ 3, which is 5, with a remainder of 2. This remainder would become the new numerator in the fractional component. *The denominator would stay the same.* Result? $5\frac{2}{3}$.

Let's solve a few examples to strengthen our understanding of the conversion between mixed numbers and improper fractions, and vice versa.

1.031 **Convert $3\frac{4}{5}$ into an improper fraction.**

$\dfrac{5 \times 3 + 4}{5}$ First, multiply 5 by 3. Then add 4 to it. Keep the denominator the same.

$\dfrac{19}{5}$ This is the final result.

1.032 **Rewrite $6\frac{3}{8}$ as an improper fraction.**

$\dfrac{8 \times 6 + 3}{8}$ Multiply 8 by 6. Then add 3. Denominator stays the same.

$$\frac{51}{8}$$ Final result.

1.033 **Change $4\frac{1}{3}$ to an improper fraction.**

$$\frac{3 \times 4 + 1}{3} \Rightarrow \frac{13}{3}$$

1.034 **Convert 9½ into an equivalent improper fraction.**

$$\frac{2 \times 9 + 1}{2} \Rightarrow \frac{19}{2}$$

1.035 **Convert 2¾ into an improper fraction.**

$$\frac{4 \times 2 + 3}{4} \Rightarrow \frac{11}{4}$$

PRACTICE EXAMPLES

1.036 Convert $2\frac{11}{12}$ into an improper fraction.

1.037 Rewrite $5\frac{4}{7}$ as an improper fraction.

1.038 Change $7\frac{3}{7}$ to an improper fraction.

1.039 Convert $1\frac{4}{5}$ into an equivalent improper fraction.

1.040 Convert $13\frac{2}{3}$ into an improper fraction.

SPACE FOR CALCULATIONS

1.041 **Convert $\frac{17}{4}$ into a mixed fraction.**

First, divide 17 by 4. That gives 4, with a remainder of 1. That 1 is now the numerator of the new fractional component, whose denominator stays 4.

Therefore, 4 and ¼, or simply 4¼.

1.042 **Rewrite $\frac{13}{6}$ as a mixed number.**

Divide 13 by 6. That gives 2, with a remainder of 1. That 1 is now the numerator of the new fractional component, whose denominator stays 6.

Therefore, 2 and $\frac{1}{6}$, or simply $2\frac{1}{6}$.

1.043 Change $\frac{49}{5}$ to a mixed fraction.

49 ÷ 5. That's 9. Remainder 4. Fractional part $\frac{4}{5}$.

Result: $9\frac{4}{5}$.

1.044 Convert $\frac{8}{7}$ into an equivalent mixed number.

8 ÷ 7. That's 1. Remainder 1. Fractional part $\frac{1}{7}$.

Result: $1\frac{1}{7}$.

1.045 Convert $\frac{18}{13}$ into a mixed fraction.

18 ÷ 13. That's 1. Remainder 5. Fractional part $\frac{5}{13}$.

Result: $1\frac{5}{13}$.

PRACTICE EXAMPLES (Convert into mixed fractions)

1.046 $\frac{51}{50}$ **1.048** $\frac{36}{11}$ **1.050** $\frac{32}{3}$

1.047 $\frac{7}{5}$ **1.049** $\frac{23}{7}$

SPACE FOR CALCULATIONS

FRACTIONS RAISED TO HIGHER TERMS

When the numerator and the denominator of a fraction are multiplied by *the same integer*, positive or negative, the resulting fraction is said to have been **raised to higher terms**. Each term—the numerator as well as the denominator—is of a greater magnitude than its original self.

For example, consider the fraction ¾. Let's multiply both its terms by 5.

$$\frac{3}{4} \Rightarrow \frac{3 \times 5}{4 \times 5} \Rightarrow \frac{15}{20}$$

Notice that the value of the original fraction was 3 ÷ 4, or 0.75. In comparison, the value of the new fraction is 15 ÷ 20, which is also 0.75. *Raising a fraction to higher terms does not change its value.*

FRACTIONS REDUCED TO LOWER TERMS

When the numerator and the denominator of a fraction are divided by *the same integer*, positive or negative, the resulting fraction is said to have been **reduced to lower terms**. Each term—the numerator as well as the denominator—is of a smaller magnitude than its original self.

This time, consider the fraction $\frac{15}{12}$. Let's divide both its terms by 3.

$$\frac{15}{12} \Rightarrow \frac{15 \div 3}{12 \div 3} \Rightarrow \frac{5}{4}$$

The value of the original fraction was 15 ÷ 12, or 1.25, and the value of the new fraction is 5 ÷ 4, which is also 1.25. Once again, *reducing a fraction to lower terms does not change its value.*

ADDITION/SUBTRACTION OF FRACTIONS

The easy rule for addition/subtraction of **like fractions** (for example $\frac{p}{q}$ and $\frac{r}{q}$) is:

$$\frac{p}{q} \pm \frac{r}{q} \Rightarrow \frac{p \pm r}{q}$$

The rule for addition/subtraction of **unlike fractions** (for example $\frac{p}{q}$ and $\frac{r}{s}$) is:

$$\frac{p}{q} \pm \frac{r}{s} \Rightarrow \frac{ps \pm qr}{qs}$$

The numerator here is *ps ± qr*, not *qr ± ps*. While adding, *ps + qr* is the same as *qr + ps*. But while subtracting, *ps – qr* is not the same as *qr – ps*. So, here's how to remember it:

$$\frac{p}{q} \pm \frac{r}{s} \Rightarrow \frac{ps \pm qr}{qs}$$

Most people are right-handed. So, first cross your arms. Then, take the *right-arm product first.*

Another approach to adding/subtracting unlike fractions is to *equalize their denominators* by taking their LCM, thus making the *unlike* fractions *like*. This is more useful when dealing with three or more fractions.

1.051 $3\frac{4}{5} + 7\frac{3}{5}$

There are two ways to go from here. You can convert the mixed fractions into improper fractions, and then add them. But I recommend not doing it, because we will be dealing with larger numbers. Instead, let's add the integer parts 3 and 7 separately, and add the fractional parts $\frac{4}{5}$ and $\frac{3}{5}$ separately. Then let's bring them together.

$$(3 + 7) + \left(\frac{4}{5} + \frac{3}{5}\right)$$

$$10 + \frac{7}{5}$$

$$10 + 1 + \frac{2}{5}$$

$$11\frac{2}{5}$$

1.052 $11\frac{5}{13} - 8\frac{2}{13}$

Same approach: Integers separately, fractions separately.

$$(11 - 8) + \left(\frac{5}{13} - \frac{2}{13}\right) \quad \Rightarrow \quad 3 + \frac{3}{13} \quad \Rightarrow \quad 3\frac{3}{13}$$

1.053 $7 - 3\frac{3}{4}$

A mixed fraction is being subtracted from an integer. First, *convert the integer into a mixed fraction by introducing an artificial fractional component.* We know that 7 is $(6 + 1)$, where 1 can be written as $\frac{10}{10}$ or $\frac{17}{17}$ or $\frac{8}{8}$ or $\frac{4}{4}$ etc. We are going to write $\frac{4}{4}$ for 1, because we need the denominator to be 4, since the other fractional component has it too.

$$6\frac{4}{4} - 3\frac{3}{4} \quad \Rightarrow \quad (6 - 3) + \left(\frac{4}{4} - \frac{3}{4}\right) \quad \Rightarrow \quad 3 + \frac{1}{4} \quad \Rightarrow \quad 3\frac{1}{4}$$

1.054 $16 - 1\frac{3}{5}$

Same approach as above: 16 is $(15 + 1)$, or $15 + \frac{5}{5}$, or $15\frac{5}{5}$

$$15\frac{5}{5} - 1\frac{3}{5} \quad \Rightarrow \quad (15 - 1) + \left(\frac{5}{5} - \frac{3}{5}\right) \quad \Rightarrow \quad 14 + \frac{2}{5} \quad \Rightarrow \quad 14\frac{2}{5}$$

1.055 $4\frac{4}{7} - 12\frac{5}{8}$

$$(4 - 12) + \left(\frac{4}{7} - \frac{5}{8}\right)$$

$$-8 + \left(\frac{4 \times 8 - 7 \times 5}{7 \times 8}\right) \qquad \text{Recall the rule of } right\ arm\ product\ first!$$

$$-8 + \left(\frac{32 - 35}{56}\right) \quad \Rightarrow \quad -8 - \frac{3}{56} \quad \Rightarrow \quad -\left(8 + \frac{3}{56}\right) \quad \Rightarrow \quad -8\frac{3}{56}$$

PRACTICE EXAMPLES

1.056 $1\frac{4}{5} + 7\frac{2}{5}$ **1.058** $8 - 2\frac{11}{12}$ **1.060** $10\frac{4}{5} - 13\frac{7}{9}$

1.057 $6\frac{5}{6} - 4\frac{1}{6}$ **1.059** $9\frac{3}{8} - 6\frac{5}{7}$

MULTIPLICATION OF FRACTIONS

When multiplying two or more fractions, convert mixed fractions (if any) into improper fractions first. Then reduce *any numerator with any denominator*, if it can be reduced. That will help you deal with smaller numbers. Finally, horizontally multiply out the numerators, and horizontally multiply out the denominators. Consider the following examples.

1.061 $\dfrac{3}{4} \times 2\dfrac{2}{5} \times \dfrac{10}{6}$

$\dfrac{3}{4} \times \dfrac{12}{5} \times \dfrac{10}{6}$ 　　Convert $2\dfrac{2}{5}$ into an improper fraction.

$\dfrac{\cancel{3}^{1}}{\cancel{4}^{1}} \times \dfrac{\cancel{12}^{3}}{\cancel{5}^{1}} \times \dfrac{\cancel{10}^{2}}{\cancel{6}^{2}}$ 　　*Reduce* 3 with 6, then 4 with 12, then 5 with 10.

3 　　Because 2 in the numerator cancelled with 2 in the denominator

1.062 $2\dfrac{1}{8} \times 1\dfrac{15}{34}$

$\dfrac{17}{8} \times \dfrac{49}{34}$ 　　Convert both the mixed numbers into improper fractions.

$\dfrac{\cancel{17}^{1}}{8} \times \dfrac{49}{\cancel{34}^{2}}$ 　　Reduce 17 with 34. Recall that 17 twice is 34.

$\dfrac{49}{16}$ 　　Horizontally multiply out the terms.

1.063 $1\dfrac{3}{4} \times 3\dfrac{2}{3} \times 1\dfrac{1}{11}$

$\dfrac{7}{4} \times \dfrac{11}{3} \times \dfrac{12}{11}$ 　　The 11s can cancel each other. The 4×3 can *together* cancel the 12.

$\dfrac{7}{\cancel{4}^{1}} \times \dfrac{\cancel{11}^{1}}{\cancel{3}^{1}} \times \dfrac{\cancel{12}^{1}}{\cancel{11}^{1}}$

7

1.064 $14 \times 2\frac{1}{7} \times 3\frac{1}{3}$

$\cancel{14}^{2} \times \dfrac{\cancel{15}^{5}}{\cancel{7}^{1}} \times \dfrac{10}{\cancel{3}^{1}}$

100

1.065 $\left(1 + \dfrac{1}{2}\right)\left(1 + \dfrac{1}{3}\right)\left(1 + \dfrac{1}{4}\right)\left(1 + \dfrac{1}{5}\right)\ldots\ldots\ldots\left(1 + \dfrac{1}{99}\right)$

Clearly, there's no way you'd be expected to perform 98 different additions, and then multiply them all! On the GRE/GMAT you only have 1½ to 2 minutes per question. So there must be a trick to this problem. Notice that each set of parentheses has a *mixed fraction* inside it.

$\left(1\frac{1}{2}\right)\left(1\frac{1}{3}\right)\left(1\frac{1}{4}\right)\left(1\frac{1}{5}\right)\ldots\ldots\ldots\left(1\frac{1}{99}\right)$ Convert each into an improper fraction. Then multiply.

$\dfrac{3}{2} \times \dfrac{4}{3} \times \dfrac{5}{4} \times \dfrac{6}{5} \times \ldots\ldots\ldots\ldots \dfrac{100}{99}$ Cancel the 3s, then the 4s, the 5s, the 6s … … … the 99s.

$\dfrac{100}{2}$ That's all that finally remains.

50   Result.

PRACTICE EXAMPLES

1.066 $\dfrac{16}{15} \times \dfrac{3}{8} \times \dfrac{5}{4}$

1.067 $7\frac{1}{3} \times 2\frac{1}{2}$

1.068 $1\frac{3}{5} \times 2\frac{2}{3} \times \dfrac{5}{16}$

1.069 $\dfrac{3}{14} \times \dfrac{2}{5} \times 21$

1.070 $\left(1 + \dfrac{1}{2}\right)\left(1 - \dfrac{1}{3}\right)\left(1 + \dfrac{1}{4}\right)\left(1 - \dfrac{1}{5}\right)\left(1 + \dfrac{1}{6}\right)\left(1 - \dfrac{1}{7}\right)\ldots\ldots\ldots\left(1 + \dfrac{1}{98}\right)\left(1 - \dfrac{1}{99}\right)$

SPACE FOR CALCULATIONS

DIVISION OF FRACTIONS

When dividing one fraction by another, simply take the *reciprocal* of the fraction in the denominator, and multiply it by the fraction in the numerator. In other words, simply flip the *second fraction*, and then multiply it by the first fraction. This rule can be written in two equivalent forms:

$$\frac{\frac{p}{q}}{\frac{r}{s}} = \frac{p}{q} \times \frac{s}{r} \qquad \text{or} \qquad \frac{p}{q} \div \frac{r}{s} = \frac{p}{q} \times \frac{s}{r}$$

If there are any mixed numbers, you would need to convert them into improper fractions first. Also, as always, look to reduce (or cancel) the terms, and deal with smaller numbers.

1.071 $\quad \dfrac{4}{3} \div \dfrac{12}{5}$

$\dfrac{\cancel{4}^{1}}{3} \times \dfrac{5}{\cancel{12}^{3}}$ Flip the second fraction. Reduce 4 with 12.

$\dfrac{5}{9}$ Horizontally multiply out the terms.

1.072 $\quad 7\dfrac{7}{8} \div \dfrac{21}{2}$

$\dfrac{8 \times 7 + 7}{8} \times \dfrac{2}{21}$ Mixed number → Improper fraction. Flip the second fraction.

$\dfrac{\cancel{63}^{3}}{\cancel{8}^{4}} \times \dfrac{\cancel{2}^{1}}{\cancel{21}^{1}}$ 21 reduces with 63, and 2 reduces with 8.

$\dfrac{3}{4}$

1.073 $\quad \dfrac{2}{3} \div 9$

$\dfrac{2}{3} \times \dfrac{1}{9}$ 9 is the same as $\dfrac{9}{1}$. Its reciprocal is $\dfrac{1}{9}$.

$\dfrac{2}{27}$ Horizontally multiply out the terms.

1.074 $\quad 2\dfrac{8}{9} \div 6\dfrac{1}{2}$

$\dfrac{9 \times 2 + 8}{9} \div \dfrac{2 \times 6 + 1}{2} \quad \Rightarrow \quad \dfrac{26}{9} \div \dfrac{13}{2} \quad \Rightarrow \quad \dfrac{\cancel{26}^{2}}{9} \times \dfrac{2}{\cancel{13}^{1}} \quad \Rightarrow \quad \dfrac{4}{9}$

PRACTICE EXAMPLES

1.075 $\quad \dfrac{11}{13} \div \dfrac{12}{39}$ **1.076** $\quad 5\dfrac{4}{5} \div \dfrac{58}{10}$ **1.077** $\quad 14 \div \dfrac{7}{8}$ **1.078** $\quad 5\dfrac{1}{2} \div 6\dfrac{4}{5}$

SPACE FOR CALCULATIONS

COMPLEX FRACTIONS

Sometimes, a fraction may contain a numerator which is already a fraction, and a denominator which is also a fraction. Such a fraction is called a **complex fraction**, meaning a fraction with fraction(s) within it. (Did you notice that I used the word 'fraction' seven times in just two lines?) Generally speaking, every complex fraction can be simplified into a proper or improper fraction.

1.079 $\dfrac{\frac{3}{5}}{\frac{24}{25}}$

$\dfrac{\cancel{3}^{1}}{\cancel{5}^{1}} \times \dfrac{\cancel{25}^{5}}{\cancel{24}^{8}}$ Flip the fraction in the denominator. Reduce the terms.

$\dfrac{5}{8}$ Horizontally multiply out the terms.

1.080 $\dfrac{12-\frac{1}{3}}{14}$

$\dfrac{\frac{12}{1}-\frac{1}{3}}{\frac{14}{1}}$ Because 12 and 14 are the same as $\dfrac{12}{1}$ and $\dfrac{14}{1}$.

$\dfrac{36-1}{3} \times \dfrac{1}{14}$ *Right arm product first* rule for numerator. Flip the denominator.

$\dfrac{\cancel{35}^{5}}{3} \times \dfrac{1}{\cancel{14}^{2}}$ Reduce 14 and 35 with 7.

$\dfrac{5}{6}$ Horizontally multiply out the terms.

1.081 $\dfrac{\frac{4}{5}+\frac{5}{6}}{\frac{7}{15}}$

$\dfrac{24+25}{30} \times \dfrac{15}{7}$ *Right arm product first* rule for numerator. Flip the denominator.

$\dfrac{\cancel{49}^{7}}{\cancel{30}^{2}} \times \dfrac{\cancel{15}^{1}}{\cancel{7}^{1}}$ Reduce the terms.

$\dfrac{7}{2}$ Horizontally multiply out the terms.

1.082 $\dfrac{\dfrac{2}{7}+\dfrac{3}{4}}{29}$

$$\dfrac{\dfrac{8+21}{28}}{\dfrac{29}{1}} \quad\Rightarrow\quad \dfrac{29^1}{28}\times\dfrac{1}{29^1} \quad\Rightarrow\quad \dfrac{1}{28}$$

1.083 $\dfrac{\dfrac{7}{8}+3}{4}$

$$\dfrac{\dfrac{7}{8}+\dfrac{3}{1}}{\dfrac{4}{1}} \quad\Rightarrow\quad \dfrac{7+24}{8}\times\dfrac{1}{4} \quad\Rightarrow\quad \dfrac{31}{8}\times\dfrac{1}{4} \quad\Rightarrow\quad \dfrac{31}{32}$$

1.084 $\dfrac{\dfrac{2}{3}+\dfrac{2}{3}+\dfrac{2}{3}+\dfrac{2}{3}}{\dfrac{2}{3}\times\dfrac{2}{3}}$

It might be tempting to add the like fractions in the numerator, and multiply out the ones in the denominator. But a quick look at the numerator will tell you that $\frac{2}{3}$ is being added to itself 4 times, i.e. it is equivalent to $4\times\dfrac{2}{3}$.

$\dfrac{4\times\dfrac{2}{3}}{\dfrac{2}{3}\times\dfrac{2}{3}}$ Now you see why I chose this path?

$\dfrac{4\times\dfrac{2}{\cancel{3}}}{\dfrac{2}{3}\times\dfrac{\cancel{2}}{\cancel{3}}}$ Cancel the $\dfrac{2}{3}$ from the numerator with *only one* of the $\dfrac{2}{3}$ from the denominator.

$\dfrac{\cancel{4}^2}{1}\times\dfrac{3}{\cancel{2}^1}$ Flip the lower fraction. Then reduce.

6

PRACTICE EXAMPLES

1.085 $\dfrac{\dfrac{3}{8}}{\dfrac{6}{7}}$

1.086 $\dfrac{20-\dfrac{4}{5}}{12}$

1.087 $\dfrac{\dfrac{2}{3}+\dfrac{1}{4}}{\dfrac{11}{6}}$

1.088 $\dfrac{\dfrac{1}{9}+\dfrac{2}{7}}{5}$

1.089 $\dfrac{4+\dfrac{5}{7}}{\dfrac{11}{14}}$

1.090 $\dfrac{3\times\dfrac{1}{6}\times\dfrac{1}{6}}{\dfrac{1}{6}+\dfrac{1}{6}+\dfrac{1}{6}+\dfrac{1}{6}}$

SPACE FOR
CALCULATIONS

COMPARISON OF FRACTIONS

To find the greater of two fractions, e.g. $\frac{a}{b}$ and $\frac{c}{d}$, multiply the numerator of the first fraction by the denominator of the second ($a \times d$), and write it above the first fraction. Multiply the numerator of the second fraction by the denominator of the first ($b \times c$), and write it above the second fraction. Whichever product looks greater, the fraction below it is the greater fraction. The process is depicted below.

$$ad \qquad\qquad bc$$

$$\frac{a}{b} \qquad\qquad \frac{c}{d}$$

The fingers being positioned upwards (not downwards) means the products (ad and bc) need to be placed *above* the fractions, where the fingers point.

$ad > bc$?   Then $\frac{a}{b} > \frac{c}{d}$

$bc > ad$?   Then $\frac{c}{d} > \frac{a}{b}$

$ad = bc$?   Then $\frac{a}{b} = \frac{c}{d}$

1.091 **Which fraction is greater, $\frac{5}{4}$ or $\frac{10}{9}$?**

$$45 \qquad\qquad 40$$

$$\frac{5}{4} \qquad\qquad \frac{10}{9}$$

Since 45 > 40, it follows that $\frac{5}{4} > \frac{10}{9}$.

1.093 **Is $\frac{11}{13}$ less than $\frac{3}{4}$?**

$$44 \qquad\qquad 39$$

$$\frac{11}{13} \qquad\qquad \frac{3}{4}$$

$\frac{11}{13}$ is not less than, but greater than $\frac{3}{4}$.

1.092 **Find the greater of $\frac{2}{7}$ and $\frac{6}{17}$.**

$$34 \qquad\qquad 42$$

$$\frac{2}{7} \qquad\qquad \frac{6}{17}$$

Therefore, $\frac{6}{17}$ is the greater fraction.

1.094 Which one of the fractions $\frac{2}{3}, \frac{4}{7}$ and $\frac{6}{11}$ is the greatest in value?

Compare the first two fractions. Take the greater of them, and then compare it with the third one. That would yield the fraction with the greatest value. **Remember:** *During the second comparison, discard the original products that you placed on top of the first two fractions.*

It appears that $\frac{2}{3}$ turns out to be greater each time. So $\frac{2}{3}$ is the greatest in value among them.

1.095 Arrange the fractions $\frac{5}{9}, \frac{3}{8}$ and $\frac{10}{17}$ in ascending (increasing) order.

The ascending order is: $\frac{3}{8}, \frac{5}{9}, \frac{10}{17}$.

1.096 Arrange the fractions $\frac{6}{7}, \frac{5}{8}$ and $\frac{8}{9}$ in descending (decreasing) order.

The descending order is: $\frac{8}{9}, \frac{6}{7}, \frac{5}{8}$.

PRACTICE EXAMPLES

1.097 Which fraction is greater, $\frac{3}{5}$ or $\frac{23}{30}$?

1.098 Is $\frac{13}{17}$ greater than $\frac{2}{3}$?

1.099 Find the smaller of $\frac{21}{16}$ and $\frac{4}{3}$.

1.100 Find the smallest of $\frac{5}{4}, \frac{8}{6}$ and $\frac{7}{5}$.

1.101 Write in descending order: $\frac{3}{13}, \frac{4}{17}, \frac{1}{4}$

1.102 Arrange in ascending order: $\frac{9}{4}, \frac{5}{2}, \frac{12}{7}$

SOME FACTS ABOUT FRACTIONS

Consider the **proper fraction** $\frac{1}{2}$. Its value is 0.5. Let's add 1 to the numerator as well as to the denominator. Let's do this repeatedly and observe what happens to the value (highlighted in gray) of the original fraction.

$$\frac{1}{2} \Rightarrow \frac{1+1}{2+1} \Rightarrow \frac{2}{3} \Rightarrow \frac{2+1}{3+1} \Rightarrow \frac{3}{4} \Rightarrow \frac{3+1}{4+1} \Rightarrow \frac{4}{5} \Rightarrow \frac{4+1}{5+1} \Rightarrow \frac{5}{6} \text{ and so on.}$$

0.50 $\qquad$ 0.67 $\qquad$ 0.75 $\qquad$ 0.80 $\qquad$ 0.83 and so on.

The value seems to be going up, but the *rate of increase* seems to be slowing down. If we continue with this process, a time will come when we will have added a billion (or something of that magnitude) to both the numerator and the denominator. At that point, the numerator as well as the denominator will be very close to a billion, with the denominator being slightly greater. Since they both will be almost the same, the value of the fraction will be very close to 1, but slightly less than 1.

Generally speaking, in a *proper fraction* $\frac{x}{y}$ where x and y are both positive, if we add the same positive number (say c), to x as well as to y, the value of the resulting fraction *increases*. The larger the number c added to both x and y, the larger the value of the resulting fraction. But it does not reach infinity; it only gets closer and closer to 1, but never quite makes it to 1.

Now consider the **improper fraction** $\frac{3}{2}$. Its value is 1.5. Let's run an analysis similar to that above.

$$\frac{3}{2} \Rightarrow \frac{3+1}{2+1} \Rightarrow \frac{4}{3} \Rightarrow \frac{4+1}{3+1} \Rightarrow \frac{5}{4} \Rightarrow \frac{5+1}{4+1} \Rightarrow \frac{6}{5} \Rightarrow \frac{6+1}{5+1} \Rightarrow \frac{7}{6} \text{ and so on.}$$

1.50 $\qquad$ 1.33 $\qquad$ 1.25 $\qquad$ 1.20 $\qquad$ 1.17 and so on.

This time, the value of the fraction *decreases* as we keep adding the same positive number to the numerator as well as to the denominator. If we continue on, then the value of the fraction will get closer and closer to 1, but will never quite make it to 1. It will always be slightly greater than 1.

Generally speaking, in an *improper fraction* $\frac{p}{q}$ where p and q are both positive, if we add the same positive number (say c), to p as well as to q, the value of the resulting fraction *decreases*. The larger the number c added to both x and y, the smaller the value of the resulting fraction. But it does not drop to negative infinity; it only gets closer and closer to 1, but stays slightly greater than 1.

This analysis is depicted in the graph below.

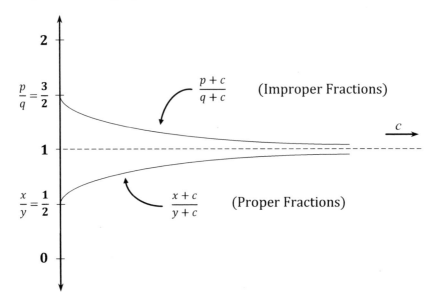

1.103 **If $\frac{a}{b} = \frac{3}{7}$ where a and b are positive numbers, which of the following is greater than $\frac{a}{b}$?**

$$\text{(i)}\,\frac{a+1}{b+1} \qquad \text{(ii)}\,\frac{a+2}{b+1} \qquad \text{(iii)}\,\frac{a}{b+1} \qquad \text{(iv)}\,\frac{b}{a} \qquad \text{(v)}\,\frac{a+b}{a-b}$$

Please note that just because $\frac{a}{b} = \frac{3}{7}$, it does not automatically mean that $a = 3$, and $b = 7$. Maybe $a = 300$, and $b = 700$. Or maybe $a = 3000$, and $b = 7000$. The value of $\frac{a}{b}$ would still be $\frac{3}{7}$.

(i) $\frac{a}{b}$ is a *proper* fraction. The same positive number (+1) is added to a as well as to b. Therefore, the resulting fraction has a greater value than $\frac{a}{b}$.

(ii) Adding 1 to the numerator and 1 to the denominator will result in a greater value of the fraction. Adding 1 more to the numerator will result in an even greater value than $\frac{a}{b}$.

(iii) Numerator is unchanged. Denominator has increased. The resulting fraction is less than $\frac{a}{b}$.

(iv) Original fraction is *proper*. Its value is less than 1. The flipped fraction is *improper*. Its value is greater than 1.

(v) Regardless of the individual values of *a* and *b*, the value of *a* + *b* will always be positive, whereas the value of *a* – *b* will always be negative. Therefore, $\frac{a+b}{a-b}$ is negative. *Any negative number is always less than any positive number.*

Notice that in none of the above five parts did we have to perform the actual calculations to tell the result. Such is the nature of *mathematical reasoning* type questions on the GRE/GMAT.

1.104 If $\frac{m}{n} = \frac{5}{4}$ where *m* and *n* are positive numbers, which of the following is less than $\frac{m}{n}$?

(i) $\dfrac{m+76}{n+76}$ (ii) $\dfrac{m+\frac{5}{6}}{n+\frac{4}{5}}$ (iii) $\dfrac{m}{n+\frac{1}{87}}$ (iv) $\dfrac{n}{m}$ (v) $\dfrac{m-(-n)}{m-n}$

Once again, keep in mind that *m* and *n* are not necessarily 5 and 4, respectively.

(i) $\frac{m}{n}$ is an *improper* fraction. The same positive number (+76) is added to *m* as well as to *n*. Therefore, the resulting fraction has a lesser value than $\frac{m}{n}$.

(ii) Adding $\frac{4}{5}$ to the numerator and to the denominator will increase the value of the fraction. Instead, adding $\frac{5}{6}$ $(>\frac{4}{5})$ to the numerator will further increase the value of the fraction.

(iii) Numerator is unchanged. Denominator has increased. The result is less than $\frac{m}{n}$.

(iv) The *improper* fraction is now a *proper* fraction. Its value has dropped below 1 from its original value above 1.

(v) $m-(-n)$ is the same as $m + n$. Both $m + n$ and $m - n$ are positive. The numerator has increased, whereas the denominator has decreased. The result is greater than $\frac{m}{n}$.

PRACTICE EXAMPLES

1.105 If $\frac{x}{y} = \frac{5}{6}$ where *x* and *y* are positive numbers, which of the following is (are) greater than $\frac{x}{y}$?

(i) $\dfrac{x+8\frac{2}{7}}{y+8\frac{2}{7}}$ (ii) $\dfrac{x+3.2}{y+3.1}$ (iii) $\dfrac{x+\frac{1}{5}}{y+\frac{1}{6}}$ (iv) $\dfrac{y}{x}$ (v) $\dfrac{x-y}{x+y}$

1.106 If $\frac{p}{q} = \frac{29}{7}$ where *p* and *q* are positive numbers, which of the following is (are) less than $\frac{p}{q}$?

(i) $\dfrac{q}{p}$ (ii) $\dfrac{p-q}{p-(-q)}$ (iii) $\dfrac{\frac{6}{5}+p}{\frac{6}{5}+q}$ (iv) $\dfrac{4.8+p}{5.3+q}$ (v) $\dfrac{p+\frac{11}{12}}{q+\frac{12}{13}}$

TWO EQUIVALENT FRACTIONS

Consider two fractions $\frac{a}{b}$ and $\frac{c}{d}$ whose terms have the values: $a = 1$, $b = 2$, $c = 100$ and $d = 200$.

The two fractions would have the values $\quad \frac{a}{b} = \frac{1}{2} = 0.5 \quad$ and $\quad \frac{c}{d} = \frac{100}{200} = 0.5$

Therefore, $\quad \frac{a}{b} = \frac{c}{d}$

The following properties hold true for such **equivalent fractions**.

1) $ad = bc \qquad$ Their cross products are equal.

2) $\dfrac{b}{a} = \dfrac{d}{c} \qquad$ Inverses of equivalent fractions are equivalent too.

3) $\dfrac{a}{c} = \dfrac{b}{d} \qquad$ Quotient of their numerators = Quotient of their corresponding denominators

4) $\left(\dfrac{a}{b}\right)^2 = \left(\dfrac{c}{d}\right)^2 \qquad$ Their squares are equal. (So are their cubes, or any higher powers.)

5) $\dfrac{a \pm b}{b} = \dfrac{c \pm d}{d}$

6) $\dfrac{a+b}{a-b} = \dfrac{c+d}{c-d}$

Let's verify each of these properties using the above values $a = 1$, $b = 2$, $c = 100$ and $d = 200$.

1) $1 \times 200 = 2 \times 100 \quad \Rightarrow \quad 200 = 200$

2) $\dfrac{2}{1} = \dfrac{200}{100} \quad \Rightarrow \quad 2 = 2$

3) $\dfrac{1}{100} = \dfrac{2}{200}$

4) $\left(\dfrac{1}{2}\right)^2 = \left(\dfrac{100}{200}\right)^2 \quad \Rightarrow \quad 0.5^2 = 0.5^2$

5) $\dfrac{1+2}{2} = \dfrac{100+200}{200} \quad \Rightarrow \quad \dfrac{3}{2} = \dfrac{300}{200} \quad$ and $\quad \dfrac{1-2}{2} = \dfrac{100-200}{200} \quad \Rightarrow \quad \dfrac{-1}{2} = \dfrac{-100}{200}$

6) $\dfrac{1+2}{1-2} = \dfrac{100+200}{100-200} \quad \Rightarrow \quad \dfrac{3}{-1} = \dfrac{300}{-100}$

RULE OF THREE

Consider two equivalent fractions: $\frac{a}{b} = \frac{c}{d}$. If the values of three of the four terms are known, then the fourth term can be calculated by equating the cross products of the terms, and then dividing both the sides of the equation by the term associated with the unknown term. **Three known terms** help determine the fourth unknown term. Therefore, rule of **three**. The following examples make it clear.

1.107 $\dfrac{3}{4} = \dfrac{x}{16}$

$3 \times 16 = 4 \times x$ Equate the cross products of the terms.

$\dfrac{3 \times 16}{4} = \dfrac{4 \times x}{4}$ Divide both sides by the term associated with the unknown term.

$\dfrac{3 \times \cancel{16}^{\,4}}{\cancel{4}^{\,1}} = \dfrac{\cancel{4}^{\,1} \times x}{\cancel{4}^{\,1}}$ Cancel and reduce.

$12 = x$ Multiply out the terms.

1.108 $\dfrac{7}{3} = \dfrac{49}{m}$

$7 \times m = 3 \times 49$ Equate the cross products of the terms.

$\dfrac{7 \times m}{7} = \dfrac{3 \times 49}{7}$ Divide both sides by the term associated with the unknown term.

$\dfrac{\cancel{7}^{\,1} \times m}{\cancel{7}^{\,1}} = \dfrac{3 \times \cancel{49}^{\,7}}{\cancel{7}^{\,1}}$ Cancel and reduce.

$m = 21$ Multiply out the terms.

1.109 $\dfrac{25}{a} = \dfrac{5}{2}$

$25 \times 2 = a \times 5$ $\Rightarrow$ $\dfrac{25 \times 2}{5} = \dfrac{a \times 5}{5}$ $\Rightarrow$ $\dfrac{\cancel{25}^{\,5} \times 2}{\cancel{5}^{\,1}} = \dfrac{a \times \cancel{5}^{\,1}}{\cancel{5}^{\,1}}$ $\Rightarrow$ $10 = a$

1.110 $\dfrac{c}{12} = \dfrac{5}{15}$

$c \times 15 = 12 \times 5$ $\Rightarrow$ $\dfrac{c \times \cancel{15}^{\,1}}{\cancel{15}^{\,1}} = \dfrac{\cancel{12}^{\,4} \times \cancel{5}^{\,1}}{\cancel{15}^{\,\cancel{3}\,1}}$ $\Rightarrow$ $c = 4$

1.111 $\dfrac{49}{42} = \dfrac{7}{f}$

$49 \times f = 42 \times 7$ $\Rightarrow$ $\dfrac{\cancel{49}^{\,1} \times f}{\cancel{49}^{\,1}} = \dfrac{\cancel{42}^{\,6} \times \cancel{7}^{\,1}}{\cancel{49}^{\,\cancel{7}\,1}}$ $\Rightarrow$ $f = 6$

PRACTICE EXAMPLES (Find the unknown term)

1.112 $\dfrac{6}{7} = \dfrac{b}{14}$

1.113 $\dfrac{4}{5} = \dfrac{8}{n}$

1.114 $\dfrac{16}{y} = \dfrac{8}{5}$

1.115 $\dfrac{e}{17} = \dfrac{5}{51}$

1.116 $\dfrac{39}{12} = \dfrac{13}{d}$

SPACE FOR CALCULATIONS

DECIMALS

Decimals are numbers that have an integer, and a decimal fraction. A dot "." also known as a **decimal point** separates the integer from the decimal fraction. For example, the number 3.14 has 3 as the integer, and .14 as the decimal fraction. The decimal fraction is always positive, and less than one. The integer can be positive, negative, or zero. Thus, 4.08, 0.92, and -5.661 are all decimals.

Consider the decimal 47,128.635. In expanded notation, it can be written as:

$$(4 \times 10,000) + (7 \times 1,000) + (1 \times 100) + (2 \times 10) + (8) + \left(6 \times \dfrac{1}{10}\right) + \left(3 \times \dfrac{1}{100}\right) + \left(5 \times \dfrac{1}{1,000}\right)$$

i.e. $40,000 + 7,000 + 100 + 20 + 8 + 0.6 + 0.03 + 0.005$

Observe that every digit, except the units digit, in 47,128.635 can be expressed as a multiple of a power of 10. Therefore, the name 'decimal.' Also, note that:

➤ 6 is in the **tenths place** in 47,128.635
➤ 3 is in the **hundredths place** in 47,128.635
➤ 5 is in the **thousandths place** in 47, 128.635

And so on...

ROUNDING OFF DECIMALS

When decimals are to be rounded off to a certain digit, look at the digit to the immediate right of that digit. If it is 5 or more, then round off the digit one unit higher. If the digit to the immediate right is less than 5, then round off the digit to itself.

For example, if you were to round off 3.04816 to the *hundredths place*, you would look at the digit in the *thousandths place*. That's the digit 8. Is 8 greater than or equal to 5? Yes, it is. So, you would round off the hundredths place to one digit higher, i.e. 4 would be rounded off to 5. The result would be 3.05.

Alternatively, if you were to round off 3.04816 to the *thousandths place*, you would now look at the digit in the *ten thousandths place*. That's the digit 1. Is 1 greater than or equal to 5? No, it's not. So, you would round off the thousandths place to itself, i.e. 8 would remain 8. The result would be 3.048.

ADDITION/SUBTRACTION OF DECIMALS

While adding or subtracting decimals, you would only need to vertically align the decimal point. After that, simply follow the basic rules of addition/subtraction. If one of the numbers being added or subtracted is purely an integer (for example, 31), it would need to be expressed as a decimal, i.e. 31.0 or 31.00 depending upon the number of digits to be filled in.

1.117   3.25 + 0.172 + 19

$$
\begin{array}{r}
3.\ 2\ 5\ \ \\
0.\ 1\ 7\ 2 \\
+\ \ 1\ 9\ \ \ \ \ \ \ \\
\hline
\end{array}
$$

First, align the decimal point.

$$
\begin{array}{r}
\quad\ \ 1\quad\ \ \ \\
3.\ 2\ 5\ 0 \\
{}_1\ \ 0.\ 1\ 7\ 2 \\
+\ \ 1\ 9.\ 0\ 0\ 0 \\
\hline
2\ 2.\ 4\ 2\ 2 \\
\end{array}
$$

Fill in any extra zeros as necessary.
Then perform the addition.

1.118   9.46728 – 3.178

$$
\begin{array}{r}
3\ \ 15\ \ 17\ \ \ \ \ \\
9.\ \cancel{4}\ \cancel{6}\ \cancel{7}\ 2\ 8 \\
-\ \ 3.\ 1\ 7\ 8\ 0\ 0 \\
\hline
6.\ 2\ 8\ 9\ 2\ 8 \\
\end{array}
$$

Align the decimal point. Fill in the zeros.
Perform the subtraction.

1.119   6.4 – 2.117

$$
\begin{array}{r}
3\ \ 9\ \ 10\ \\
6.\ \cancel{4}\ \cancel{0}\ \cancel{0} \\
-\ \ 2.\ 1\ 1\ 7 \\
\hline
4.\ 2\ 8\ 3 \\
\end{array}
$$

1.120   78 – 0.43

$$
\begin{array}{r}
7\ \ 9\ \ 10\ \\
7\ \cancel{8}.\ \cancel{0}\ \cancel{0} \\
-\ \ \ \ \ 0.\ 4\ 3 \\
\hline
7\ 7.\ 5\ 7 \\
\end{array}
$$

PRACTICE EXAMPLES

1.121 $1.791 + 13 + 0.843$ **1.123** $7.5 - 3.283$

1.122 $8.76482 - 8.129$ **1.124** $31 - 0.54$

SPACE FOR CALCULATIONS

MULTIPLICATION OF DECIMALS

When multiplying a decimal by another decimal, first carry out the multiplication by ignoring the decimal points. Then, count the *total number of digits to the right of the decimal points in both the numbers*. Finally, in the product, introduce the decimal point *to the left side of* those many digits.

When multiplying a decimal by a power of 10, for example 10, or 100, or 1000 etc. simply move the decimal point *to the right by as many digits as there are zeros in the power of 10.*

1.125 **8.93 × 5.4**

```
        4   1
        3   1
    8.  9   3
  ×     5.  4
  ───────────
    1   1
    3   5   7   2
 4  4   6   5  +
 ───────────────
 4  8.  2   2   2
```

Ignore the decimal points, and multiply the two numbers. Then, add the decimal point before the 3rd digit from the right, because there are a total of 3 digits after the decimal points in the two numbers (2 digits in 8.93, and 1 digit in 5.4).

1.126 **1.34 × 0.006**

```
         2   2
     1.  3   4
 ×  0.   0   0   6
 ─────────────────
 0.  0   0   8   0   4
```

Ignore the decimal points, and multiply. Then introduce the decimal point before the 5th digit from the right. There were only 3 digits in the product. So we added 2 extra zeros on the left to make up for the digits.

1.127 **5.03 × 1,000**

We need to move the decimal point three digits to the right. But there aren't three digits after the decimal point. So let's introduce a few extra zeros *after the existing digits*. Simply adding zeros won't change the value of 5.03, so let's be generous.

5.03000000 × 1,000   Now, move the decimal point 3 digits to the right.

5030.00000   Discard the extra zeros. 5030.00000 is the same as 5030.

5030   Final answer

1.128 **25.3 × 100**

25.3000 × 100   The decimal point moves 2 digits to the right.

2530   After discarding the extra zeros

PRACTICE EXAMPLES

1.129 4.93 × 10.5   **1.131** 4.14 × 10,000
1.130 0.071 × 4.75   **1.132** 4.98 × 100

SPACE FOR CALCULATIONS

DIVISION OF DECIMALS

We know that if we divide a number (be it an integer, a fraction, or a decimal) by the number 1, its value will not change. For example, $\frac{14}{1}$ is the same as just 14. Similarly, $\frac{5.619}{1}$ is the same as 5.619. Also, if we multiply both the numerator and the denominator of $\frac{4.287}{3.1}$ by 1000, we will get $\frac{4287}{3100}$. Its original value will still be the same, because multiplying and dividing by the same number does nothing to the original value. This technique helps eliminate the decimal points. We now have two integers. Dividing one integer by another is much easier than dividing one decimal by another.

When dividing a decimal by a power of 10, for example 10, or 100, or 1000 etc. simply move the decimal point *to the left by as many digits as there are zeros in the power of 10*. If there aren't enough digits, then introduce a few extra zeros to facilitate the shifting.

1.133 **0.529 ÷ 1.7**

$\frac{0.529}{1.7} \times \frac{1000}{1000}$ Multiplying and dividing by 10 will eliminate the decimal in the denominator, but not in the numerator. Use 1000 instead.

$$\frac{529}{17} \times \frac{1}{100}$$

Think about convenience. Would you rather divide 529 by 17 or 1700? Clearly, 17. So, use only as many zeros from the 1000 in the denominator as you need to eliminate the decimal point. Leave the rest as a power of 10. Here, we have used the 10, and left behind 100. ($1000 = 10 \times 100$)

$$
\begin{array}{r}
31.1 \\
17 \overline{)\,529} \\
-51 \\
\hline
19 \\
-17 \\
\hline
20 \\
-17 \\
\hline
3
\end{array}
$$

Carry out the actual division. Just one or two digits after the decimal point are good enough. You don't need to keep calculating for all eternity!

$$31.1 \times \frac{1}{100}$$

$\dfrac{529}{17}$ gives us 31.1. But we need to bring back the $\dfrac{1}{100}$ that we had temporarily isolated.

0.311

After shifting the decimal point 2 digits to the left

1.134 **$3.7 \div 9$**

$$\frac{3.7}{9} \times \frac{10}{10}$$

3.7 has 1 digit after the decimal point. 9 has none. So use 10.

$$\frac{37}{9} \times \frac{1}{10}$$

It is easier to divide 37 by 9 than by 90. So, use up only one 10.

$$4.11 \times \frac{1}{10}$$

Carry out the division. Then bring back the $\dfrac{1}{10}$.

0.411

After shifting the decimal point 1 digit to the left

1.135 **$7 \div 29$**

$$\frac{7}{29} \times \frac{10}{10} \quad \Rightarrow \quad \frac{70}{29} \times \frac{1}{10} \quad \Rightarrow \quad 2.41 \times \frac{1}{10} \quad \Rightarrow \quad 0.241 \qquad \text{Remember } convenience.$$

1.136 **$3.79 \div 1000$**

$000003.79 \div 1000$ Add zeros to create space. Move the decimal point 3 digits to the left.

0.00379   After discarding the unwanted zeros

1.137 **6.753 ÷ 0.14**

$$\frac{6.753}{0.14} \times \frac{1000}{1000} \quad \Rightarrow \quad \frac{6753}{14} \times \frac{1}{10} \quad \Rightarrow \quad 482.3 \times \frac{1}{10} \quad \Rightarrow \quad 48.23$$

1.138 **9 ÷ 0.16**

$$\frac{9}{0.16} \times \frac{100}{100} \quad \Rightarrow \quad \frac{900}{16} \quad \Rightarrow \quad 56.25$$

This is a very systematic method that will always get you the right answer. The more you practice, the faster you will get. On the GRE/GMAT speed is as important as accuracy.

PRACTICE EXAMPLES (Solve many more until you're fast. Use any two numbers you can think of.)

1.139 0.688 ÷ 1.2	**1.141** 4 ÷ 25	**1.143** 5.491 ÷ 0.13
1.140 3.1 ÷ 7	**1.142** 8.76 ÷ 100	**1.144** 77 ÷ 0.15

SPACE FOR CALCULATIONS

RECURRING DIGITS IN DECIMALS

4 divided by 3 gives 1.333333………… The question is: how many digits should we write? Well, that depends upon however many digits are expected to be mentioned. For example, if you buy 3 apples for $4, then each apple costs $\frac{\$4}{3}$ which is $1.33333……… We only need two digits, since a dollar can only be split into 100 cents. So $1.33 will do. Notice that we rounded off the number to the *hundredths* place. But this rounding-off resulted in a portion of the number being discarded, known as *error*. You can eliminate this error by telling the reader that the digit was repeating itself infinitely; simply place a horizontal bar on top of that digit. For example, $1.\overline{3}$ means that the digit 3 forever repeats itself. If several digits repeat a sequence, then place a bar covering the entire sequence.

For example, $2.45\overline{791}$ would mean the digits 7, 9 and 1 forever recur in that order, i.e. 2.45791791791791.........

FRACTIONS WITHIN DECIMALS

Sometimes, a number may be of the form 6.308¼. That looks strange! To rewrite the number as a decimal, first find the value of the ¼ in the number, i.e. ¼ = 0.25. Then pick the digits *after the decimal point* (2 and 5), and put them in place of ¼ in the original number. So, the decimal equivalent of 6.308¼ would be 6.30825. Similarly, the decimal equivalent of $27.49\frac{2}{3}$ would be $27.49\overline{6}$, and that of 0.61¾ would be 0.6175. What is the decimal equivalent of $3.711\frac{5}{6}$? It's $3.7118\overline{3}$. What about $0.25\frac{2}{5}$? It's 0.254. Quite easy, isn't it?

DECIMAL FRACTIONS   →   PROPER FRACTIONS

The *decimal fraction part* in a decimal can be converted into a proper fraction by multiplying and dividing it by the power of 10 necessary to eliminate the decimal point. The resulting fraction may then be reduced to its lowest terms. The *integer part* in the decimal stays the same.

1.145 **Convert 0.026 into a proper fraction.**

$$0.026 \times \frac{1000}{1000} \quad \Rightarrow \quad \frac{26}{1000} \quad \Rightarrow \quad \frac{\overset{13}{\cancel{26}}}{\underset{500}{\cancel{1000}}} \quad \Rightarrow \quad \frac{13}{500}$$

1.146 **Change 17.32 into a mixed fraction.**

$$17 + 0.32 \quad \Rightarrow \quad 17 + \left(0.32 \times \frac{100}{100}\right) \quad \Rightarrow \quad 17 + \frac{32}{100} \quad \Rightarrow \quad 17 + \frac{\overset{8}{\cancel{32}}}{\underset{25}{\cancel{100}}} \quad \Rightarrow \quad 17\frac{8}{25}$$

1.147 **Rewrite $0.6\frac{4}{5}$ as a proper fraction.**

$$0.6\frac{4}{5} \quad \Rightarrow \quad 0.68 \quad \Rightarrow \quad 0.68 \times \frac{100}{100} \quad \Rightarrow \quad \frac{68}{100} \quad \Rightarrow \quad \frac{\overset{17}{\cancel{68}}}{\underset{25}{\cancel{100}}} \quad \Rightarrow \quad \frac{17}{25}$$

1.148 **Convert 11.7½ into a mixed number.**

$$11.7\frac{1}{2} \quad \Rightarrow \quad 11.75 \quad \Rightarrow \quad 11 + \left(0.75 \times \frac{100}{100}\right) \quad \Rightarrow \quad 11 + \frac{75}{100} \quad \Rightarrow \quad 11 + \frac{\overset{3}{\cancel{75}}}{\underset{4}{\cancel{100}}} \quad \Rightarrow \quad 11\frac{3}{4}$$

PRACTICE EXAMPLES

1.149 Convert 0.475 into a proper fraction.

1.150 Change 21.25 into a mixed fraction.

1.151 Rewrite 0.1½ as a proper fraction.

1.152 Convert 26.0¼ into a mixed number.

PERCENTAGES

A percentage is a quantity expressed out of a hundred. It is followed by the % sign. For example, 36% means $\frac{36}{100}$ or 0.36. It is essentially a decimal, only expressed differently. The % sign is the equivalent of ($\div$ 100). If you delete the sign, then you would need to immediately replace it with ($\div$ 100) or $\left(\times \frac{1}{100}\right)$. On the other hand, if there is a number without the % sign, and you want to add the sign, then in addition to that, you would also need to multiply it by 100 (because % means $\div$ 100). The idea is to not disturb the integrity of the original number when converting it from one form to another.

PERCENT $\rightarrow$ **DECIMAL**

A percent can be converted into a decimal simply by replacing the % sign with ($\div$ 100) or $\left(\times \frac{1}{100}\right)$. Let's convert the following percents into decimals.

1.153 **10.47%**

$$10.47 \times \frac{1}{100}$$ Replace the % sign with $\left(\times \frac{1}{100}\right)$.

$$\frac{10.47}{100}$$ $\left(\times \frac{1}{100}\right)$ really means *divide by* 100.

$$0.1047$$ Shift the decimal 2 places to the *left.* (Remember?)

1.154 **0.935%**

$$0.935 \times \frac{1}{100} \quad \Rightarrow \quad \frac{0.935}{100} \quad \Rightarrow \quad 0.00935$$

1.155 **87½%**

$$87\frac{1}{2}\% \quad \Rightarrow \quad 87.5\% \quad \Rightarrow \quad \frac{87.5}{100} \quad \Rightarrow \quad 0.875$$

1.156 $12.8\frac{2}{3}\%$

$$12.8\frac{2}{3}\% \quad \Rightarrow \quad 12.8\overline{6}\% \quad \Rightarrow \quad \frac{12.8\overline{6}}{100} \quad \Rightarrow \quad 0.1286\overline{}$$

1.157 **0.75¼%**

$$0.75\frac{1}{4}\% \quad \Rightarrow \quad 0.7525\% \quad \Rightarrow \quad \frac{0.7525}{100} \quad \Rightarrow \quad 0.007525$$

PRACTICE EXAMPLES (Convert into decimals)

1.158	10.71%	**1.160**	$97\frac{1}{3}\%$	**1.162**	0.84¾%
1.159	0.897%	**1.161**	$11.4\frac{5}{6}\%$		

SPACE FOR CALCULATIONS

DECIMAL   →   PERCENT

A decimal can be expressed as a percent by multiplying it by 100, and then attaching the % sign. Remember, we're not altering the *value* of the number; we're only expressing it differently. Let's convert the following into percents.

1.163 **0.412**

$0.412 \times 100\%$ Multiply by 100. Attach the % sign.

41.2%   Shift the decimal 2 places to the *right*.

1.164 **0.034**

$0.034 \quad \Rightarrow \quad 0.034 \times 100\% \quad \Rightarrow \quad 3.4\%$

1.165 **8.07**

$8.07 \quad \Rightarrow \quad 8.07 \times 100\% \quad \Rightarrow \quad 807\%$

1.166 **46**

$46 \quad \Rightarrow \quad 46 \times 100\% \quad \Rightarrow \quad 4600\%$

1.167 **1.91**

$1.91 \quad \Rightarrow \quad 1.91 \times 100\% \quad \Rightarrow \quad 191\%$

PRACTICE EXAMPLES (Convert into percents)

| **1.168** | 0.706 | **1.170** | 4.11 | **1.172** | 5.45 |
| **1.169** | 0.085 | **1.171** | 9 | | |

SPACE FOR CALCULATIONS

PERCENT → FRACTION

A percent can be converted into a fraction by replacing the % sign with $\left(\times \frac{1}{100}\right)$, or simply dividing by 100. The resulting fraction may be *reduced* if necessary. Convert these into fractions:

1.173 **34%**

$$34 \times \frac{1}{100}$$ Replace the % sign with $\left(\times \frac{1}{100}\right)$.

$$\frac{34}{100}$$ $\left(\times \frac{1}{100}\right)$ really means *divide by* 100.

$$\frac{34}{100} \, \frac{17}{50}$$ Reduce by 2.

1.174 **120%**

$$120 \times \frac{1}{100} \quad \Rightarrow \quad \frac{120}{100} \, \frac{6}{5} \quad \Rightarrow \quad \frac{6}{5} \quad \Rightarrow \quad 1\frac{1}{5}$$ Show as a mixed fraction *if asked for.*

1.175 **12¾%**

$$12\frac{3}{4}\% \quad \Rightarrow \quad 12.75\% \quad \Rightarrow \quad \frac{12.75}{100} \quad \Rightarrow \quad \frac{12.75}{100} \times \frac{100}{100} \quad \Rightarrow \quad \frac{1275}{10,000} \, \frac{51}{400} \quad \Rightarrow \quad \frac{51}{400}$$

Here's an easier way: First convert into an improper fraction. Then replace the % sign.

$$12\frac{3}{4}\% \quad \Rightarrow \quad \frac{4 \times 12 + 3}{4}\% \quad \Rightarrow \quad \frac{51}{4}\% \quad \Rightarrow \quad \frac{51}{4} \times \frac{1}{100} \quad \Rightarrow \quad \frac{51}{400}$$

1.176 **8.4%**

$$8.4 \times \frac{1}{100} \quad \Rightarrow \quad \frac{8.4}{100} \quad \Rightarrow \quad \frac{8.4}{100} \times \frac{10}{10} \quad \Rightarrow \quad \frac{84}{1000} \quad \Rightarrow \quad \frac{84}{1000} \, \frac{21}{250} \quad \Rightarrow \quad \frac{21}{250}$$

1.177 **0.5%**

$$0.5 \times \frac{1}{100} \quad \Rightarrow \quad \frac{0.5}{100} \quad \Rightarrow \quad \frac{0.5}{100} \times \frac{10}{10} \quad \Rightarrow \quad \frac{5}{1000} \quad \Rightarrow \quad \frac{\cancel{5}^1}{\cancel{1000}^{200}} \quad \Rightarrow \quad \frac{1}{200}$$

PRACTICE EXAMPLES (Convert into fractions)

1.178 75%   **1.180** $5\frac{4}{5}\%$ **1.182** 0.61%

1.179 250%   **1.181** 8.7%

SPACE FOR CALCULATIONS

FRACTION → PERCENT

A fraction can be converted into a percent by multiplying by 100, and attaching the % sign. Once again, remember that we're not changing the *value* of the fraction; we're only changing its *form*. Let's try converting the following into percents.

1.183 $\frac{3}{4}$

$\frac{3}{4} \times 100\%$ Multiply by 100. Attach the % sign.

$\frac{3}{4} \times \frac{100}{1}\%$ Remember multiplication of fractions by integers?

$\frac{3 \times \cancel{100}^{25}}{\cancel{4}^1}\%$ Please, always *reduce* first. Deal with smaller numbers.

75%

1.184 $\frac{12}{15}$

$\frac{12}{15} \times 100\% \quad \Rightarrow \quad \frac{12}{15} \times \frac{100}{1}\% \quad \Rightarrow \quad \frac{\cancel{12}^4 \times \cancel{100}^{20}}{\cancel{15}^{3}{}^1}\% \quad \Rightarrow \quad 80\%$

1.185 $\frac{5}{8}$

$\frac{5}{8} \times 100\% \quad \Rightarrow \quad \frac{5 \times \cancel{100}^{25}}{\cancel{8}^2}\% \quad \Rightarrow \quad \frac{125}{2}\% \quad \Rightarrow \quad 62.5\%$

1.186 $2\dfrac{3}{5}$

$$2\dfrac{3}{5} \quad \Rightarrow \quad \dfrac{5 \times 2 + 3}{5} \quad \Rightarrow \quad \dfrac{13}{5} \quad \Rightarrow \quad \dfrac{13}{5} \times 100\% \quad \Rightarrow \quad \dfrac{13 \times \cancel{100}^{20}}{\cancel{5}^{1}}\% \quad \Rightarrow \quad 260\%$$

1.187 $\dfrac{51}{68}$

$$\dfrac{51}{68} \times 100\% \quad \Rightarrow \quad \dfrac{\cancel{51}^{3} \times \cancel{100}^{25}}{\cancel{68}^{4} \,^{1}}\% \quad \Rightarrow \quad 75\% \qquad\qquad \text{Reduce 51 and 68 by 17.}$$

PRACTICE EXAMPLES (Convert into percents)

1.188 $\dfrac{7}{8}$ **1.190** $\dfrac{15}{18}$ **1.192** $\dfrac{39}{65}$

1.189 $\dfrac{19}{25}$ **1.191** $3\tfrac{1}{2}$

SPACE FOR CALCULATIONS

COMMON EQUIVALENT FRACTIONS, DECIMALS AND PERCENTS

Here's a list of some fractions we regularly encounter, and their equivalent decimals and percents. Memorizing them will considerably speed up your calculations. Remember, you only have 1½ minutes per question on the GRE, and 2 minutes per question on the GMAT. Time is of the essence!

FRACTION	DECIMAL	PERCENT
$\frac{1}{4}$	0.25	25%
$\frac{1}{3}$	$0.\bar{3}$	$33.\bar{3}\%$
$\frac{2}{3}$	$0.\bar{6}$	$66.\bar{6}\%$
$\frac{3}{4}$	0.75	75%
$\frac{4}{3}$	$1.\bar{3}$	$133.\bar{3}\%$
$\frac{5}{6}$	$0.8\bar{3}$	$83.\bar{3}\%$

FRACTION	DECIMAL	PERCENT
$\frac{1}{5}$	0.2	20%
$\frac{2}{5}$	0.4	40%
$\frac{3}{5}$	0.6	60%
$\frac{4}{5}$	0.8	80%
$\frac{5}{4}$	1.25	125%
$\frac{6}{5}$	1.2	120%

RATIO AND PROPORTION

A **ratio** is a quotient of two quantities. Does that sound familiar? Yes, it's the same as a fraction. The ratio of x to y is written as $x : y$. It can also be written just like the fraction $\frac{x}{y}$. What makes a ratio more useful than a fraction is that a ratio can be used to compare *any* number of quantities, not just two. For example, let's say that Sam, Pam and Tam inherit \$5M, \$6M and \$7M from their respective parents. The ratio of the amounts inherited by the three of them can be written as $5 : 6 : 7$. Clearly, this could not have been expressed purely in terms of a fraction.

A **proportion** is a statement expressing the equivalence of two ratios. The symbol :: is used to denote the equivalence. For example, $a : b :: c : d$ means the ratio $a : b$ is equal to the ratio $c : d$. The :: symbol is rather outdated. On the GRE/GMAT, you are more likely to see a proportion written as $a : b = c : d$, or something like $p : q : r = 4 : 7 : 9$. Let's imagine a proportion expressed in the form of several numbers, e.g. $w : x : y : z = 3 : 12 : 27 : 9$. It means w represents 3 parts; x, 12 parts; y, 27 parts; and z, 9 parts. So there must be a total of $3 + 12 + 27 + 9$ parts, i.e. 51 parts. The proportion (or share) of w is 3 parts out of 51 parts, or $\frac{3}{51}$; the proportion of x is $\frac{12}{51}$; of y is $\frac{27}{51}$; of z is $\frac{9}{51}$.

A proportion can be *reduced* to its lowest terms by *using the same factor*. In the above example, the proportion $3 : 12 : 27 : 9$ has all of its terms divisible by 3. It can therefore be reduced by 3, and rewritten as $1 : 4 : 9 : 3$. Similarly, a proportion can also be *raised* to higher terms, e.g. $3 : 12 : 27 : 9$ can be *raised* by multiplying all terms by 2, and rewritten as $6 : 24 : 54 : 18$.

1.193   **Express 2½ : 3 as a fraction.**

$\dfrac{2\frac{1}{2}}{3}$ That's a complex fraction. We need to simplify it.

$\dfrac{\frac{2 \times 2 + 1}{2}}{\frac{3}{1}}$ Convert numerator into an improper fraction.

$\dfrac{5}{2} \times \dfrac{1}{3}$ Flip the denominator and multiply.

$\dfrac{5}{6}$

1.194   **Express $2\frac{1}{3} : 5\frac{1}{3}$ as a ratio of two whole numbers.**

$\dfrac{2\frac{1}{3}}{5\frac{1}{3}} \quad \Rightarrow \quad \dfrac{\frac{3 \times 2 + 1}{3}}{\frac{3 \times 5 + 1}{3}} \quad \Rightarrow \quad \dfrac{\frac{7}{3}}{\frac{16}{3}} \quad \Rightarrow \quad \dfrac{7}{3^1} \times \dfrac{3^1}{16} \quad \Rightarrow \quad \dfrac{7}{16}$

1.195   Rewrite $2 : 7\frac{5}{6}$ as a ratio of two integers.

$$\frac{2}{7\frac{5}{6}} \quad \Rightarrow \quad \frac{2}{\frac{6 \times 7 + 5}{6}} \quad \Rightarrow \quad \frac{2}{\frac{47}{6}} \quad \Rightarrow \quad 2 \times \frac{6}{47} \quad \Rightarrow \quad \frac{12}{47}$$

1.196   Ratio of b to c is $8 : 3$. What is the ratio of b to $2c$?

We're given that $b : c = 8 : 3$. Clearly, c represents 3. The question is, in the ratio $b : 2c$, what would $2c$ represent? Obviously, 2×3, or 6. Could you imagine it would be *that* easy?

1.197   What is the ratio of the number of days in January to that in a regular calendar year?

Well, January has 31 days; a year has 365. We're considering a regular year, not a leap year. Therefore, the ratio is $31 : 365$. It can't be *reduced* by any number. So that's the final answer.

1.198   Rewrite the proportion $2 : 3 :: 12 : x$ as an equality between two fractions, and find x.

$$\frac{2}{3} = \frac{12}{x} \quad \Rightarrow \quad 2 \times x = 3 \times 12 \quad \Rightarrow \quad \frac{2^1 \times x}{2^1} = \frac{3 \times 12^6}{2^1} \quad \Rightarrow \quad x = 18$$

Remember the *rule of three*? That's what we used here.

PRACTICE EXAMPLES

1.199   Rewrite the ratio $4\frac{2}{3} : 21$ as a fraction.

1.200   Express $1\frac{2}{3} : 4\frac{5}{6}$ as a ratio of two integers.

1.201   Rewrite $\frac{7}{6} : 14$ as a ratio of two whole numbers.

1.202   Ratio of x to y is $3 : 4$. What is the ratio of $2x : 3y$?

1.203   What is the ratio of the value of a dime to that of a quarter?

1.204   Express $4 : 7 :: c : 28$ as an equality between two fractions, and find c.

SPACE FOR CALCULATIONS

COMBINING PROPORTIONS

Let's look at two different ratios $k : m = 2 : 5$ and $m : n = 1 : 4$. They can both be combined into a single ratio $k : m : n$. To do that, find a common number that would be represented by the common letter (m) in both the proportions. In the first one, m represents 5, and in the second one, m

represents 1. If we try to reduce the first ratio by 5, we will get 1 for m, but then the 2 will become $\frac{2}{5}$ which is not an integer. We need all three terms for k, m, and n to be integers. So let's raise the second ratio by 5 to get $m : n = (1 \times 5) : (4 \times 5)$, or simply $m : n = 5 : 20$. Now the proportion can be written in terms of the numbers that are represented by each letter as $k : m : n = 2 : 5 : 20$.

1.205 **Combine the proportions $e : f = 7 : 4$, and $g : f = 3 : 8$ into a single proportion.**

The letter f is common in both proportions. In the first one, it represents 4, and in the second one, it represents 8. So we need to raise the first proportion by 2 to get $e : f = (7 \times 2) : (4 \times 2)$, or $e : f = 14 : 8$. Now the common letter f represents the same number in both the proportions. They can be combined into $e : f : g = 14 : 8 : 3$.

1.206 $s : t = 3 : 5$, and $t : v = 3 : 5$. Find $s : t : v$.

t represents 5 in the first proportion, and 3 in the second. Simply raising or reducing either ratio won't help. The common number for t would be the LCM of 5 and 3, which is 15. So, we should raise *both* the ratios: the first one by 3, and the second one by 5.

We now have $s : t = (3 \times 3) : (5 \times 3)$, and $t : v = (3 \times 5) : (5 \times 5)$.

That is, $s : t = 9 : 15$, and $t : v = 15 : 25$

Finally, $s : t : v = 9 : 15 : 25$.

1.207 $p : q = 3 : 7$, and $q : r = 1 : 2$. Find $p : q : r$.

q represents 7 in the first proportion, and 1 in the second. We can raise the second proportion by 7, to get $q : r = (1 \times 7) : (2 \times 7)$, that is $q : r = 7 : 14$.

Combining the two proportions, we get $p : q : r = 3 : 7 : 14$.

PRACTICE EXAMPLES

1.208 $a : b = 7 : 2$, and $c : b = 5 : 3$. $a : b : c = ?$ **1.209** $w : x = 2 : 3$, and $x : y = 4 : 9$. $w : x : y = ?$

SPACE FOR CALCULATIONS

AVERAGE

An **average** of a set of elements is the sum of all those elements divided by the number of elements. For example, the average of 5, 11, 7 and 9 is $\left(\frac{5+11+7+9}{4}\right)$ which is equal to 8. The average of p, q and r is $\left(\frac{p+q+r}{3}\right)$. An average is also referred to as the **arithmetic mean**, or just **mean**.

The average of a set of numbers that are *uniformly spaced out* is the *midpoint* of those numbers. For example 2, 5, 8, 11 and 14 are spaced out by 3, i.e. 5 is 3 greater than 2; 8 is 3 greater than 5, and so on. The midpoint is 8, which is the average. There's no need to perform the actual calculations!

So, what is the average of 6, 0, 8, −2, 4 and 2? Should you do the actual math? Maybe not. First, arrange the numbers in ascending (increasing) order: −2, 0, 2, 4, 6, 8. They are uniformly spaced out, but there's no midpoint; 2 and 4 are *both* at the center of the list. Take *their average*, which is 3. This is the mean of the entire list. Cut short your calculations, and save time on the GRE/GMAT!

1.210 **Find the average (arithmetic mean) of 3, 5, 7 and 9.**

The numbers are arranged in increasing order, and are evenly spaced out. 5 and 7 are both at the center of the list. Their average is $\frac{5+7}{2}$ or simply 6. This is the average of 3, 5, 7 and 9.

1.211 **The arithmetic mean of 5, 11, 2 and x is 6. Find x.**

$\frac{5 + 11 + 2 + x}{4} = 6$ Add them up. Divide by 4. That is equal to 6, given.

$\frac{18 + x}{4} = \frac{6}{1}$ $\frac{6}{1}$ is the same as 6. This now looks like $\frac{a}{b} = \frac{c}{d}$

$18 + x = 6 \times 4$ Equate cross products of equivalent fractions.

$18 + x = 24$

$x = 6$

1.212 **What is the average of the first nine consecutive positive integers?**

That would be $\left(\frac{1+2+3+4+5+6+7+8+9}{9}\right)$. Should we add them up and divide by 9? No, because it's a list of *uniformly spaced out numbers*. Their midpoint is 5. That's the average.

1.213 **$p + q = 12$. What is the arithmetic mean of p and q?**

The arithmetic mean of p and q would be $\left(\frac{p+q}{2}\right)$, or $\left(\frac{12}{2}\right)$, or 6.

PRACTICE EXAMPLES

1.214 What is the arithmetic mean of 12, 0, 3, 10 and 5?

1.215 If the mean of 3, y, 19 and 1 is 5, find y.

1.216 Find the arithmetic mean of the positive integers from 1 to 15.

1.217 $m + n = 8$. Find the average of m and n.

EXPONENTS

Consider any number, say 7, that is multiplied by itself 5 times, i.e. $7 \times 7 \times 7 \times 7 \times 7$. The result may be rather large. It can be expressed in exponential notation as 7^5, where 7 is called the **base**, and 5 is called the **exponent**, or **power**, or **index**. In simple English, an exponent is the number of times an element is multiplied by itself. If x is multiplied by itself 12 times, then the result would be x^{12}. The following are the rules when performing calculations involving exponents.

➤ $a^b \times a^c = a^{b+c}$ As an extension of this, $a^b \times a^c \times a^d \times a^e \ldots = a^{b+c+d+e\ldots}$

➤ $\dfrac{a^b}{a^c} = a^{b-c}$

➤ $\left(a^b\right)^c = a^{bc} = \left(a^c\right)^b$ As an extension of this, $\left(\left(\left(\left(a^b\right)^c\right)^d\right)^e\right)^{\ldots} = a^{bcde\ldots}$

➤ $a^{-b} = \dfrac{1}{a^b}$ This rule helps rewrite negative exponents in positive terms.

➤ $a^b \times c^b = (ac)^b$ As an extension of this, $a^b \times c^b \times d^b \times e^b \ldots = (acde \ldots)^b$

➤ $\dfrac{a^b}{c^b} = \left(\dfrac{a}{c}\right)^b$

➤ $a^0 = 1$ where $a \neq 0$, because 0^0 is undefined.

PROOF OF $a^0 = 1$: $\dfrac{a^x}{a^x} = a^{x-x} = a^0$ But $\dfrac{a^x}{a^x} = \dfrac{a^{\cancel{x}}}{a^{\cancel{x}}} = 1$ Therefore, $a^0 = 1$

It is generally a good practice to express the final result in terms of positive exponents.

1.218 $\dfrac{3^4 \times 3^5 \times 7^{11}}{7^{20}}$

     $3^{4+5} \times 7^{11-20}$ Because $a^b \times a^c = a^{b+c}$ and $\dfrac{a^b}{a^c} = a^{b-c}$

     $3^9 \times 7^{-9}$ or $\dfrac{3^9}{7^9}$ or $\left(\dfrac{3}{7}\right)^9$ Because $a^{-b} = \dfrac{1}{a^b}$ and $\dfrac{a^b}{c^b} = \left(\dfrac{a}{c}\right)^b$

1.219 $\dfrac{-2^4 \times 2^{-4}}{(-2)^4}$

     $\dfrac{-(2 \times 2 \times 2 \times 2)}{(-2) \times (-2) \times (-2) \times (-2)} \times \dfrac{1}{2^4}$ Because $a^{-b} = \dfrac{1}{a^b}$ $\left(\text{i.e. } 2^{-4} = \dfrac{1}{2^4}\right)$

     $-1 \times \dfrac{1}{16}$ or $-\dfrac{1}{16}$

1.220 $\left(4^2\right)^5$

     $4^{2 \times 5}$ Because $\left(a^b\right)^c = a^{bc}$

     4^{10}

1.221 $\dfrac{\left(5^2\right)^3}{25^4}$

Notice that 25 is 5 to the 2nd power, i.e. $25 = 5^2$. When faced with examples involving different bases being powers of one another, express all bases in terms of the *smallest base*.

$\dfrac{(5^2)^3}{(5^2)^4}$

$(5^2)^{3-4}$ Because $\dfrac{a^b}{a^c} = a^{b-c}$

5^{-2} or $\dfrac{1}{5^2}$ or $\dfrac{1}{25}$ Because $a^{-b} = \dfrac{1}{a^b}$

1.222 $\dfrac{27 \times 3^8}{\left(9^2\right)^3}$

$\dfrac{3^3 \times 3^8}{((3^2)^2)^3}$ Express all bases in terms of 3 (the *smallest base*).

$\dfrac{3^{11}}{3^{12}}$ Because $a^b \times a^c = a^{b+c}$ and $\left(\left(a^b\right)^c\right)^d = a^{bcd}$

3^{-1} or $\dfrac{1}{3}$ Because $a^{-b} = \dfrac{1}{a^b}$

1.223 $p^4 = 5$. **Find** p^8.

$p^8 = (p^4)^2$ Because $a^{bc} = \left(a^b\right)^c$ Think backwards!

$p^8 = (5)^2$ Because we are given that $p^4 = 5$

$p^8 = 25$

1.224 $27^x = 9^2$. **Find** x.

$(3^3)^x = (3^2)^2$ Express all bases in terms of 3 (the smallest base).

$3^{3x} = 3^4$ Because $\left(a^b\right)^c = a^{bc}$

$3x = 4$ Same bases. Therefore, the exponents *must be equal*.

$x = \dfrac{4}{3}$

1.225 $16 \times 2^8 \times 4 = 4^a$. **Find** a.

$2^4 \times 2^8 \times 2^2 = (2^2)^a$ Express all bases in terms of 2 (the smallest base).

$2^{4+8+2} = 2^{2a}$ Because $a^b \times a^c \times a^d = a^{b+c+d}$, and $\left(a^b\right)^c = a^{bc}$

$2^{14} = 2^{2a}$

$14 = 2a$,   or $7 = a$ Same bases. Therefore equal exponents.

PRACTICE EXAMPLES

1.226 $\dfrac{-5^2 \times 5^{-4}}{5^7}$

1.227 $((2^3)^4)^5$

1.228 $\dfrac{64^{13}}{(4^3)^{12}}$

1.229 $\dfrac{8^{-12} \times 2^{40}}{(4^2)^2}$

1.230 $x^5 = 3$. Find x^{15}.

1.231 $25^m = 125$. Find m.

1.232 $9 \times 3^{11} \times 27 = 3^b$. Find b.

SPACE FOR CALCULATIONS

LCM OF THREE OR MORE INTEGERS

Knowledge of exponents can help us determine the LCM of more than two integers. Here's how:

1) First, prime factorize each integer.
2) From the prime factorizations, select all *distinct primes* having the greatest exponents.
3) Multiply the selected prime numbers having the greatest exponents together. That's the LCM.

The following examples should make the method clear.

1.233 **Find the LCM of 24, 45 and 20.**

Prime factorization of 24:	$2 \times 2 \times 2 \times 3$	or	$2^3 \times 3$
Prime factorization of 45:	$3 \times 3 \times 5$	or	$3^2 \times 5$
Prime factorization of 20:	$2 \times 2 \times 5$	or	$2^2 \times 5$

The unique primes are 2, 3 and 5. The greatest power of 2 is 2^3; of 3 is 3^2; and of 5 is just **5**. Multiply them together.

$2^3 \times 3^2 \times 5$ or $8 \times 9 \times 5$ or   360   This is the LCM of 24, 45 and 20.

1.234 **Find the LCM of 25, 8, 12 and 22.**

Prime factorization of 25:	5×5	or	5^2
Prime factorization of 8:	$2 \times 2 \times 2$	or	2^3
Prime factorization of 12:	$2 \times 2 \times 3$	or	$2^2 \times 3$
Prime factorization of 22:	2×11		

The unique primes are 5, 2, 3 and 11. The greatest power of 5 is 5^2; of 2 is 2^3; of 3 is **3**; and of 11 is **11**. Multiply them together to get the LCM.

$5^2 \times 2^3 \times 3 \times 11$ or $25 \times 8 \times 3 \times 11$ or   6600

PRACTICE EXAMPLES

| 1.235 | Find the LCM of 6, 7, 8 and 18. | | 1.236 | Find the LCM of 12, 18, 24, 30 and 36. |

SPACE FOR CALCULATIONS

RADICALS

In its simplest form, a **radical** (also referred to as a **radical expression**) is a root. For example, $\sqrt[3]{7}$ is 3rd root (or cube root) of 7; $\sqrt[5]{11}$ is 5th root of 11; $\sqrt{16}$ is 2nd root (or square root) of 16 etc. If the index is absent in the radical sign ($\sqrt{\ }$), then it is assumed to be a square root. That's the standardized notation. Generally speaking, a radical of the form $\sqrt[n]{x}$ is the nth root of x. Radicals are essentially exponents, with the exponent being a fraction, not an integer. Mathematically, $\sqrt[3]{7} = 7^{\frac{1}{3}}$, $\sqrt[5]{11} = 11^{\frac{1}{5}}$, $\sqrt{16} = 16^{\frac{1}{2}}$, and so on. Generally, $\sqrt[n]{x} = x^{\frac{1}{n}}$.

$\sqrt[n]{x}$ would be the number which, when multiplied by itself n times, yields x. So, what would be $\sqrt{16}$ (or $\sqrt[2]{16}$) ? Could it be 4? Yes, because 4 when multiplied by itself gives 16. What about -4? It is too, because $(-4) \times (-4)$ is also 16. Therefore, *a square always has a positive and a negative root.* What is $\sqrt[3]{8}$? It is 2, because $2 \times 2 \times 2 = 8$. There's no negative root here. What is $\sqrt[3]{-7}$? We can't tell it without using a calculator (which is NOT allowed on the GMAT), but we do know that it is negative, because only $(-) \times (-) \times (-) = (-)$. *The odd root of any negative number is always a negative number.* Also, $\sqrt{5} \times \sqrt{5} = 5$; $\sqrt[3]{7} \times \sqrt[3]{7} \times \sqrt[3]{7} = 7$; and so on.

Radical expressions obey all the operating rules of exponents that we've studied. Here are some of them (and some others) in radical notation, which you will encounter quite commonly.

➢ $\sqrt[n]{a} = a^{\frac{1}{n}}$

    As an extension, $\sqrt[n]{\sqrt[m]{a}} = \left(a^{\frac{1}{m}}\right)^{\frac{1}{n}} = a^{\left(\frac{1}{m}\right)\cdot\left(\frac{1}{n}\right)} = a^{\frac{1}{mn}} = \sqrt[mn]{a}$

➢ $\sqrt{a \times b} = \sqrt{a} \times \sqrt{b}$

    As an extension, $\sqrt{a \times b \times c} = \sqrt{a} \times \sqrt{b} \times \sqrt{c}$

➢ $\sqrt{\dfrac{a}{b}} = \dfrac{\sqrt{a}}{\sqrt{b}}$

➢ $\sqrt{a \pm b} \neq \sqrt{a} \pm \sqrt{b}$

➢ $x\sqrt{a} \pm y\sqrt{a} = (x \pm y)\sqrt{a}$

    Please note that $x\sqrt{a}$ means $x \times \sqrt{a}$, not $\sqrt[x]{a}$.

Quite often, you will come across integers under the square root sign, that do not have square roots, e.g. $\sqrt{108}$. *Most of such radicals can be rewritten in terms of smaller radicands.* Recall that a *radicand* is the number underneath the root sign. Simply follow these steps:

1) Prime factorize the radicand: $\sqrt{108}$ $\sqrt{2 \times 2 \times 3 \times 3 \times 3}$
2) Pair-up the repeating primes *with their own kind*: $\sqrt{(2 \times 2) \times (3 \times 3) \times 3}$
3) From each pair, bring *one* prime out of the radical sign: $2 \times 3 \times \sqrt{3}$
4) Multiply out the terms, and rewrite if necessary: $6\sqrt{3}$

1.237 $\sqrt{98} + 2\sqrt{2} - \sqrt{72}$

$\sqrt{2 \times (7 \times 7)} + 2\sqrt{2} - \sqrt{2 \times (2 \times 2) \times (3 \times 3)}$ Prime factorize the radicands. Pair-up.

$7\sqrt{2} + 2\sqrt{2} - 2 \times 3 \times \sqrt{2}$ Bring out *one* of each repeating prime.

$3\sqrt{2}$ Because $x\sqrt{a} \pm y\sqrt{a} = (x \pm y)\sqrt{a}$

1.238 $\sqrt{\dfrac{5}{3}} \times \dfrac{\sqrt{72}}{2\sqrt{45}}$

$\dfrac{\sqrt{5}}{\sqrt{3}} \times \dfrac{\sqrt{(3\times3)\times(2\times2)\times2}}{2\sqrt{(3\times3)\times5}}$ Prime factorize. Pair-up. Also, use $\sqrt{\dfrac{a}{b}} = \dfrac{\sqrt{a}}{\sqrt{b}}$

$\dfrac{\sqrt{5}}{\sqrt{3}} \times \dfrac{3\times2\times\sqrt{2}}{2\times3\times\sqrt{5}}$ Bring out *one* of each repeating prime.

$\dfrac{\cancel{\sqrt{5}}}{\sqrt{3}} \times \dfrac{\cancel{6}\sqrt{2}}{\cancel{6}\cancel{\sqrt{5}}}$ Reduce the terms. Remember fractions?

$\dfrac{\sqrt{2}}{\sqrt{3}}$ or $\sqrt{\dfrac{2}{3}}$

PRACTICE EXAMPLES

1.239 $\sqrt{5} - 4\sqrt{45} + 3\sqrt{20}$

1.240 $\dfrac{3\sqrt{10}}{\sqrt{28}} \times \dfrac{\sqrt{2}}{\sqrt{15}}$

ORDER OF OPERATIONS

When performing calculations involving several types of operations as well as parentheses and exponents together, how to decide which ones are to be done first? The answer lies in the acronym *PEMDAS*: *P*arenthesis (or brackets), *E*xponents (also radicals), *M*ultiplication, *D*ivision, *A*ddition, *S*ubtraction. That's the order in which calculations should be performed.

The two exceptions to *PEMDAS* are: In case of operations involving only multiplication and division, execute the operations from left to right. Same with operations involving only addition and subtraction.

1.241 $3 + 16 \div (9 - 5)^2$

According to *PEMDAS*, first parentheses, next exponents, then division, and finally addition.

$3 + 16 \div 4^2$	Parentheses
$3 + 16 \div 16$	Exponents
$3 + 1$	Division
4	Addition

1.242 $3^3 - 2\sqrt{(25 - 21)^3}$

$3^3 - 2\sqrt{4^3}$	Parentheses
$27 - 2\sqrt{64}$	Exponents
$27 - 2 \times 8$	Radicals
$27 - 16$	Multiplication
11	Subtraction

1.243 $8 \div 4 \times 5 \div 2 \div 5 \times 3$

3	Multiplication and division only. Hence, go straight from left to right.

1.244 $10 + 6 - 5 - 5 - 5 + 8$

9	Addition and subtraction only. Go from left to right uninterrupted.

PRACTICE EXAMPLES

1.245 $8 - \dfrac{(3+4)^2}{7}$

1.246 $12 \div 24 \times 4 \div 5 \times 10$

1.247 $\dfrac{8 - 3\sqrt{10^2 - 96}}{4}$

1.248 $-6 + 11 - 4 + 7 + 2 + 2$

SPACE FOR CALCULATIONS

ABSOLUTE VALUES

An absolute value (or modulus) of a number is, as the name suggests, absolutely the value of the number without regard to its sign. Therefore, modulus is always positive or zero, but never negative. Absolute value of a number is denoted by placing a vertical bar before as well as after the number. For example, modulus of -4.19 is written as $|-4.19|$, and is equal to 4.19. Modulus of 25 is written as $|25|$ and is equal to 25.

1.249 $|-13.24|$

13.24   Take the value, discard the sign.

1.250 $|4.187|$

4.187   The number itself.

1.251 $|-2| - |-6|$

$2 - 6$ or -4 Take individual absolute values. Then subtract.

1.252 $\dfrac{-|4| \times 3}{|-12|}$

$\dfrac{-4 \times 3}{12}$ The negative sign was outside the modulus sign. So it stays.

$\dfrac{-12}{12}$ or -1

1.253 $\dfrac{|-25|}{(-5) \times |-5|}$

$\dfrac{25}{-25}$ Please don't confuse between parentheses and modulus sign.

-1

PRACTICE EXAMPLES

1.254 $|-3.78|$

1.255 $|14| - |-23|$

1.256 $|-11| + |-1|$

1.257 $\dfrac{-|-37| + |-20|}{|-12| + |-22|}$

1.258 $\dfrac{|-8| \times |-3|}{(-6) \times (-2)}$

PROPERTIES OF ZERO

From a mathematical reasoning point of view, the important properties of zero are:

➤ Zero is neither positive nor negative.
➤ Zero is neither prime nor composite.
➤ Zero is an even integer.
➤ $\frac{x}{0} = \infty$, where $x \neq 0$. It's fair to say that $\frac{x}{0}$ is undefined, because ∞ has an undefined value.
➤ $\frac{0}{0}$ is a meaningless expression. It has no value at all. It is NOT infinity (∞).
➤ $0^x = 0$, where $x \neq 0$.
➤ $x^0 = 1$, where $x \neq 0$.
➤ 0^0 is a meaningless expression. No value. NOT infinity.

PROPERTIES OF ONE

➤ One is neither prime nor composite.
➤ One is an odd integer.
➤ $1^x = 1$, where x is any number.
➤ $x^1 = x$, where x is any number.

PROPERTIES OF NUMBERS BETWEEN ZERO AND ONE

All numbers x between zero and one, i.e. $0 < x < 1$, are proper fractions or decimal fractions.

➤ $\frac{1}{x} > x$ because $\frac{1}{x}$ would be an improper fraction (flipped!), which is greater than 1.
➤ $+\sqrt{x} > x$ and $-\sqrt{x} < x$ (positive root is greater, negative root is smaller)
➤ $\sqrt[n]{x} > x$ where n is a positive integer greater than one (positive root only)
➤ $x^n < x$ where n is a positive integer (i.e. fraction $\times$ fraction $\times$ ... ... = even smaller fraction)
➤ $x^{-n} > x$, i.e. $\frac{1}{x^n} > x$ where n is a positive integer
➤ $nx < n$ where n is a positive integer
➤ $nx > n$ where n is a negative integer

Please note that between $\frac{1}{x}$, $+\sqrt{x}$, $\sqrt[n]{x}$ and x^{-n}, only $\frac{1}{x}$ is greater than one. All others, while being greater than x, are still less than one.

1.259 $x = 0.62$. **Which of the following is (are) greater than x?** $+\sqrt{x}$, x^2, $\frac{1}{x}$, $-6x$

From the properties, we can conclude that $+\sqrt{x}$ is greater than x, x^2 is less than x, $\frac{1}{x}$ is greater than x, and $-6x$ is less than x. We need not perform any calculations with 0.62.

1.260 $p = 36\%$. Arrange the following in descending (decreasing) order: $5p$, p^3, $+\sqrt{p}$, p, $\dfrac{1}{p}$

$\dfrac{1}{p}$ and $5p$ are both greater than one. Let's find out the greater of them.

$$5p = 5 \times 36\% = \cancel{5}^1 \times \frac{36}{\cancel{100}^{20}} = 1.8$$

$$\frac{1}{p} = \frac{\cancel{100}^{25}}{\cancel{36}^{9}} = \frac{25}{9} = 2.\overline{7}$$

$+\sqrt{p}$ is greater than p, but less than one, whereas p^3 is less than p.

Therefore, the descending order is: $\dfrac{1}{p}$, $5p$, $+\sqrt{p}$, p, p^3

PRACTICE EXAMPLES

1.261 $w = \dfrac{47}{74}$. Which of the following is (are) less than w? $-3w$, $+\sqrt{w}$, w^3, $2w$

1.262 $z = \dfrac{2}{9}$. Arrange the following in ascending (increasing) order: $\dfrac{1}{z}$, z, z^4

You won't need any *space for calculations* here; there aren't any calculations to be done.

SCIENTIFIC NOTATION

When calculations result in very large or very small numbers, they can be expressed in the form $\pm N.NNN \times 10^{\pm NN}$, where N simply represents a digit. This format involves expressing the number as a multiple of a power of 10. The number can have *only one digit* before the decimal point, and any number of digits after the decimal point. For example, 5.143×10^6 is in scientific notation, whereas 21.754×10^{-33} is not. Let's express the following examples in scientific notation.

1.263 0.317

3.17×10^{-1}

1.264 $-1,475,000$

-1.475×10^6

1.265 86×10^{-12}

8.6×10^{-11}

1.266 $\dfrac{59}{10,000}$

5.9×10^{-3}

PRACTICE EXAMPLES

1.267 0.882

1.268 9,469,000,000

1.269 -4×200^2

1.270 $\dfrac{60,000}{300}$

SPACE FOR CALCULATIONS

CHAPTER 2

ALGEBRA

THE ORIGIN OF VARIABLES

Thousands of years ago, when the numbering system was first formulated by a civilization in remote antiquity, they must have assigned each number to represent a quantity. For example, the number 5 back then must have been used to denote the number of fingers on the human hand. Today, the number 5 still denotes the same thing. If my guess is correct, thousands of years from now, it will still denote the same quantity! That holds true for the number 8, and 1, and 476, and any other number you may imagine. Meaning, numbers do not change; they have *constant* values. Therefore, in the language of algebra, they are referred to as **constants**.

In the last chapter, we used some letters to represent certain quantities. But from example to example, each letter (for instance x) had a different value. It did not have any permanent value; it *varied* depending upon the context or the application. Such entities are called **variables**. Arithmetic was mostly about constants (numbers), whereas algebra is mostly about variables ($x, y, p, c, n...$).

The laws of addition, subtraction, multiplication, division, exponents, reductions etc. pertaining to numbers can be applied to variables too. After all, a variable only represents an *unknown* number.

ALGEBRAIC NOTATIONS

Consider two variables x and y. The following are the notations for various algebraic operations.

Addition: $\qquad\qquad x + y$

Subtraction: $\qquad\quad x - y$

Multiplication: $\qquad x \times y \quad$ or $\quad x \cdot y \quad$ or $\quad xy \quad$ or $\quad x(y) \quad$ or $\quad (x)y \quad$ or $\quad (x)(y)$

Division: $\qquad\qquad \dfrac{x}{y} \quad$ or $\quad {}^{x}/_{y} \quad$ or $\quad x \div y \quad$ or $\quad y\overline{)x}$

POLYNOMIALS

When an algebraic term includes a constant multiplied by a variable, it is called a **monomial**. For example, in the term $3x^2$, 3 is the constant, multiplied by x^2, the variable. The constant in a monomial is also referred to as the **coefficient**. When only a variable exists by itself, it is still a monomial. For example, d can be rewritten as $1d$, where 1 is the constant. When two monomials are added to, or subtracted from, each other, they're collectively called a **binomial**. Examples would be $7a + 4b$, $c - 5.17k$, etc. Three monomials form a **trinomial**, and so on. The general term for two or more monomials added or subtracted is **polynomial**. Hence, one may say that a binomial, a trinomial, a quadrinomial etc. are all polynomials.

Two or more monomials when multiplied by one another, do not form a polynomial. They would still form a monomial. For example, $(4a)(2c)(x)$ yields $8acx$, which is a monomial, whereas $4a + 2c + x$ is a trinomial (a polynomial).

Using the values $a = 3$, $t = -4$ and $y = 5$, let's evaluate the following algebraic expressions.

2.001 $a^2t - t^2y$

$3^2(-4) - (-4)^2(5)$ Simply plug in (substitute) the values for a, t and y.

$9(-4) - 16(5)$

$-36 - 80$

-116

2.002 $ay^2 - aty$

$3(5)^2 - 3(-4)(5)$ $\Rightarrow$ $75 + 60$ $\Rightarrow$ 135

2.003 $(t + a)(t - y)$

$(-4 + 3)(-4 - 5)$ $\Rightarrow$ $(-1)(-9)$ $\Rightarrow$ 9

2.004 $\sqrt{\dfrac{-ty^2}{t^2 + ay + 50}}$

$\sqrt{\dfrac{-(-4)(5)^2}{(-4)^2 + (3)(5) + 50}}$ $\Rightarrow$ $\sqrt{\dfrac{4 \times 25}{16 + 15 + 50}}$ $\Rightarrow$ $\sqrt{\dfrac{100}{81}}$ $\Rightarrow$ $\pm\dfrac{10}{9}$

In example 2.004 above here, please note that the square root of a number can be positive or negative. For example, the square root of 25 can be 5 or -5, because $5 \times 5 = 25$ and $-5 \times -5 = 25$ as well. In other words, $\sqrt{25} = \pm5$. We need to consider both the values. Hence, the $\pm$ sign.

PRACTICE EXAMPLES (Evaluate for $a = 3$, $t = -4$ and $y = 5$)

2.005 $\dfrac{20 + y - t}{a^2 - ty}$

2.006 $\dfrac{a + t}{a - t - y}$

2.007 The average (arithmetic mean) of a, t and y.

2.008 $\dfrac{1}{\sqrt{4ay - t^3}}$

SPACE FOR CALCULATIONS

ADDITION/SUBTRACTION OF POLYNOMIALS

As we know, a polynomial comprises two or more monomials being added or subtracted. The monomials that have the same *variable part* are called **like terms**. For example, $24c^2by$, $-7bc^2y$ and $4\frac{1}{2}byc^2$ have the same variable part. The order of c^2, b and y in the monomials doesn't matter because they are all multiplied together. On the other hand, the monomials having different *variable parts* are called **unlike terms**.

When adding monomials, group the like terms together. Then simply add their coefficients (or constant parts) while keeping the variable part as it is. In the above example, adding the three coefficients, we get $24 - 7 + 4\frac{1}{2} = 21\frac{1}{2}$. You can choose to write the variable part in any way, say byc^2. Therefore, the sum of the above three like terms is $21\frac{1}{2}byc^2$.

Also, note that a monomial $\frac{12xy}{z}$ would be the same as the monomial $12\frac{xy}{z}$. When we multiply $\frac{xy}{z}$ by 12, we're essentially multiplying it by $\frac{12}{1}$. Here's what happens:

$$12 \times \frac{xy}{z} \Rightarrow \frac{12}{1} \times \frac{xy}{z} \Rightarrow \frac{12 \times xy}{1 \times z} \Rightarrow \frac{12xy}{z}.$$

2.009 $3xy - \dfrac{12zc}{a} + 4xy + \dfrac{15cz}{a}$

$(3xy + 4xy) + \left(-\dfrac{12zc}{a} + \dfrac{15cz}{a}\right)$ Group the *like terms*.

$7xy + \dfrac{3zc}{a}$ Add the coefficients of the like terms. Rewrite.

2.010 $8pxy + 2ab + xpy - 12ab - 3ypx$

$(8pxy + xpy - 3ypx) + (2ab - 12ab)$ Group the like terms.

$6pxy - 10ab$ Add the coefficients. Rewrite.

2.011 $-7ab - \dfrac{12ab}{y} + (-ab) - \left(-\dfrac{20ba}{y}\right)$

$-7ab - ab - \dfrac{12ab}{y} + \dfrac{20ab}{y} \Rightarrow -8ab + \dfrac{8ab}{y}$

2.012 $4\dfrac{c^2x^4}{y^3} + 6\dfrac{a}{y^2} - \dfrac{8c^2x^4}{y^3} - \dfrac{2a}{y^2}$

$4\dfrac{c^2x^4}{y^3} - \dfrac{8c^2x^4}{y^3} + 6\dfrac{a}{y^2} - \dfrac{2a}{y^2} \Rightarrow -4\dfrac{c^2x^4}{y^3} + 4\dfrac{a}{y^2}$

2.013 **Find the average of** $3x + 8y - 7$, $4y - 4x + 5$, **and** $7x - 12y + 2$.

We know that the average of the above three polynomials would be their sum divided by 3.

$(3x - 4x + 7x) + (8y + 4y - 12y) + (-7 + 5 + 2)$ Group the like terms when adding.

$6x + 0 + 0$

$6x$

Now divide $6x$ by 3, which gives $2x$, which is the average.

PRACTICE EXAMPLES

$\boxed{2.014}$ $3mp + 12ma - 4mp + 13ma$

$\boxed{2.015}$ $bdy + \dfrac{caz}{7} - 12byd + \dfrac{cza}{21}$

$\boxed{2.016}$ $\dfrac{p^2(4x)}{n^3} + \dfrac{11ac}{b} + \dfrac{5xp^2}{n^3} - \dfrac{7ac}{b}$

$\boxed{2.017}$ $3ax + x^2y^2 - 4ax + 2y^2x^2 + 2xa - 3x^2y^2$

$\boxed{2.018}$ If the sum of $9a - 6b + c$, $5b - 4a - 3c$, and $2b - 5a + 2c$ is 8, find $b^2 + 3$.

SPACE FOR CALCULATIONS

MULTIPLICATION OF POLYNOMIALS

When multiplying two or more *monomials*, simply multiply their coefficients separately, and their variables separately. Then multiply them together in the form of one monomial. For example, $2x^5(11c)(3x^2c^3)$ $\Rightarrow$ $(2 \times 11 \times 3)(x^5 \times c \times x^2 \times c^3)$ $\Rightarrow$ $66x^{5+2}c^{1+3}$ $\Rightarrow$ $66x^7c^4$. Notice that we used the laws of exponents (from page 44) to multiply the variables.

When multiplying a *monomial* with a *polynomial*, or two polynomials with each other, the following are the rules that need to be adhered to.

$a(b + c) = ab + ac$ In general, $a(b \pm c \pm d \pm \cdots) = ab \pm ac \pm ad \pm \cdots$

Multiply the first term (a) with each of the terms in the parentheses (b, c, d, e etc.), and add or subtract them in the sequence in which they occur.

$ab + ac = a(b + c)$ $ab \pm ac \pm ad \pm \cdots = a(b \pm c \pm d \pm \cdots)$

Extract (*factor out*) the common element (a) from each of the terms (ab, ac, ad, ae etc.); quite the opposite of the above rule.

$(a + b)(x + y) = ax + ay + bx + by$ $(a + b)(x + y + z \dots) = (ax + ay \dots) + (bx + by \dots)$

Multiply the first term (a) from the first set of brackets with each term from the second set of brackets (x, y, z etc.). Then proceed to multiply the second term (b) from the first set of brackets with each term from the second set of brackets. If there were more terms in the first set of parenthesis, then you would proceed in a similar manner to the next term.

Before proceeding beyond this point, please review the laws of exponents from page 44. You will be using them very frequently.

$\boxed{2.019}$ $(7x^2)(-x^4)$

$-7x^{2+4}$ Multiply the coefficients. For variables, use $a^b \times a^c = a^{b+c}$

$-7x^6$

2.020 $(-2ac^2)(c^2xa^3)(-27x^3)$

$54a^{1+3}c^{2+2}x^{1+3}$ Final sign is positive, because $(-) \times (-) = (+)$

$54a^4c^4x^4$

$54(acx)^4$ Because $x^a y^a z^a = (xyz)^a$. But this step is optional.

2.021 $-4mn(m^2 + 2nm)$

$-4mn \cdot m^2 - 4mn \cdot 2nm$ Because $a(b + c) = ab + ac$. Be mindful of the $(-)$ sign.

$-4m^{1+2}n - 8m^{1+1}n^{1+1}$

$-4m^3n - 8m^2n^2$

2.022 $(a + bx)(3 - cy)$

$a \cdot 3 - a \cdot cy + bx \cdot 3 - bx \cdot cy$ Be mindful of the $(-)$ sign.

$3a - acy + 3bx - bcxy$ It's okay to rearrange the variables within a monomial.

2.023 $(8 - 5c)(c^2 + c - 8)$

$8c^2 + 8c - 64 - 5c^3 - 5c^2 + 40c$ On similar lines as the previous example.

$3c^2 + 48c - 64 - 5c^3$ Add the *like terms*.

$-5c^3 + 3c^2 + 48c - 64$ Rearrange exponents in decreasing order.

The final step in the above example is optional. But it is generally a good practice to arrange the terms in decreasing order of their exponents (c^3, c^2, c^1, c^0). Keep in mind that $c^0=1$, i.e. the number 64 in the above example can be rewritten as $64c^0$. Therefore, it is in the last place.

PRACTICE EXAMPLES

2.024 $(lm^2n)(-3n^3m^2)$

2.025 $(-p^2q)(3aq)(-2pa^2)(-q)$

2.026 $(ic - cu)(4uc)$

2.027 $(f + yi)(8 - ay)$

2.028 $(a + b - x)(a^2 + x)$

SPACE FOR CALCULATIONS

DIVISION OF POLYNOMIALS

Division of a *monomial* by another *monomial* is done on similar lines as multiplication. The constants may be reduced by a common factor, and the variables may be reduced using the laws of exponents.

For example, $\dfrac{35x^2b^5}{14b^3x^3} \Rightarrow \dfrac{35^5x^{2-3}b^{5-3}}{14^2} \Rightarrow \dfrac{5x^{-1}b^2}{2} \Rightarrow \dfrac{5b^2}{2x}$. The 35 and 14 get reduced by 7. Also, it is generally a good practice to express the exponents as positive numbers. Therefore, you may notice that I rewrote the x^{-1} from the numerator as just x in the denominator, in the final step.

On similar lines, when a *polynomial* is divided by a *monomial*, the above method can be used again, except that the terms in the polynomial may each be separately divided by the denominator. For example, $\dfrac{2p^3r+r^2s^2}{2p^2rs} \Rightarrow \dfrac{2p^3r}{2p^2rs} + \dfrac{r^2s^2}{2p^2rs} \Rightarrow \dfrac{p}{s} + \dfrac{rs}{2p^2}$

2.029 $\dfrac{-12ab^2}{8ab}$

$\dfrac{-12^3ab^2}{8^2ab}$ Reduce 12 and 8 by 4. Cancel the a's. Reduce b^2 with b.

$\dfrac{-3b}{2}$

2.030 $\dfrac{25x^2pc^3}{-5x^3c^4}$

$\dfrac{25^5x^2pc^3}{-5^1x^3c^4} \Rightarrow -\dfrac{5p}{xc}$

2.031 $\dfrac{-3m^2n^2x^4}{-9m^2n^2x^3}$

$\dfrac{-3^1m^2n^2x^4}{-9^3m^2n^2x^3} \Rightarrow \dfrac{x}{3}$

2.032 $\dfrac{2x^2y^4+3y^3zx}{x^2y^3}$

$\dfrac{2x^2y^4}{x^2y^3}+\dfrac{3y^3zx}{x^2y^3} \Rightarrow 2y+\dfrac{3z}{x}$

2.033 $\dfrac{6a^2cy+9y^2ca-3c^2ay}{-3acy}$

$\dfrac{6^2a^2cy}{-3^1acy}+\dfrac{9^3y^2ca}{-3^1acy}-\dfrac{3c^2ay}{-3acy} \Rightarrow -2a-3y+c$

PRACTICE EXAMPLES

2.034 $\dfrac{-7p^2q^3}{p^3q^2}$

2.035 $\dfrac{6w^2xy^3}{-29x^2wy^3}$

2.036 $\dfrac{-4b^{12}xy^7}{-12x^3b^{11}}$

2.037 $\dfrac{k^3b^4-c^3k^2b^3}{-k^2b^3}$

2.038 $\dfrac{mx^2n+3xn^2m-8m^2xn}{-2m^2n^2x^2}$

MONOMIALS RAISED TO A POWER

When a monomial is raised to a power, simply multiply the power with each of the exponents within the monomial. Also, please be alert about whether the final sign will be positive or negative. For

example, $\dfrac{-(a^2x^5)^2}{(-a^3x)^4} \Rightarrow \dfrac{-a^{2\times2}x^{5\times2}}{a^{(3\times4)}x^4} \Rightarrow \dfrac{-a^4x^{10}}{a^{12}x^4} \Rightarrow -a^{4-12}x^{10-4} \Rightarrow -a^{-8}x^6 \Rightarrow -\dfrac{x^6}{a^8}$

Notice that the $(-)$ sign in the numerator was outside the brackets. So, it was unaffected by the squaring up. But the $(-)$ sign in the denominator disappeared because it was raised to an *even* power. Also, let's maintain the practice of expressing the final result in terms of positive exponents.

2.039 $(-a^2cb^5)^3$

$\quad -a^{2\times3}c^3b^{5\times3}$ The $(-)$ sign stays because it's multiplied by itself *odd* number of times.

$\quad -a^6c^3b^{15}$

2.040 $\left(\dfrac{3c^2y^{11}z^4}{4c^3y^{10}z^4}\right)^{-2}$

$\quad \left(\dfrac{4c^3y^{10}z^4}{3c^2y^{11}z^4}\right)^{2}$ Because $a^{-b} = \frac{1}{a^b}$.   Flip the numerator and denominator.

$\quad \left(\dfrac{4c^{3-2}y^{10-11}}{3}\right)^{2}$ Rewrite in terms of exponents, or simply reduce. It's your choice.

$\quad \left(\dfrac{4cy^{-1}}{3}\right)^{2}$

$\quad \left(\dfrac{4c}{3y}\right)^{2}$

$\quad \dfrac{16c^2}{9y^2}$ Because $\left(\dfrac{ab}{cd}\right)^2 = \dfrac{a^2b^2}{c^2d^2}$

2.041 $\left(\dfrac{xyz}{a^3bc^2}\right)^{14} \div \left(\dfrac{y^4z^5}{c^9a^{14}}\right)^{3}$

$\quad \left(\dfrac{xyz}{a^3bc^2}\right)^{14} \times \left(\dfrac{c^9a^{14}}{y^4z^5}\right)^{3}$ Because $\dfrac{a^{14}}{b^{14}} \div \dfrac{c^3}{d^3} = \dfrac{a^{14}}{b^{14}} \times \dfrac{d^3}{c^3}$.   Recall fractions?

$$\frac{x^{14}y^{14}z^{14}}{a^{42}b^{14}c^{28}} \times \frac{c^{27}a^{42}}{y^{12}z^{15}} \Rightarrow \frac{x^{14}y^{14}z^{14}c^{27}a^{42}}{a^{42}b^{14}c^{28}y^{12}z^{15}} \Rightarrow \frac{x^{14}y^{14-12}z^{14-15}c^{27-28}}{b^{14}} \Rightarrow \frac{x^{14}y^{-2}z^{-1}c^{-1}}{b^{14}} \Rightarrow \frac{x^{14}}{y^2zcb^{14}}$$

PRACTICE EXAMPLES

2.042 $(4x^{-2}y^7e^5)^{-3}$

2.043 $\dfrac{-\left(p^3q^2r^{12}\right)^2}{\left(-p^4qr^8\right)^3}$

2.044 $\left(\dfrac{a^3b^2}{d^4e^4f^3}\right)^2 \div \left(\dfrac{a^2bc^2}{e^3f^2}\right)^3$

SPECIAL POLYNOMIAL PRODUCTS

The following are three polynomial products quite frequently tested on the GRE/GMAT, the third one being the *most* common.

1) $(a+b)^2$ $\Rightarrow$ $(a+b)(a+b)$
   $\Rightarrow$ $a \cdot a + a \cdot b + b \cdot a + b \cdot b$
   $\Rightarrow$ $a^2 + ab + ba + b^2$
   $\Rightarrow$ $a^2 + 2ab + b^2$ $\qquad$ Therefore, $(a+b)^2 = a^2 + 2ab + b^2$

2) $(a-b)^2$ $\Rightarrow$ $(a-b)(a-b)$
   $\Rightarrow$ $a \cdot a - a \cdot b - b \cdot a + b \cdot b$
   $\Rightarrow$ $a^2 - ab - ba + b^2$
   $\Rightarrow$ $a^2 - 2ab + b^2$ $\qquad\qquad\qquad$ $(a-b)^2 = a^2 - 2ab + b^2$

3) $(a+b)(a-b) \Rightarrow$ $a \cdot a - a \cdot b + b \cdot a - b \cdot b$
   $\Rightarrow$ $a^2 - ab + ba - b^2$
   $\Rightarrow$ $a^2 - b^2$ $\qquad\qquad\qquad$ $(a+b)(a-b) = a^2 - b^2$

The third type of polynomial product above is often tested in a rather *hidden* way on the tests. For example, $25 - c^2$, where you have to recognize that 25 is a square (5^2); so is c^2. The expression is of the form $a^2 - b^2$, which can be factorized into $(a+b)(a-b)$. Hence, $25 - c^2 = (5+c)(5-c)$. In a similar fashion, $x^2 - 1$ would be $(x+1)(x-1)$, and $m^2n^2 - k^2$ would be $(mn+k)(mn-k)$.

The following examples are of the form $a^2 - b^2$. Let's factorize them into the form $(a+b)(a-b)$.

2.045 $p^2 - 4q^2$

$\qquad p^2 - (2q)^2$ $\qquad\qquad$ Observe that 4 is 2^2, therefore $4q^2$ is $(2q)^2$.

$\qquad (p+2q)(p-2q)$

2.046 $64m^2 - y^2$

$(8m)^2 - y^2 \Rightarrow (8m + y)(8m - y)$

2.047 $49a^2 - 25$

$(7a)^2 - 5^2 \Rightarrow (7a + 5)(7a - 5)$

2.048 $\dfrac{a^2 b^2}{c^2} - d^2$

$\left(\dfrac{ab}{c}\right)^2 - d^2 \Rightarrow \left(\dfrac{ab}{c} + d\right)\left(\dfrac{ab}{c} - d\right)$

2.049 $(a + m)^2 - (a - m)^2$

In comparing this to the form $a^2 - b^2$, consider what's inside the first set of brackets as a, and what's inside the second set of brackets as b. Then factorize into the form $(a + b)(a - b)$.

$$\left(\underbrace{a + m}_{a}\right)^2 - \left(\underbrace{a - m}_{b}\right)^2 \Rightarrow \left\{\left(\underbrace{a + m}_{a}\right) + \left(\underbrace{a - m}_{b}\right)\right\} \cdot \left\{\left(\underbrace{a + m}_{a}\right) - \left(\underbrace{a - m}_{b}\right)\right\}$$

Now, simplify what's inside the curled brackets { }. Be mindful of the $(-)$ sign when simplifying the second set of curled brackets.

$\{a + m + a - m\} \cdot \{a + m - a + m\}$

$\{2a\} \cdot \{2m\}$

$4am$

2.050 $\left(\dfrac{1}{c} + \dfrac{1}{q}\right)^2 - \left(\dfrac{1}{c} - \dfrac{1}{q}\right)^2$

Exactly on the same lines as the above example, consider the contents of the first brackets as a, and those of the second brackets as b. Then factorize into the form $(a + b)(a - b)$.

$$\left(\underbrace{\dfrac{1}{c} + \dfrac{1}{q}}_{a}\right)^2 - \left(\underbrace{\dfrac{1}{c} - \dfrac{1}{q}}_{b}\right)^2 \Rightarrow \left\{\left(\underbrace{\dfrac{1}{c} + \dfrac{1}{q}}_{a}\right) + \left(\underbrace{\dfrac{1}{c} - \dfrac{1}{q}}_{b}\right)\right\} \cdot \left\{\left(\underbrace{\dfrac{1}{c} + \dfrac{1}{q}}_{a}\right) - \left(\underbrace{\dfrac{1}{c} - \dfrac{1}{q}}_{b}\right)\right\}$$

Simplify the curled brackets. Be cautious while handling the $(-)$ sign.

$\left\{\dfrac{1}{c} + \dfrac{1}{q} + \dfrac{1}{c} - \dfrac{1}{q}\right\} \cdot \left\{\dfrac{1}{c} + \dfrac{1}{q} - \dfrac{1}{c} + \dfrac{1}{q}\right\}$

$\left\{\dfrac{2}{c}\right\} \cdot \left\{\dfrac{2}{q}\right\}$

$\dfrac{4}{cq}$

2.051 $81x^4 - 16y^4$

$(9x^2)^2 - (4y^2)^2 \qquad \qquad$ Because $x^4 = (x^2)^2$ and $y^4 = (y^2)^2$

$(9x^2 + 4y^2)(9x^2 - 4y^2)$

The second set of parentheses $(9x^2 - 4y^2)$ look like the form $a^2 - b^2$ yet again. Factorize it, leaving the first set of parentheses $(9x^2 + 4y^2)$ as it is.

$$(9x^2 + 4y^2)((3x)^2 - (2y)^2) \quad \Rightarrow \quad (9x^2 + 4y^2)(3x + 2y)(3x - 2y)$$

It cannot be factorized any further. So this is the final answer.

2.052 If $\dfrac{x^2 - y^2}{x - y} = 14$, find the average of x and y.

We can factorize $x^2 - y^2$ into $(x + y)(x - y)$, and then rewrite the expression as:

$$\dfrac{(x+y)(x-y)}{x-y} = 14$$

Just like fractions involving pure numbers, fractions involving algebraic terms can also be reduced. Meaning, the term $(x - y)$ cancels out from the numerator and the denominator.

We're left with $x + y = 14$.

The average of x and y would be $\dfrac{x+y}{2}$, i.e. $\dfrac{14}{2}$, i.e. 7.

PRACTICE EXAMPLES

2.053 $36x^2 - y^2$
2.054 $c^2 - 16d^2$
2.055 $1 - 81z^2$
2.056 $p^2 - \dfrac{q^2}{r^2 s^2}$

2.057 $\left(\dfrac{1}{32} + 8\right)^2 - \left(\dfrac{1}{32} - 8\right)^2$

2.058 $\left(\dfrac{1}{4} + \dfrac{1}{x}\right)^2 - \left(\dfrac{1}{4} - \dfrac{1}{x}\right)^2$

2.059 $t^8 - w^8$

2.060 If $d^2 - c^2 = 11$, and $d - c = 4$, find the arithmetic mean of d and c.

SPACE FOR CALCULATIONS

POLYNOMIALS OF THE FORM $ax^2 + bx + c$

Consider the polynomial $2x^2 + 7x + 3$. It is of the general form $ax^2 + bx + c$, in which $a = 2$, $b = 7$ and $c = 3$. Such polynomials can be factorized (factored) by following the method outlined below:

Think of two numbers which, when multiplied by each other, will yield the product ac (here, $ac = 2 \times 3$, or 6), and when added to each other, will yield b (in this case, $b = 7$).

The product ac (or 6) can be obtained by : 6×1 -6×-1 2×3 -2×-3

By *adding* which of those pairs of numbers can we get b (or 7)? Clearly, it's $6 + 1$. (None of the other pairs yields 7. For example, $-6 + -1$ gives -7; $2 + 3$ gives 5; $-2 + -3$ gives -5.)

Split the middle term $(7x)$ using the two numbers we selected $(6 + 1)$ into $6x + 1x$, or simply $6x + x$.

We can rewrite the polynomial as $2x^2 + 6x + x + 3$.

$2x^2 + \underline{6x + x} + 3$ Now, factorize the first two terms. Meaning, extract *the most you can* from the first two terms (the ones with a line underneath). Then do the same thing to the next two terms (the ones with a line above).

$2x(x + 3) + 1(x + 3)$ Notice that you can't extract anything from the second two terms. So, simply write 1. *Please don't write a zero, or the entire term will disappear!*

Observe that the term $(x + 3)$ repeats itself in the first as well as the second term. Factor it out from the two terms, just the way we extracted $2x$ and 1 in the previous step.

$(x + 3)(2x + 1)$ We've now fully factorized the polynomial $2x^2 + 7x + 3$.   End of story.

Using this method, let's factorize the following polynomials.

2.061 $x^2 + 5x + 6$

The product ac is 1×6, or 6. It can be obtained by: 1×6 -1×-6 (2×3) -2×-3
This pair yields the sum b, or 5. $- - - - - - - - - - - - - - - - - - \nearrow$

Split the middle term as the sum of the pair of numbers selected $(2x + 3x)$.

$x^2 + 2x + \overline{3x + 6}$ Factorize the first two terms; then factorize the next two terms.

$x(x + 2) + 3(x + 2)$ Now, factorize the repeating term again.

$(x + 2)(x + 3)$

2.062 $3y^2 + 4y + 1$

The product ac is 3×1, or 3. It can be obtained by: (3×1) -3×-1
This pair yields the sum b, or 4. $- - - - - - - - - \nearrow$

Split the middle term as the sum of the pair of numbers selected $(3y + y)$.

$3y^2 + 3y + \overline{y + 1}$ Factorize the first two terms; then the next two terms.

$3y(y + 1) + 1(y + 1)$ Now, factorize the repeating term again.

$(y + 1)(3y + 1)$

2.063 $p^2 - 10p + 21$

The product ac is 1×21, or 21. It can be obtained by: 21×1 -21×-1 7×3 (-7×-3)
This pair yields the sum b, or -10. $- - - - - - - - - - - - - - - - \nearrow$

Split the middle term as the sum of the pair of numbers selected $(-7p -3p)$.

$p^2 - 7p - \overline{3p + 21}$ Factorize the first two terms; then the next two terms.

$p(p - 7) - 3(p - 7)$ Factorize the repeating term again. Be mindful of the ($-$) sign.

$(p - 7)(p - 3)$

As a precaution in this step, simply repeat the contents of the first brackets, i.e. $(p - 7)$ in the second term. Then match up the term outside the second brackets (-3, not $+3$). If you do this for every single example of the form $ax^2 + bx + c$, then you will never go wrong. Also, think of the time savings!

2.064 $5z^2 - 12z + 4$

The product ac is 5×4, or 20. It can be obtained by: 20×1 -20×-1 5×4 -5×-4
10×2 -10×-2 ← — — — — — — This pair yields the sum b, or -12.

$5z^2 - 10z - 2z + 4$ The middle term is split. Now factorize the terms.

$5z(z - 2) - 2(z - 2)$ Extract the common term. Rewrite.

$(z - 2)(5z - 2)$

2.065 $g^2 + 16g - 36$

Product ac is -36. It can be obtained by: 1×-36 -1×36 2×-18 -2×18 3×-12
-3×12 4×-9 -4×9 6×-6 This pair yields the sum b, or 16.

$g^2 - 2g + 18g - 36$ The middle term is split. Factorize the terms.

$g(g - 2) + 18(g - 2)$ Extract the common term. Rewrite.

$(g - 2)(g + 18)$

2.066 $4p^2 + pq - 3q^2$

This time, two variables are involved: p and q. Do not worry. Stick to the numbers, and everything else will fall in its proper place. Here's how:

Product ac is -12. It can be obtained by: 1×-12 -1×12 2×-6 -2×6 3×-4
-3×4 ← — — — — — — — — — — This pair yields the sum b, or 1.

$4p^2 - 3pq + 4pq - 3q^2$ The middle term is split. Factorize the terms.

$p(4p - 3q) + q(4p - 3q)$ Extract the common term. Rewrite.

$(4p - 3q)(p + q)$

So you see, the number of variables doesn't matter; it's all about the coefficients (constants). Let's solve a few more such examples.

2.067 $e^2 - 8ef - 33f^2$

Product ac is -33. It can be obtained by: 11×-3 -11×3 1×-33 -1×33

This pair yields the sum b, or -8. — — — — —

$e^2 - 11ef + 3ef - 33f^2$ The middle term is split. Factorize the terms.

$e(e - 11f) + 3f(e - 11f)$ Extract the common term. Rewrite.

$(e - 11f)(e + 3f)$

2.068 $2m^2 - 13mn - 7n^2$

Product ac is -14. It can be obtained by: 14×-1 $\;\;\boxed{-14 \times 1}\;\;$ 7×-2 $\;\;-7 \times 2$

This pair yields the sum b, or -13.

$\underline{2m^2 - 14mn} + \overline{mn - 7n^2}$ The middle term is split. Factorize the terms.

$2m(m - 7n) + n(m - 7n)$ Extract the common term. Rewrite.

$(m - 7n)(2m + n)$

2.069 **If $a = 95$, evaluate $a^2 - 5a - 50$.**

Substituting the value of a in the polynomial could give you a headache while performing the calculations. So why not factorize instead?

Product ac is -50. It can be obtained by: 50×-1 $\;\;-50 \times 1$ $\;\;25 \times -2$ $\;\;-25 \times 2$ $\;\;10 \times -5$
$\boxed{-10 \times 5}$ This pair yields the sum b, or -5.

$\underline{a^2 - 10a} + \overline{5a - 50}$ The middle term is split. Factorize the terms.

$a(a - 10) + 5(a - 10)$ Extract the common term. Rewrite.

$(a - 10)(a + 5)$ Now substitute $a = 95$, and calculate.

$(95 - 10)(95 + 5)$ $\Rightarrow$ $(85)(100)$ $\Rightarrow$ 8500

PRACTICE EXAMPLES (Factorize)

2.070 $d^2 + 12d + 11$	**2.074** $r^2 + 48r - 100$	**2.078** If $n = 13$, evaluate
2.071 $7h^2 + 12h + 5$	**2.075** $12s^2 + 5st - 2t^2$	$\sqrt{n^2 - 9n - 36}$.
2.072 $j^2 - 16j + 39$	**2.076** $u^2 - 23uv - 50v^2$	
2.073 $9k^2 - 15k + 4$	**2.077** $15w^2 - 2wx - x^2$	

SPACE FOR CALCULATIONS

ALGEBRAIC FRACTIONS

A fraction in which the numerator and/or the denominator comprise a monomial or a polynomial is called an **algebraic fraction**. The rules of addition, subtraction, multiplication and division for algebraic fractions are the same as they are for numerical fractions. Particular care should be taken while reducing algebraic fractions. For example, it is not right to *reduce* this way: $\frac{\cancel{2x}+1}{\cancel{2x}+3}$. But it is correct to reduce this way: $\frac{(3y+7)\cancel{(5x-4)}}{\cancel{5x-4}}$. As a rule, you can only reduce an entire term, not just a part of it.

2.079 $\dfrac{5a+7}{xy} + \dfrac{3a-12}{yx}$

$\dfrac{5a+7+3a-12}{xy}$ — They're *like fractions*. Simply add numerators.

$\dfrac{8a-5}{xy}$

2.080 $\dfrac{4+3n}{n-4} - \dfrac{12+n}{n-4}$

$\dfrac{4+3n-12-n}{n-4}$ — Be careful with the $(-)$ sign: $-(12+n) = -12-n$

$\dfrac{2n-8}{n-4}$ — Look forward to factorizing (factoring) terms, if possible.

$\dfrac{2\cancel{(n-4)}}{\cancel{n-4}}$ — Cancel and *reduce*. Rewrite.

2

2.081 $\dfrac{8c}{p} - \dfrac{7c}{p^2}$

$\dfrac{8c\cdot p^2 - 7c\cdot p}{p\cdot p^2}$ — *Unlike fractions*; use the *right-arm product first* rule (page 14).

$\dfrac{8cp^2-7cp}{p^3}$ — Look forward to factorizing. cp appears common. Extract it.

$\dfrac{cp(8p-7)}{p^3}$ — Reduce with p.

$\dfrac{c(8p-7)}{p^2}$

2.082 $\dfrac{1}{1-r} + \dfrac{2}{r+1}$

$\dfrac{r+1+2(1-r)}{(1-r)(1+r)}$ — *Right-arm product first.* Also, rewrite $r+1$ as $1+r$ in denominator.

$\dfrac{r+1+2-2r}{1-r^2}$ — Because $(a-b)(a+b) = a^2 - b^2$

$\dfrac{3-r}{1-r^2}$

2.083 $\dfrac{\left(b^2-1\right)-3}{c^2} \cdot \dfrac{2c}{3b+6}$

Two algebraic fractions are being multiplied. Simplify the numerator on the left. Factorize the denominator on the right. Also, reduce the c and the c^2.

$\dfrac{b^2-4}{c} \cdot \dfrac{2}{3(b+2)}$ Is that it? Is something familiar about $b^2 - 4$?

$\dfrac{(b+2)(b-2)}{c} \cdot \dfrac{2}{3(b+2)}$ Because $a^2 - b^2 = (a + b)(a - b)$

$\dfrac{2(b-2)}{3c}$ Because $(b + 2)$ got cancelled out (reduced).

2.083 $\dfrac{7m-21}{m} \div \dfrac{9-m^2}{km}$

$\dfrac{7m-21}{m} \times \dfrac{km}{9-m^2}$ Because $\dfrac{a}{b} \div \dfrac{c}{d} = \dfrac{a}{b} \times \dfrac{d}{c}$

$\dfrac{7(m-3)}{1} \times \dfrac{k}{(3-m)(3+m)}$ m got cancelled out. The terms got factorized.

$\dfrac{-7(3-m)k}{(3-m)(3+m)}$ Because $(m - 3) = -(3 - m)$

$\dfrac{-7k}{3+m}$ Because $(3 - m)$ got cancelled out.

2.084 $\dfrac{j^2+5j-6}{18+3j} \div \dfrac{1-j}{9q}$

$\dfrac{j^2+5j-6}{18+3j} \times \dfrac{9q}{1-j}$ Because $\dfrac{a}{b} \div \dfrac{c}{d} = \dfrac{a}{b} \times \dfrac{d}{c}$

$\dfrac{(j+6)(j-1)}{3(6+j)} \times \dfrac{9q}{-(j-1)}$ Because $j^2 + 5j - 6 = (j + 6)(j - 1)$, and $1 - j = -(j - 1)$

$-3q$ $j + 6$ was the same as $6 + j$. So they cancelled out too.

2.085 $\dfrac{\frac{7}{d} - \frac{16}{3d}}{\frac{5}{d^2}}$

This is a complex *algebraic* fraction, similar to a complex fraction (page 19). Simplify the numerator using the *right-arm product first* rule. Also, bring the denominator up, and multiply its reciprocal (flip) by the numerator.

$\dfrac{21d - 16d}{3d^2} \times \dfrac{d^2}{5} \quad \Rightarrow \quad \dfrac{5d}{3} \times \dfrac{1}{5} \quad \Rightarrow \quad \dfrac{d}{3}$

2.086 $\dfrac{\frac{t}{v} - 2}{\frac{2v+t}{v}}$

$\dfrac{\frac{t}{v} - \frac{2}{1}}{\frac{2v + t}{v}} \quad \Rightarrow \quad \dfrac{\frac{t - 2v}{v}}{\frac{2v + t}{v}} \quad \Rightarrow \quad \dfrac{t - 2v}{v} \times \dfrac{v}{2v + t} \quad \Rightarrow \quad \dfrac{t - 2v}{t + 2v}$

2.087 $\dfrac{\dfrac{k}{12} - \dfrac{3}{k}}{\dfrac{1}{6} + \dfrac{1}{k}}$

$$\dfrac{\dfrac{k^2 - 36}{12k}}{\dfrac{k + 6}{6k}} \quad \Rightarrow \quad \dfrac{\dfrac{(k + 6)(k - 6)}{12k}}{\dfrac{k + 6}{6k}} \quad \Rightarrow \quad \dfrac{(k + 6)(k - 6)}{12k} \times \dfrac{6k}{k + 6} \quad \Rightarrow \quad \dfrac{k - 6}{2}$$

PRACTICE EXAMPLES

2.088 $\dfrac{7g+1}{3xc} - \dfrac{4g-5}{3cx}$

2.089 $\dfrac{p+3}{p-q} + \dfrac{3-p}{p-q}$

2.090 $\dfrac{g-2}{5} + \dfrac{1}{g+2}$

2.091 $\dfrac{k}{18-2k} \times \dfrac{81-k^2}{3k^2}$

2.092 $\dfrac{c+d}{ab} \div \dfrac{c^2-d^2}{ab^2}$

2.093 $\dfrac{6t-9}{14+2s} \cdot \dfrac{s^2+4s-21}{5-2(t+1)}$

2.094 $\dfrac{n}{a^2} - \dfrac{b}{a}$

2.095 $\dfrac{f - \dfrac{1}{f}}{\dfrac{1+f}{f}}$

2.096 $\dfrac{h - \dfrac{1}{e}}{\dfrac{1-eh}{3e^2}}$

2.097 $\dfrac{\dfrac{c}{4a} - \dfrac{a}{c}}{\dfrac{c}{a} - 2}$

SPACE FOR CALCULATIONS

RADICAL EXPRESSIONS

In the previous chapter (page 47), we studied laws pertaining to radicals involving numbers. In this section, we will apply the same laws to *variables*. For example, $x \cdot x = x^2$. Also, $(-x) \cdot (-x) = x^2$. Therefore, $(\pm x)^2 = x^2$. If we take the square root of x^2, it could be x or $-x$, i.e. $\sqrt{x^2} = \pm x$.

Similarly, $d \cdot d \cdot d = d^3$, or $\sqrt[3]{d^3} = d$. There's no $\pm$ here, because $(-d) \cdot (-d) \cdot (-d) = -d^3$, not d^3.

2.098 $\sqrt{25x^2y^2}$

$\sqrt{5 \cdot 5 \cdot x \cdot x \cdot y \cdot y}$

$\pm 5xy$ Add the $\pm$ sign only once, not $(\pm 5)(\pm x)(\pm y)$.

2.099 $\sqrt{50c^4a^2}$

We know that 50 doesn't have a square root. So we need to prime factorize it. That would be $5 \times 5 \times 2$, or simply $5^2 \times 2$. Also, c^4 is $(c^2)^2$.

$\sqrt{5^2 \times 2 \times (c^2)^2 \times a^2}$

$5c^2a\sqrt{2}$ The $\pm$ sign comes *only when the square root sign goes*, i.e. when you actually write the value of $\sqrt{2}$.

2.100 $\sqrt[3]{8b^3d^6}$

Since we are finding the cube root here, let's express the *radicand* in terms of cubes, not squares. Meaning, express d^6 as $(d^2)^3$, and not as $(d^3)^2$.

$\sqrt[3]{2^3b^3(d^2)^3}$

$2bd^2$

2.101 $\dfrac{p^2}{r} \cdot \sqrt{\dfrac{q^2r^3}{p^4}}$

$\dfrac{p^2}{r} \cdot \sqrt{\dfrac{q^2r^2r}{(p^2)^2}}$ Express r^3 as r^2r to extract part of it out of the radical sign.

$\dfrac{p^2}{r} \cdot \dfrac{qr\sqrt{r}}{p^2}$ Now, cancel and *reduce.*

$q\sqrt{r}$ The $\pm$ sign comes *only when the square root sign goes.*

2.102 $\dfrac{3 \cdot \sqrt{64m^2n^3k^7}}{2 \cdot \sqrt[3]{64m^6n^3k^9}}$

$\dfrac{3 \cdot \sqrt{8^2m^2n^2n(k^3)^2k}}{2 \cdot \sqrt[3]{4^3(m^2)^3n^3(k^3)^3}}$ Note: Numerator is square root; denominator is cube root.

$\dfrac{3 \cdot 8 \cdot m \cdot n \cdot k^3\sqrt{nk}}{2 \cdot 4 \cdot m^2 \cdot n \cdot k^3}$ Now, cancel and reduce.

$$\frac{3\sqrt{nk}}{m}$$

The $\pm$ sign comes *only when the square root sign goes*.

2.103 $\sqrt{\dfrac{\sqrt[3]{b^6 c^9}}{4a^2}}$

$$\sqrt{\frac{\sqrt[3]{(b^2)^3(c^3)^3}}{2^2 a^2}} \;\Rightarrow\; \sqrt{\frac{b^2 c^3}{2^2 a^2}} \;\Rightarrow\; \sqrt{\frac{b^2 c^2 c}{2^2 a^2}} \;\Rightarrow\; \frac{bc\sqrt{c}}{2a}$$

2.104 $\dfrac{t-v}{\sqrt{t^2-v^2}}$

$\dfrac{\sqrt{t-v}\cdot\sqrt{t-v}}{\sqrt{(t+v)(t-v)}}$ Because $t^2 - v^2 = (t+v)(t-v)$, and $t-v = \sqrt{t-v}\cdot\sqrt{t-v}$

$\dfrac{\sqrt{t-v}\cdot\sqrt{t-v}}{\sqrt{t+v}\cdot\sqrt{t-v}}$ Because $\sqrt{(t+v)(t-v)} = \sqrt{t+v}\cdot\sqrt{t-v}$

$\dfrac{\sqrt{t-v}}{\sqrt{t+v}}$ After cancelling out the $\sqrt{t-v}$ term

$\sqrt{\dfrac{t-v}{t+v}}$

You might be thinking, "What was the point of all this algebraic manipulation? Wasn't the original expression simple enough?"   Well, on the GRE/GMAT, some questions may be along the lines of "Which of the following expressions is equivalent to … …" To answer such questions, you should be able to work your way around and come up with equivalent expressions. Hence, all this *manipulation*.

2.105 $\dfrac{\dfrac{\sqrt{u}}{\sqrt{v}} - \dfrac{1}{\sqrt{uv}}}{\sqrt{u}\sqrt{v}}$

$\dfrac{\dfrac{u\sqrt{v} - \sqrt{v}}{v\sqrt{u}}}{\sqrt{uv}}$ Use *right-arm product first* rule for numerator.

$\dfrac{\sqrt{v}(u-1)}{v\sqrt{u}} \times \dfrac{1}{\sqrt{uv}}$ Flip the denominator, and multiply it with the numerator.

$\dfrac{\sqrt{v}(u-1)}{uv\sqrt{v}}$ Now, reduce the $\sqrt{v}$.

$\dfrac{u-1}{uv}$ There won't be a $\pm$ sign, because $\sqrt{v}$ got cancelled out.

2.106 $4y\sqrt{y^3} - \sqrt{9y^5}$

$4y\sqrt{y^2 y} - \sqrt{3^2 (y^2)^2 y}$

$4y^2\sqrt{y} - 3y^2\sqrt{y}$

$y^2\sqrt{y}$ The $\pm$ sign comes *only when the square root sign goes*.

PRACTICE EXAMPLES

2.107 $\sqrt{49a^2c^2b^2}$

2.108 $\sqrt{98d^2e^4f^4}$

2.109 $\sqrt[3]{\dfrac{125g^3}{h^3}}$

2.110 $\dfrac{x\sqrt{3x^2y^5}}{3\sqrt{y}}$

2.111 $\dfrac{3\sqrt{700f}}{\sqrt{63e^2f^3}}$

2.112 $\dfrac{\sqrt[3]{16p^4qr^4}}{\sqrt{4r^2p^2}}$

2.113 $\dfrac{\sqrt[3]{8\cdot\sqrt[5]{\sqrt[7]{-1}}}}{\sqrt{-2\cdot\sqrt[3]{-8a^6}}}$

2.114 $\dfrac{\sqrt{\dfrac{h}{s}}-\sqrt{\dfrac{s}{h}}}{\sqrt{\dfrac{h}{s}-\dfrac{s}{h}}}$

2.115 $12a\sqrt{3a}-\sqrt{48a^3}-4\sqrt{12a}$

SPACE FOR CALCULATIONS

EQUATIONS

In its simplest form, any **equation** expresses the equality between two sets of terms. The terms could be monomials, polynomials, or constants. For example, $4-7a=12$, $x^2=3-4c$ etc. are equations. If the equality sign $(=)$ is absent, then it is only a polynomial, not an equation. An equation essentially has two sides, both being equal. Here are the rules that apply to all types of equations:

➢ It is permissible to *add (or subtract) the same quantity to both sides of the = sign*. It doesn't alter the original equality. But adding (or subtracting) to only one side does.

➢ It is permissible to *multiply (or divide) both sides of the = sign by the same quantity*. But multiplying (or dividing) only one side disturbs the original equality.

➢ It is permissible to *square up both sides of the = sign*. Same with raising it to other *integer* powers, i.e. cube, 4th power, −7th power, 5th power, −11th power etc. But doing it to only one side invalidates the original equality. Also, be careful while taking any *even roots* of both sides. For example, if $x^2=a^2$, then taking square roots, we would get $x=\pm a$, not just $x=a$.

LINEAR EQUATIONS

A **linear equation** is an equation having one, two or three variables *in the first degree only*, with no variables *multiplied by each other*. When graphed, a linear equation represents a *line* in a 2-D or a 3-D coordinate system. Therefore, the name 'linear.' For example, $2x + y = 8$ is a linear equation. So is $x - 5y + 11 = 3z$. But $x^2 + 3 = 4y$ is not, because of the 2nd degree term x^2. Nor is $3 + xy = x - 1$, because two variables (x and y) are multiplied by each other. We will not be plotting any graphs right now; we will consider that part when we study coordinate geometry in Chapter 4.

2.116 $3x + 7 = 13$ **Find** x.

To find x, we need to isolate x step by step. Meaning, first get rid of the 7, then the 3.

$3x = 13 - 7$ Subtracting 7 from both sides of the = sign

$3x = 6$

$. x = 2$ Dividing both sides by 3

2.117 $1 + \dfrac{g}{7} = -4$

$\dfrac{g}{7} = -4 - 1$ Subtracting 1 from both sides

$\dfrac{g}{7} = -5$

$g = -35$ Multiplying both sides by 7

2.118 $\dfrac{3}{z} - 2 = 5 + \dfrac{2}{3z}$

$\dfrac{3}{z} - \dfrac{2}{3z} = 5 + 2$ Bring the z terms onto one side, and the constants onto the

other, i.e. add 2 to both sides, and subtract $\dfrac{2}{3z}$ from both sides.

$\dfrac{9z - 2z}{3z^2} = 7$ *Unlike* fractions. *Right-arm product first* rule.

$\dfrac{7z}{3z^2} = 7$ $\Rightarrow$ $\dfrac{7}{3z} = \dfrac{7}{1}$ $\Rightarrow$ $3z \times 7 = 7 \times 1$ $\Rightarrow$ $z = \dfrac{7}{7 \times 3}$ $\Rightarrow$ $z = \dfrac{1}{3}$

2.119 $2x - c = 10 - 3c$ **Find the sum of** x **and** c.

To find $x + c$, get x and c onto the same side of the equation by adding $3c$ to both sides.

$2x + 2c = 10$

$2(x + c) = 10$ Factoring out the 2

$x + c = 5$ Dividing both sides by 2

2.120 $m + 7 = \dfrac{13 - 2b}{2}$ **What is the arithmetic mean (average) of** m **and** b?

The average of m and b is $\dfrac{m+b}{2}$. We need to get m and b on the same side. Then divide by 2.

$2m + 14 = 13 - 2b$ Multiplying both sides by 2

$2m + 2b = 13 - 14$ Subtracting 14 from both sides; adding $2b$ to both sides

$2(m + b) = -1$ $\Rightarrow$ $m + b = -\dfrac{1}{2}$ $\Rightarrow$ $\dfrac{m + b}{2} = -\dfrac{1}{4}$

2.121 $5q + 6 = 20$ **What is $5q + 16$?**

Be on the lookout for a *trend*. On the GRE/GMAT, such questions are quite common. If you find the value of q using the regular method, it may take you some time. Instead, a quick glance at what's given and what's asked should give you the hint:

$5q + 6$ and $5q + 16$.   They're only 10 apart from each other, i.e. add 10 to $5q + 6$, and you end up getting $5q + 16$. That means, $5q + 16$ must be 20 + 10, or simply 30.

2.122 $3y - 8 = -1$ **What is $\sqrt{3y + 18}$?**

Same as the last example. Add 26 to $3y - 8$, and you end up with $3y + 18$. Then take the root.

$3y - 8 + 26 = -1 + 26$

$3y + 18 = 25$

$\sqrt{3y + 18} = \sqrt{25}$

$\sqrt{3y + 18} = \pm 5$ The $\pm$ sign comes *only when the square root sign goes.*

2.123 $\frac{4}{5}d - \frac{2}{3} = 17$ **What is $\frac{2}{3} - \frac{4}{5}d$?**

We know that $a - b = -(b - a)$.

Similarly, $\frac{2}{3} - \frac{4}{5}d$ is simply $-\left(\frac{4}{5}d - \frac{2}{3}\right)$, or -17.

2.124 $\frac{2c}{d} = 7$ **What is $\frac{4c}{d}$?**

Notice something between $\frac{2c}{d}$ and $\frac{4c}{d}$?

$\frac{4c}{d}$ is simply 2 times $\frac{2c}{d}$, or 2 times 7, or 14.

2.125 $\frac{a}{b} = 15$ **What is $\frac{3a}{5b}$?**

Once again, notice the *trend*.

$\frac{3a}{5b}$ is simply $\frac{3}{5} \times \frac{a}{b}$, or $\frac{3}{5} \times 15$, or 9.

2.126 $2p + 6 = 24$ **Find $\frac{p+3}{2}$.**

Look for the *trend*. There's no need to find p and then substitute it in $\frac{p+3}{2}$.

$2(p + 3) = 24$ Factoring out the 2

$p + 3 = 12$ Dividing both sides by 2

$\frac{p+3}{2} = 6$ Dividing both sides by 2 again

2.127 $\sqrt{7n} = \sqrt{3 - 8n}$ **Find n.**

$7n = 3 - 8n$ It's okay to square both sides. The root sign goes away.

$7n + 8n = 3$ Adding $8n$ to both sides

$15n = 3$

$$n = \frac{3}{15}$$ Dividing both sides by 15

$$n = \frac{1}{5}$$ Reducing by 3

2.128 **If e is positive, and $\sqrt{e+2} = \dfrac{2\sqrt{3}}{\sqrt{e-2}}$, find e.**

$$e + 2 = \frac{\left(2\sqrt{3}\right)^2}{e-2}$$ Squaring up both sides

$$\frac{e+2}{1} = \frac{12}{e-2}$$

$$(e+2)(e-2) = 12$$ Equating cross products

$$e^2 - 4 = 12$$ Because $(a+b)(a-b) = a^2 - b^2$

$$e^2 = 16$$ Adding 4 to both sides

$$e = 4$$ We can ignore $e = -4$, because e is positive.

2.129 $\dfrac{a}{2} + \dfrac{a}{3} + \dfrac{a}{4} = 13$ **Find a.**

$$a\left(\frac{1}{2} + \frac{1}{3} + \frac{1}{4}\right) = 13$$ Factoring out the a

$$a\left(\frac{1\times6}{2\times6} + \frac{1\times4}{3\times4} + \frac{1\times3}{4\times3}\right) = 13$$ The LCM of 2, 3, and 4 is 12. Accordingly, raise each fraction to get 12 in the denominator.

$$a\left(\frac{6}{12} + \frac{4}{12} + \frac{3}{12}\right) = 13 \quad \Rightarrow \quad a\left(\frac{13}{12}\right) = 13 \quad \Rightarrow \quad a = 13 \times \left(\frac{12}{13}\right) \quad \Rightarrow \quad \mathbf{a = 12}$$

2.130 $\dfrac{2q+1}{3q-2} = \dfrac{4}{5}$ **Find q.**

$$5(2q+1) = 4(3q-2)$$ Equating cross products

$$10q + 5 = 12q - 8$$

$$10q - 12q = -8 - 5$$ Subtracting $12q$ and 5 from both sides

$$-2q = -13 \quad \Rightarrow \quad q = \frac{-13}{-2} \quad \Rightarrow \quad q = 6\frac{1}{2}$$

PRACTICE EXAMPLES

2.131 $12 - 5b = 2$ Find b.

2.132 $\dfrac{x}{4} - 11 = 1$ Find x.

2.133 $1 + \dfrac{7}{5z} = 2 - \dfrac{7}{z}$ Find z.

2.134 What is $a + c$ if $2a - c = 8 + 3a$?

2.135 Find the arithmetic mean of p and q if $\dfrac{p+4}{11-2q} = \dfrac{1}{2}$.

2.136 If $11f + 4 = -3$, find $(11f - 4)^2$.

2.137 If $\dfrac{a}{b} - \dfrac{c}{d} = -12$, what is $\dfrac{c}{d} - \dfrac{a}{b}$?

2.138 If $\dfrac{3k}{2m} = 18$, what is $\dfrac{2k}{3m}$?

2.139 $3g - 51 = 9$ What is $g - 17$?

2.140 Find h if $6 \cdot \sqrt{\dfrac{8h+21}{h-3}} = 18$.

2.141 $3 + \sqrt{6y+1} = 10$ Find y.

SYSTEM OF EQUATIONS

In the previous section, we solved linear equations in *one variable*, i.e. either x, or y, or a, or q etc. In this section, we will consider two linear equations at a time, both in *two variables*. A set of such equations is called a **system of equations**. We can solve them simultaneously to figure out the value of both the variables. There are two methods to do this. Let's consider each with examples.

METHOD OF ADDITION/SUBTRACTION

Recall that adding (or subtracting) the same quantity to both sides of the equation does not void the validity of the original equation. Similarly, adding (or subtracting) one equation from another is valid too. For example, consider two equations: $2x + 5y = 16$ and $2x + 3y = 10$

To find x and y, we need to first eliminate one of them, and find the other. The term $2x$ appears in both the equations. If we subtract one equation from the other, $2x$ would go away.

$$(2x + 5y) - (2x + 3y) = 16 - 10$$

Observe that we are subtracting $2x + 3y$ from one side, whereas 10 from the other side. But if you look at the second equation, you will realize that $2x + 3y$ is the same as 10. So, subtracting one thing is the same as subtracting the other.

$$2x + 5y - 2x - 3y = 6 \quad \Rightarrow \quad 2y = 6 \quad \Rightarrow \quad y = 3$$

Substituting this value of y in any of the two equations (say the first one), we get:

$$2x + 5(3) = 16 \quad \Rightarrow \quad 2x + 15 = 16 \quad \Rightarrow \quad 2x = 1 \quad \Rightarrow \quad x = \frac{1}{2}$$

2.142 **Find x and y if $x + y = 2$ and $3x + y = 10$.**

Upon subtracting the first equation from the second, y would go. We can subtract the second equation from the first too, but that would yield negative numbers. As far as possible, let's try to avoid them. But please remember, regardless of whichever way we go, the final result will still be the same.

$(3x + y) - (x + y) = 10 - 2 \quad \Rightarrow \quad 3x + y - x - y = 8 \quad \Rightarrow \quad 2x = 8 \quad \Rightarrow \quad x = 4$

Substituting this value of x in the first equation, we get:

$4 + y = 2 \quad \Rightarrow \quad y = -2$

2.143 **Find k and c using $2k - c = 1$ and $c - 5k = -7$.**

If we add the two equations, we can eliminate c right away.

$(2k - c) + (c - 5k) = 1 - 7 \quad \Rightarrow \quad 2k - c + c - 5k = -6 \quad \Rightarrow \quad -3k = -6 \quad \Rightarrow \quad k = 2$

Substituting $k = 2$ in the first equation, we get:

$2(2) - c = 1 \quad \Rightarrow \quad 4 - c = 1 \quad \Rightarrow \quad -c = -3 \quad \Rightarrow \quad c = 3$

2.144 **Find a and b if $2a - 3b = 6$ and $a + 2b = 1$.**

Simply adding or subtracting one equation from the other will not help. But we can multiply (both sides of) the second equation by 2 to get $2a$ in it. Then we can eliminate a.

$2(a + 2b) = 2(1) \quad \Rightarrow \quad 2a + 4b = 2$

Now let's subtract the first equation from this one.

$(2a + 4b) - (2a - 3b) = 2 - 6 \quad \Rightarrow \quad 2a + 4b - 2a + 3b = -4 \quad \Rightarrow \quad 7b = -4 \quad \Rightarrow \quad b = -\dfrac{4}{7}$

Substituting $b = -\dfrac{4}{7}$ in the second equation, we get:

$a + 2\left(-\dfrac{4}{7}\right) = 1 \quad \Rightarrow \quad a - \dfrac{8}{7} = 1 \quad \Rightarrow \quad a = \dfrac{1}{1} + \dfrac{8}{7} \quad \Rightarrow \quad a = \dfrac{15}{7} \quad \Rightarrow \quad a = 2\dfrac{1}{7}$

2.145 **Find p and q using $2p + 7q = -1$ and $17 + 3p = 5q$.**

To put the second equation in perspective, let's first rearrange its terms.

$3p - 5q = -17$

Now, we can do one of two things: Either multiply the first equation by 5, and the second by 7 to eliminate q, or multiply the first equation by 3, and the second by 2 to eliminate p. Let's choose the second option, and *deal with smaller numbers*.

$3(2p + 7q) = 3(-1) \quad \Rightarrow \quad 6p + 21q = -3 \qquad \qquad \text{(Equation 3)}$

$2(3p - 5q) = 2(-17) \quad \Rightarrow \quad 6p - 10q = -34 \qquad \qquad \text{(Equation 4)}$

Subtracting the 4th equation from the 3rd, we get:

$(6p + 21q) - (6p - 10q) = -3 + 34 \quad \Rightarrow \quad 6p + 21q - 6p + 10q = 31 \quad \Rightarrow \quad 31q = 31$

Or, $q = 1$ \quad Substituting this in the very first equation, we get:

$2p + 7(1) = -1 \quad \Rightarrow \quad 2p + 7 = -1 \quad \Rightarrow \quad 2p = -8 \quad \Rightarrow \quad p = -4$

PRACTICE EXAMPLES (Find the unknown variables)

2.146 $g + 2b = 7$ and $g + b = 5$

2.147 $4n + m = 39$ and $m - 6n = -31$

2.148 $5y + x = 26$ and $2x - 5y = 7$

2.149 $5s + 6t = 32$ and $3s + 4t = 22$

METHOD OF SUBSTITUTION

When one of the two equations has a variable expressed in the form of other terms, simply substitute those terms for that same variable in the second equation. That immediately eliminates one variable. The other variable can be found out as in the previous method. For example, consider two equations:

$m = 2 + 3n$ and $2m + n = -3$

The value of m from the first equation (i.e. $2 + 3n$) can be directly substituted for m in the second equation.

$2(2 + 3n) + n = -3$ $\Rightarrow$ $4 + 6n + n = -3$ $\Rightarrow$ $7n = -7$ $\Rightarrow$ $n = -1$

Substitute this value of n in the first equation.

$m = 2 + 3(-1)$ $\Rightarrow$ $m = 2 - 3$ $\Rightarrow$ $m = -1$

2.150 $c = 3d$ and $2c + 5d = 11$ **Find c and d.**

Substitute the value of c from the first equation into the second.

$2(3d) + 5d = 11$ $\Rightarrow$ $6d + 5d = 11$ $\Rightarrow$ $11d = 11$ $\Rightarrow$ $d = 1$

Substitute this value of d in the first equation.

$c = 3(1)$ $\Rightarrow$ $c = 3$

2.151 $2f - 3h = 10$ and $4h = f$ **Find f and h.**

Substitute the value of f from the second equation into the first.

$2(4h) - 3h = 10$ $\Rightarrow$ $8h - 3h = 10$ $\Rightarrow$ $5h = 10$ $\Rightarrow$ $h = 2$

Substitute $h = 2$ in the second equation.

$4(2) = f$ $\Rightarrow$ $f = 8$

PRACTICE EXAMPLES (Find the unknown variables)

2.152 $6 - 2v = u$ and $2u + v = 15$

2.153 $3s + 2t = 12$ and $s = 3t - 7$

QUADRATIC EQUATIONS

We have studied polynomials of the form $ax^2 + bx + c$. Quadratic equations are simply equations containing such polynomials. The standard form of a quadratic equation is $ax^2 + bx + c = 0$. When factorized, the polynomial yields two terms, the product of which is zero.

Consider the quadratic equation $5x^2 - 12x + 4 = 0$. We can factorize the polynomial using the method we know:

The product ac is 5×4, or 20. It can be obtained by: $20 \times 1 \quad -20 \times -1 \quad 5 \times 4 \quad -5 \times -4 \quad 10 \times 2$
$-10 \times -2 \longleftarrow - - - - - - -$ This pair yields the sum b, or -12.

$5x^2 - 10x - \overline{2x + 4} = 0$ $\qquad$ The middle term is split. Now factorize the terms.

$5x(x - 2) - 2(x - 2) = 0$ $\qquad$ Extract the common term. Rewrite.

$(x - 2)(5x - 2) = 0$

In order to understand the next step, let's consider a concept: If I told you that the product of an unknown number and 14 is zero, what does that unknown number *have to be*? Zero. It's the only way the product would be zero, right? If I told you that the product of two unknown numbers is zero, what could the numbers be? Well, one of them could be zero, while the other could be anything. Or they could *both* be zero. Meaning, *at least one of them would need to be zero.*

Now look at the above two terms. Their product is zero. Meaning, either $(x - 2)$ is zero, or $(5x - 2)$ is zero, or both are zero. If the two terms were identical, they would both be zero. But in this particular case, they're not. So let's set each one of them equal to zero, and find the *possible values* of x.

$x - 2 = 0 \quad \Rightarrow \quad x = 2$ $\qquad$ This is one possible value of x.

$5x - 2 = 0 \quad \Rightarrow \quad 5x = 2 \quad \Rightarrow \quad x = \dfrac{2}{5}$ $\qquad$ This is the other possible value of x.

If x takes on either of these two values at any given time, only then the above product will be zero. No other value of x will yield a zero product above.

Quadratic equations *always* yield two roots (two values of the unknown variable). Whether the two roots would be different or identical depends upon the terms in the equation.

2.154 **Find the roots of the quadratic equation $x^2 + 13x + 12 = 0$.**

The product ac is 1×12, or 12. It can be obtained by: $12 \times 1 \quad -12 \times -1 \quad 6 \times 2 \quad -6 \times -2$
$4 \times 3 \quad -4 \times -3$ $\qquad\qquad\qquad$ This pair yields the sum b, or 13.

$x^2 + 12x + \overline{x + 12} = 0$ $\qquad$ The middle term is split. Factorize the terms.

$x(x + 12) + 1(x + 12) = 0$ $\qquad$ Extract the common term. Rewrite.

$(x + 12)(x + 1) = 0$

$x + 12 = 0 \quad \Rightarrow \quad x = -12$ $\qquad$ This is one possible value of x.

$x + 1 = 0 \quad \Rightarrow \quad x = -1$ $\qquad$ This is the other possible value of x.

2.155 **Solve the equation $11c^2 + 32c - 3 = 0$.**

The product ac is 11×-3, or -33. It can be obtained by: 11×-3 -11×3 $\overparen{33 \times -1}$
-33×1 This pair yields the sum b, or 32. -- -->

$\underline{11c^2 + 33c} - \overline{c - 3} = 0$ The middle term is split. Factorize the terms.

$11c(c + 3) - 1(c + 3) = 0$ Extract the common term. Rewrite.

$(c + 3)(11c - 1) = 0$

$c + 3 = 0 \quad \Rightarrow \quad c = -3$ This is one possible value of c.

$11c - 1 = 0 \quad \Rightarrow \quad c = \dfrac{1}{11}$ This is the other possible value of c.

2.156   **Find the roots of $p^2 - 11p + 28 = 0$.**

The product ac is 1×28, or 28. It can be obtained by: 1×28 -1×-28 2×14 -2×-14
4×7 $\overparen{-4 \times -7}$ ‹‐ ‐ ‐ ‐ ‐ ‐ ‐ ‐ This pair yields the sum b, or -11.

$\underline{p^2 - 4p} - \overline{7p + 28} = 0$ The middle term is split. Factorize the terms.

$p(p - 4) - 7(p - 4) = 0$ Extract the common term. Rewrite.

$(p - 4)(p - 7) = 0$

$p - 4 = 0 \quad \Rightarrow \quad p = 4$ This is one possible value of p.

$p - 7 = 0 \quad \Rightarrow \quad p = 7$ This is the other possible value of p.

2.157   **Solve $13a^2 - 11a - 2 = 0$.**

The product ac is 13×-2, or -26. It can be obtained by: 1×-26 -1×26 $\overparen{2 \times -13}$
-2×13 This pair yields the sum b, or -11. -- ->

$\underline{13a^2 - 13a} + \overline{2a - 2} = 0$ The middle term is split. Factorize the terms.

$13a(a - 1) + 2(a - 1) = 0$ Extract the common term. Rewrite.

$(a - 1)(13a + 2) = 0$

$a - 1 = 0 \quad \Rightarrow \quad a = 1$ This is one possible value of a.

$13a + 2 = 0 \quad \Rightarrow \quad a = -\dfrac{2}{13}$ This is the other possible value of a.

PRACTICE EXAMPLES (Solve these equations)

2.158 $m^2 + 12m + 27 = 0$ **2.160** $f^2 - 7f + 12 = 0$

2.159 $6u^2 + 5u - 4 = 0$ **2.161** $10z^2 - z - 3 = 0$

SPACE FOR CALCULATIONS

QUADRATIC FORMULA

While most quadratic equations can be factorized into integer or fractional roots, some cannot. But regardless, *all* quadratic equations can be solved using the **quadratic formula**, which is:

$$x = \frac{-b \pm \sqrt{b^2 - 4ac}}{2a}$$

where x is the unknown variable in the general quadratic form $ax^2 + bx + c = 0$.

In the quadratic equations that we have solved so far, replace the variable x with whichever variable is used (q, s, u, t etc.), and solve them again. You should get the same roots.

INCOMPLETE QUADRATIC EQUATIONS

Sometimes, from the general form $ax^2 + bx + c = 0$, either the bx term or the c term might go missing. In such a case, the equation is called an **incomplete quadratic equation**. Even then, it will have two roots.

For example, consider the equation $2x^2 + 6x = 0$. Notice that there's no c term. We would still factorize it by extracting a common term, which is $2x$:

$2x(x + 3) = 0$ Once again, we're looking at a zero product of two terms. Meaning, either $2x$ is zero, or $x + 3$ is zero.

$2x = 0$ $\Rightarrow$ $x = 0$ This is one possible root.

$x + 3 = 0$ $\Rightarrow$ $x = -3$ This is the other possible root.

In the examples below, some are incomplete quadratic equations, whereas others are complete but scattered on both sides of the = sign.

2.162 $y^2 + 2y = 8 + 9y$ **Find y.**

This is a *complete* quadratic equation. Rearrange the terms (by subtracting $9y$ and 8 from both sides).

$y^2 - 7y - 8 = 0$ $ac = -8$ Looks like -8×1

$y^2 - 8y + \overline{y - 8} = 0$ The middle term is split. Factorize the terms.

$y(y - 8) + 1(y - 8) = 0$ Extract the common term. Rewrite.

$(y - 8)(y + 1) = 0$

$y - 8 = 0$ $\Rightarrow$ $y = 8$ This is one possible value of y.

$y + 1 = 0$ $\Rightarrow$ $y = -1$ This is the other possible value of y.

2.163 **If $14 - 2m^2 = m^2 - 13$, and m is positive, what is the value of m?**

This is an *incomplete* quadratic equation. Rearrange the terms.

$-2m^2 - m^2 = -13 - 14$ $\Rightarrow$ $-3m^2 = -27$ $\Rightarrow$ $m^2 = 9$ $\Rightarrow$ $m = 3$

$m \neq -3$ because m is positive.

2.164 Given that $g^2 - 7g + 100 = 19 + 11g$, can g have two distinct values?

Complete quadratic equation. Rearrange the terms.

$g^2 - 7g - 11g + 100 - 19 = 0$

$g^2 - 18g + 81 = 0$ $ac = 81$ Looks like -9×-9

$g^2 - 9g - \overline{9g + 81} = 0$ The middle term is split. Factorize the terms.

$g(g - 9) - 9(g - 9) = 0$ Extract the common term. Rewrite.

$(g - 9)(g - 9) = 0$

$g - 9 = 0 \quad \Rightarrow \quad g = 9$ This is the *only* possible value of g.

2.165 If $8q^2 + 17q = 3q^2 + 21q$, find the integer value of q.

Incomplete quadratic equation. Rearrange the terms.

$8q^2 - 3q^2 + 17q - 21q = 0$

$5q^2 - 4q = 0$

$q(5q - 4) = 0$ Zero product of two terms.

$q = 0$ This is the *integer value* of q.

$5q - 4 = 0 \quad \Rightarrow \quad 5q = 4 \quad \Rightarrow \quad q = \dfrac{4}{5}$ This is the fractional (not integer) value of q.

2.166 $(x^2 - 9)(2x + 7)(3x - 15) = 0$ **Arrange the possible values of x in increasing order.**

This time, we have a zero product of three terms. The same logic applies.

$x^2 - 9 = 0 \quad \Rightarrow \quad x^2 = 9 \quad \Rightarrow \quad x = \pm 3$ Possible values.

$2x + 7 = 0 \quad \Rightarrow \quad 2x = -7 \quad \Rightarrow \quad x = -\dfrac{7}{2} \quad \Rightarrow \quad x = -3.5$ Possible value.

$3x - 15 = 0 \quad \Rightarrow \quad 3x = 15 \quad \Rightarrow \quad x = \dfrac{15}{3} \quad \Rightarrow \quad x = 5$ Possible value.

Possible values of x in increasing order: $-3.5 \quad -3 \quad 3 \quad 5$

PRACTICE EXAMPLES

2.167 If $5z + 17 = z^2 - 19$, what values can z take?

2.168 If $2d^2 - 25 = 11 - 7d^2$, and $d > 0$, find d.

2.169 $y^2 + 27 - 3y = 15 - 2y^2 + 10y$. Can y have two *different* values?

2.170 If $10 - e^2 = e^2 - 40$, what is the minimum possible value of e?

2.171 $(p - 6)(p^3 + 8)\left(p^2 - \dfrac{4}{9}\right) = 0$ Find the arithmetic mean of the possible values of p.

SPACE FOR CALCULATIONS

LITERAL EQUATIONS

The presence of *letters* (variables) characterizes an equation. Whether those letters are Latin or Greek doesn't matter. An equation that involves *letters* is called a **literal equation** (*literal* being the adjective of *letter*). By that definition, *all* equations (including the ones we've studied up to this point) are literal equations.

Until now, the equations we've discussed had just one or two variables. Using various methods, we determined the values of each of those variables. What if there were several variables in an equation? We could then use the techniques we've learned to isolate one variable at a time. *Isolating* a variable means *solving for that variable*. That is what we will discuss in this section.

Consider the formula for the velocity of a freely falling object striking the ground: $v^2 = u^2 + 2gh$, where v = striking velocity, u = initial velocity, g = gravitational acceleration, and h = the height through which it falls. There are four variables. If three of them are known, then the fourth can be calculated. But to do that, we would first need to isolate the fourth (unknown) variable.

If the height h were to be determined, while the other things were known (v, u and g), we would do:

$$v^2 - u^2 = 2gh \qquad \Rightarrow \qquad \frac{v^2 - u^2}{2g} = h$$

Instead, if the initial velocity u were to be determined, we would do:

$$v^2 - 2gh = u^2 \qquad \Rightarrow \qquad \sqrt{v^2 - 2gh} = u$$

When isolating a variable, make sure that the variable appears on *only one side* of the equation. A literal equation can have any number of variables with any power. The equation can be solved for any of those variables.

2.172 $\dfrac{1}{a} + \dfrac{2}{b} = \dfrac{3}{c}$ **Solve this equation for a, b and c, one at a time.**

Solving for a: $\dfrac{1}{a} = \dfrac{3}{c} - \dfrac{2}{b}$ $\Rightarrow$ $\dfrac{1}{a} = \dfrac{3b - 2c}{bc}$ $\Rightarrow$ $a = \dfrac{bc}{3b - 2c}$

Solving for b: $\dfrac{2}{b} = \dfrac{3}{c} - \dfrac{1}{a}$ $\Rightarrow$ $\dfrac{2}{b} = \dfrac{3a - c}{ac}$ $\Rightarrow$ $\dfrac{b}{2} = \dfrac{ac}{3a - c}$ $\Rightarrow$ $b = \dfrac{2ac}{3a - c}$

Solving for c: $\dfrac{1}{a} + \dfrac{2}{b} = \dfrac{3}{c}$ $\Rightarrow$ $\dfrac{b + 2a}{ab} = \dfrac{3}{c}$ $\Rightarrow$ $\dfrac{ab}{b + 2a} = \dfrac{c}{3}$ $\Rightarrow$ $\dfrac{3ab}{b + 2a} = c$

2.173 $mp^2 + q = 2qm$ **Solve it for p, q and m, one variable at a time.**

Solving for p: $mp^2 = 2qm - q$ $\Rightarrow$ $p^2 = \dfrac{2qm - q}{m}$ $\Rightarrow$ $p = \sqrt{\dfrac{q(2m - 1)}{m}}$

Note the factorization in the last step. Also, ± sign comes *only when the square root sign goes.*

Solving for q: $mp^2 = 2qm - q$ $\Rightarrow$ $mp^2 = q(2m - 1)$ $\Rightarrow$ $\dfrac{mp^2}{2m - 1} = q$

Solving for m: $mp^2 - 2qm = -q$ $\Rightarrow m(p^2 - 2q) = -q$ $\Rightarrow$ $m = \dfrac{-q}{p^2 - 2q}$

2.174 $\dfrac{7}{3t + 1} = \dfrac{2u^2}{v}$ **Solve it for t, u and v, one at a time.**

Solving for t: $\quad \dfrac{3t+1}{7} = \dfrac{v}{2u^2} \quad \Rightarrow \quad 3t+1 = \dfrac{7v}{2u^2} \quad \Rightarrow \quad 3t = \dfrac{7v}{2u^2} - 1 \quad \Rightarrow$

$\quad 3t = \dfrac{7v - 2u^2}{2u^2} \quad \Rightarrow \quad t = \dfrac{7v - 2u^2}{6u^2}$

Solving for u: $\quad \dfrac{7v}{3t+1} = 2u^2 \quad \Rightarrow \quad \dfrac{7v}{2(3t+1)} = u^2 \quad \Rightarrow \quad \sqrt{\dfrac{7v}{2(3t+1)}} = u$

Solving for v: $\quad \dfrac{3t+1}{7} = \dfrac{v}{2u^2} \quad \Rightarrow \quad \dfrac{2u^2(3t+1)}{7} = v$

In the above three examples, we've done some simple algebraic manipulations (adding/subtracting/multiplying/dividing both sides by the same quantity) to eliminate the unwanted terms while isolating the variable under consideration. Now let's solve some examples in which a variable or two might need to be *completely eliminated.*

2.175 **If $a = 3r^2$ and $c = 6r$, then express a in terms of c.**

To express a in terms of c, we must eliminate r.

From the first equation: $\quad a = 3r^2 \quad \Rightarrow \quad \dfrac{a}{3} = r^2$

From the second equation: $\quad c = 6r \quad \Rightarrow \quad c^2 = (6r)^2 \quad \Rightarrow \quad c^2 = 36r^2 \quad \Rightarrow \quad \dfrac{c^2}{36} = r^2$

Equate the values of r^2 from the two equations, and the result will be in terms of a and c only.

$\dfrac{a}{3} = \dfrac{c^2}{36} \quad \Rightarrow \quad a = \dfrac{3c^2}{36} \quad \Rightarrow \quad a = \dfrac{c^2}{12} \qquad$ a is now expressed in terms of c.

2.176 **If $py + 2 = ak$ and $b - 6 = 2yp$, then express k in terms of a and b.**

Clearly, the py (or yp) term needs to be eliminated.

From the first equation: $\quad py = ak - 2$

From the second equation: $\quad \dfrac{b-6}{2} = yp$

Equate the two values of py, and the py term will go away.

$ak - 2 = \dfrac{b-6}{2} \quad \Rightarrow \quad ak = \dfrac{b-6}{2} + \dfrac{2}{1} \quad \Rightarrow \quad ak = \dfrac{b-6+4}{2} \quad \Rightarrow \quad k = \dfrac{b-2}{2a}$

2.177 **If $\dfrac{2^{k+1}}{4^{m-1}} = \dfrac{16^{4f+g}}{8^x}$, then express x in terms of k, m, f and g.**

$\dfrac{2^{k+1}}{(2^2)^{m-1}} = \dfrac{(2^4)^{4f+g}}{(2^3)^x}$ $\qquad$ Express in terms of the smallest base 2.

$\dfrac{2^{k+1}}{2^{2m-2}} = \dfrac{2^{16f+4g}}{2^{3x}}$ $\qquad$ Because $\left(a^b\right)^c = a^{bc}$

$2^{k+1-(2m-2)} = 2^{16f+4g-3x}$ $\qquad$ Because $\dfrac{a^b}{a^c} = a^{b-c}$

$k + 1 - 2m + 2 = 16f + 4g - 3x$ $\qquad$ Same bases. Therefore, equate exponents.

$$k - 2m + 3 = 16f + 4g - 3x$$

$$k - 2m + 3 - 16f - 4g = -3x$$

$$\frac{k - 2m + 3 - 16f - 4g}{-3} = x$$

$$\frac{16f + 4g + 2m - k - 3}{3} = x$$

In the final step, multiply the numerator and denominator by –1 to make the –3 positive. All the signs in the numerator will reverse. Conveniently rearrange the terms in the numerator.

PRACTICE EXAMPLES

2.178 $ax^2 + by = cx^2 + dy$ Solve for x and y, one at a time.

2.179 $b - y^2 = 4ax - 3y^2$ Solve for a, b, x and y, one at a time.

2.180 $\dfrac{p}{2q-3r} = s$ Solve for p, q, r and s, one at a time.

2.181 If $m = cx + 4f$ and $n = 3cx - 1$, then express m in terms of n and f.

2.182 If $h = 9c$ and $4e = 9c^2$, then express e in terms of h.

2.183 If $\dfrac{125}{25^{a+b}} = \dfrac{5^y}{125^{c-d}}$ then express y in terms of a, b, c and d.

SPACE FOR CALCULATIONS

ABSOLUTE VALUE EQUATIONS

In the previous chapter, we defined absolute value (or modulus) of a number as the value of the number without regard to its sign. Absolute values are always positive or zero, but never negative. When dealing with variables, we would use the same definition. But before we take the modulus of a variable, how do we know whether the variable has a positive or negative value? For example, consider the modulus of x, or $|x|$. If $x = 5$, then $|x| = 5$ as well. But if $x = -12$, then $|x|$ has to be positive, regardless of the negative sign. To get that positive value, we would need to *take the negative of* -12. Mathematically, $|x| = -(-12) = 12$.

Algebraically, we can define **absolute value** as:

$|x| = x$	if x is already positive	e.g. $x = 5$ means $|x| = |5| = 5$
$|x| = -(x)$	if x is negative	e.g. $x = -12$ means $|x| = |-12| = -(-12) = 12$

What about an equation of the form $|x - 7| = 2$? Such an equation would be called an **absolute value equation**. If $x = 9$, then $|x - 7| = |9 - 7| = |2| = 2$. But if $x = 5$, then too $|x - 7| = |5 - 7| = |-2| = 2$. That was good guess work! But when trying to solve an absolute value equation, consider the *value of the modulus*, and the *negative of value of the modulus*, just the way we did in the definition. Set each of those values equal to the constant (here, 2). Solving them, you will get the two roots of the equation.

Considering *value of modulus*: $x - 7 = 2$ $\Rightarrow$ $x = 9$
Considering *negative of value of modulus*: $-(x - 7) = 2$ $\Rightarrow$ $-x + 7 = 2$ $\Rightarrow$ $-x = -5$ $\Rightarrow$ $x = 5$

2.184 **If $|x - 12| = 3$, find the sum of the possible values of x.**

Let's find the values of x considering the *value* and the *negative of value* of the modulus.

$x - 12 = 3$ $\Rightarrow$ $x = 15$

$-(x - 12) = 3$ $\Rightarrow$ $-x + 12 = 3$ $\Rightarrow$ $-x = -9$ $\Rightarrow$ $x = 9$

Thus the sum of the two possible values of x is $15 + 9$, i.e. 24.

2.185 **If $|2c + 5| - 7 = 10$, and $c < 0$, find c.**

First simplify the expression.

$|2c + 5| - 7 = 10$ $\Rightarrow$ $|2c + 5| = 10 + 7$ $\Rightarrow$ $|2c + 5| = 17$

Now consider the *value* and the *negative of value* of the modulus.

$2c + 5 = 17$ $\Rightarrow$ $2c = 12$ $\Rightarrow$ $c = 6$

$-(2c + 5) = 17$ $\Rightarrow$ $-2c - 5 = 17$ $\Rightarrow$ $-2c = 22$ $\Rightarrow$ $c = -11$

c is negative ($c < 0$). Therefore, $c = -11$, not 6.

2.186 **If $|12 - 5a| = 2$, find the average of the possible values of a.**

Consider the *value* and the *negative of value* of the modulus.

$12 - 5a = 2$ $\Rightarrow$ $-5a = -10$ $\Rightarrow$ $a = 2$

$-(12 - 5a) = 2$ $\Rightarrow$ $-12 + 5a = 2$ $\Rightarrow$ $5a = 14$ $\Rightarrow$ $a = \dfrac{14}{5}$

The average of the values of a is $\dfrac{2 + \frac{14}{5}}{2}$ $\Rightarrow$ $\dfrac{\frac{10+14}{5}}{2}$ $\Rightarrow$ $\dfrac{24}{5} \times \dfrac{1}{2}$ $\Rightarrow$ $\dfrac{12}{5}$ $\Rightarrow$ $2\dfrac{2}{5}$

2.187   **If $|6m + 7| = 1$, what is the integer value of m?**

Consider the *value* and the *negative of value* of the modulus.

$$6m + 7 = 1 \quad \Rightarrow \quad 6m = -6 \quad \Rightarrow \quad m = -1$$

$$-(6m + 7) = 1 \quad \Rightarrow \quad -6m - 7 = 1 \quad \Rightarrow \quad -6m = 8 \quad \Rightarrow \quad m = -\frac{8}{6} \quad \Rightarrow \quad m = -\frac{4}{3}$$

The integer value of m is –1, not $-\frac{4}{3}$.

2.188   **If $|2f^2 - 13| = 11$, what is the minimum value of f?**

Consider the *value* and the *negative of value* of the modulus.

$$2f^2 - 13 = 11 \quad \Rightarrow \quad 2f^2 = 24 \quad \Rightarrow \quad f^2 = 12 \quad \Rightarrow \quad f = \sqrt{12}$$

$$-(2f^2 - 13) = 11 \quad \Rightarrow \quad -2f^2 + 13 = 11 \quad \Rightarrow \quad -2f^2 = -2 \quad \Rightarrow \quad f^2 = 1 \quad \Rightarrow \quad f = \pm1$$

The number $\sqrt{12}$ is not an integer. But its positive value is a bit more than 3, and the negative value is a bit less than –3. On the GMAT, calculators are not allowed. So, the answer is just $-\sqrt{12}$, or $-2\sqrt{3}$.

2.189   **For a positive integer k, what is the minimum value of $|17 - 4k|$?**

We know that an absolute value can be positive or zero, but not negative. So, the least value of $|17 - 4k|$ would ideally be zero. Let's equate it to zero, and see what k is.

$$17 - 4k = 0 \quad \Rightarrow \quad 17 = 4k \quad \Rightarrow \quad k = \frac{17}{4} \quad \Rightarrow \quad k = 4\frac{1}{4}$$

But k is supposed to be an *integer*. So it could be either 4 or 5, not 4¼ .

If $k = 4$, then $|17 - 4k| = |17 - 4(4)| = |17 - 16| = |1| = 1$

If $k = 5$, then $|17 - 4k| = |17 - 4(5)| = |17 - 20| = |-3| = 3$

Clearly, the minimum value of $|17 - 4k|$ is 1, if k is to be a positive integer.

PRACTICE EXAMPLES

2.190   If $|p + 8| = 20$, find the average of the possible values of p.

2.191   If $14 - |8 + 3b| = 5$, and b is positive, find b.

2.192   If $|3u - 14| = 1$, find the sum of all the possible values of u.

2.193   If $|15 - 4k| = 17$, find the integer value of k.

2.194   Find the maximum value of h if $|40 - h^2| = 9$.

2.195   For the minimum value of $|50 - 6g|$, find the integer value of g?

INEQUALITIES

An **inequality** expresses the unequal relationship between two sets of terms. The terms could be a combination of monomials, polynomials or constants. One set of terms could be greater than (>), less than (<), greater than or equal to (≥), or less than or equal to (≤) the other set of terms. For example, $3x + 7 \geq 12$, $a + b < 4$, etc. are inequalities. The following rules govern inequalities:

➤ Just like in equations, it is permissible to add (or subtract) the same quantity to both sides of an inequality. But doing that to only one side invalidates the inequality.

➤ It is permissible to multiply (or divide) both sides of an inequality by the same *positive* number. But doing that to only one side invalidates it.

➤ *The inequality sign reverses* when both of its sides are multiplied (or divided) by the same *negative* number. For example, given that $a + b \leq 4$, if we multiply both sides by –2, then $-2a - 2b \geq -8$.

➤ *The inequality sign reverses* when we take *reciprocals* (or inverses, or flip) of both of its sides. For example, if $p - 5 > c + k$, then $\frac{1}{p-5} < \frac{1}{c+k}$

➤ Taking squares (or *even* powers) or square roots (or *even* roots) of both sides does not always keep the inequality intact. For example, $-4 < -3$. Squaring both sides, we get $16 > 9$. Inequality sign *reverses*. Instead, if we start with $4 > 3$, then squaring both sides, we get $16 > 9$. Inequality sign *does not reverse*.

2.196 **If $a + 12 > 130$, what is the minimum integer value of a?**

As a general rule, first simplify the inequality by isolating the variable.

Subtracting 12 from both sides, we get $a > 118$. Clearly, a cannot be 118, because it has to be *greater than* 118. So, the very next integer value would be 119. That's the minimum integer value of a.

2.197 **Given that $2 - g \geq 18 + 3g$, what is the maximum integer value of g?**

$-4g \geq 16$ Subtracting $3g$ and 2 from both sides.

$g \leq -4$ Dividing both sides by –4 (a *negative* number). Inequality *reverses*.

The maximum integer value of g is –4 (because g can be –4 too).

2.198 **If $7 + 2q < 3$ and $3q + 16 > 4$, find the integer value of q.**

Simplify the inequalities by isolating the variable.

$7 + 2q < 3$ ⇒ $2q < -4$ ⇒ $q < -2$

$3q + 16 > 4$ ⇒ $3q > -12$ ⇒ $q > -4$

There's only one *integer* value between –2 and –4 that q can take, which is –3.

2.199 **Given that $9b \geq 2b^2$, and b is positive, how many different whole number values can b take?**

Divide both sides of the inequality by b. The sign remains intact because b is positive.

$9 \geq 2b$ ⇒ $4.5 \geq b$ ⇒ $b \leq 4.5$

So b can take the values 1, 2, 3 and 4, but not 0 because 0 is not positive (nor negative).

2.200 **Given that $3d^3 \geq 2d^2$, what is the minimum integer value of d?**

We don't know whether d is positive or negative. But d^2 would certainly be positive.

$3d \geq 2$ Dividing both sides by d^2 (*positive*). Inequality *does not reverse.*

$d \geq \dfrac{2}{3}$

The minimum *integer* value of d would be 1.

2.201 **$11 \leq m < 17$ and $23 < n \leq 300$, m and n being whole numbers. What is the minimum value of: (i) $(m + n)$,   (ii) $(m - n)$,   (iii) $(n - m)$?**

This is a mathematical reasoning type question. Notice that m can assume integer values from 11 to 16 (not 17), and n can assume integer values from 24 (not 23) to 300.

(i)   To find the minimum value of $(m + n)$, we would need the *least* values of both m and n, which would be $(11 + 24)$, or 35.

(ii) To find the minimum value of $(m - n)$, we would need the *least* value of m, and the *greatest* value of n. The idea is to subtract the *most* from the *least*, i.e. $(11 - 300)$, or -289.

(iii) To find the minimum value of $(n - m)$, we would need to subtract the *greatest* value of m from the *least* value of n, i.e. $(24 - 16)$, or 8.

2.202 **$-16 \leq 5e + 10 < 100$.   How many integer values of e are possible?**

First, simplify the inequality.

$-26 \leq 5e < 90$ Subtracting 10 from *all three sides.*

$-\dfrac{26}{5} \leq e < 18$ Dividing all three sides by 5.

$-5\dfrac{1}{5} \leq e < 18$

The possible integer values are from -5 all the way to 17, which is a total of 23 values. Don't forget to count zero; zero *is* an integer.

2.203 **If $a + b + c + d + e + 15 \geq \dfrac{a}{2} + \dfrac{b}{2} + \dfrac{c}{2} + \dfrac{d}{2} + \dfrac{e}{2} + \dfrac{15}{2}$,   what is the *least* value of the average of a, b, c, d and e?**

First, simplify the inequality.

$\dfrac{a}{2} + \dfrac{b}{2} + \dfrac{c}{2} + \dfrac{d}{2} + \dfrac{e}{2} + 15 \geq \dfrac{15}{2}$ Subtracting $\dfrac{a}{2}, \dfrac{b}{2}, \dfrac{c}{2}, \dfrac{d}{2}$ and $\dfrac{e}{2}$ from both sides.

$\dfrac{a}{2} + \dfrac{b}{2} + \dfrac{c}{2} + \dfrac{d}{2} + \dfrac{e}{2} \geq -\dfrac{15}{2}$ Subtracting 15 from both sides.

$\dfrac{a+b+c+d+e}{2} \geq -\dfrac{15}{2}$

$a + b + c + d + e \geq -15$ Multiplying both sides by 2.

To find the average of a, b, c, d and e, divide both sides by 5 (the number of terms).

$\dfrac{a+b+c+d+e}{5} \geq -3$ Therefore, -3 is the *least* possible value of the average.

2.204 **If $3x + 2y \leq 8$, what is the maximum integer value of y for which: (i) x is a positive integer, (ii) x is a positive number?**

The question is about y. So let's first isolate y.

$$3x + 2y \le 8 \quad \Rightarrow \quad 2y \le 8 - 3x \quad \Rightarrow \quad y \le \frac{8-3x}{2} \quad \Rightarrow \quad y \le 4 - 1.5x$$

(i) To find the maximum integer value of y, we should subtract the *least* from 4, i.e. x should assume its least possible positive integer value, which is 1 (not 0). That would give us: $y \le 4 - 1.5(1) \quad \Rightarrow \quad y \le 2.5 \qquad$ The maximum *integer* value of y would then be 2.

(ii) This time, x is a positive *number*, not necessarily an *integer*. It could be a really small positive number, like 0.0000000000000000001, in which case the inequality would yield:

$y \le 4 - 1.5x$
$y \le 4 - 1.5(0.0000 \ldots \ldots \ldots 001)$
$y \le 4 - 0.00000 \ldots \ldots \ldots \ldots 0015$
$y \le 3.999999 \ldots \ldots \ldots \ldots \ldots 99985$

Therefore, the maximum integer value of y would now be 3.

2.205 Given that $p > q > 0 > r$, verify if the following statements always hold true:

(i) $q - p > 0$

(ii) $\frac{1}{q} > \frac{1}{p}$

(iii) $\frac{1}{r} > \frac{1}{q}$

(iv) $rq < rp$

(v) $\frac{q+p}{r} < \frac{q-p}{r}$

(vi) $\frac{p+q}{p-q} > \frac{q+r}{q-r}$

It is always a good practice to arrange the given inequality in increasing order from left to right, simply because that's the way the *number line* is. That way, it's easier to imagine the placement of variables (p, q, r, a, b, x, y etc.) along the number line.

The given inequality can be rewritten as: $\qquad r < 0 < q < p$

Now it's clear that r is negative, while q and p are positive, p being the largest. While solving such questions, start with the *statement under verification*, and use the permissible rules (from page 87) to reach the *given inequality* (or a part of the given inequality).

(i) Under verification is $q - p > 0$. $\qquad$ Adding p to both sides, we get $q > p$. Is this in line with the given inequality? No, because it's given that q is not greater, but less than p. Therefore, this statement does not hold true.

(ii) $\frac{1}{q} > \frac{1}{p} \quad \Rightarrow \quad q < p \qquad$ Inequality sign reverses upon taking reciprocals. The result is in line with (is a part of) the given inequality. Hence, this statement is valid.

(iii) $\frac{1}{r} > \frac{1}{q} \quad \Rightarrow \quad r < q \qquad$ This result is consistent with the given inequality. Hence, this statement is true.

(iv) $rq < rp \quad \Rightarrow \quad q > p \qquad$ Dividing both sides by r, the sign reverses, since $r < 0$. The result is contrary to the given inequality. Thus, this statement is not true.

(v) $\frac{q+p}{r} < \frac{q-p}{r} \quad \Rightarrow \quad q + p > q - p \quad$ Multiplying both sides by r (negative).
$p > -p \qquad\qquad\qquad$ Subtracting q from both sides.
$2p > 0 \qquad\qquad\qquad$ Adding p to both sides.
$p > 0 \qquad\qquad\qquad$ Dividing both sides by 2.
This result matches with the given inequality. So, the statement is valid.

(vi) $\frac{p+q}{p-q} > \frac{q+r}{q-r} \qquad$ There's no way to eliminate any term here. So look to see which terms are positive, and which ones are negative. $(p + q), (p - q)$ and $(q - r)$ are all positive, regardless of the values of p, q and r. But $(q + r)$ will be positive if q is farther from 0 than is r, whereas negative if r is farther from 0 than is q. Meaning, the left side of the inequality may be greater or less than the right side depending upon the actual values of q and r. Thus, the statement does not always hold true.

2.206 $0 < x^2 < y < 1$ **Verify if the following statements are always true:**

(i) $x < y$

(ii) $x^2 y < y$

(iii) $xy > x$

(iv) $\dfrac{x^2}{y} > 1$

(v) $\dfrac{x^2 + y}{x^2 - y} < 0$

Observe that in the given inequality, $x^2 < y$, where x and y are both between 0 and 1 (proper fractions or decimal fractions). Review the properties of such numbers from page 51.

(i) $x < y$.   Notice that x^2 is positive. Meaning, x could be positive or negative. If x is negative, then $x < y$. But if x is positive, then x could be less than or greater than y. For example, x could be $\dfrac{1}{2}$ and y could be $\dfrac{1}{3}$. In that case, $x > y$, while $x^2 < y$ $\left(x^2 \text{ being } \dfrac{1}{4}\right)$. Therefore, this statement is not always true.

(ii) $x^2 y < y$.   Dividing both sides by y (positive), we get $x^2 < 1$. This is in line with the given inequality. Hence, this statement is valid.

(iii) $xy > x$.   Since x may be positive or negative, dividing both sides by x may or may not reverse the sign. Therefore, we would be left wondering whether $y > 1$ or $y < 1$. This statement is therefore not always true.

(iv) $\dfrac{x^2}{y} > 1$.   Multiplying both sides by y (positive), we get $x^2 > y$, which does not match up with the given inequality. Therefore, not a valid statement.

(v) $\dfrac{x^2 + y}{x^2 - y} < 0$ Multiplying both sides by $(x^2 - y)$, a negative quantity, we get $x^2 + y > 0$. This result is valid since x^2 and y are both positive in the given inequality.

PRACTICE EXAMPLES

2.207 If $3f + 7 \geq 25$ and $3f - 19 < 23$, what is the arithmetic mean of the maximum and minimum integer values of f?

2.208 If $3c^2 \leq \dfrac{4c^3}{5}$ and $d^3 < 5d^2$, what integer value is common to both c and d?

2.209 $-8 < j \leq 20$ and $14 \leq k < 70$, where j and k are whole numbers. What is the maximum value of: (i) $(j + k)$, (ii) $(j - k)$, (iii) $(k - j)$?

2.210 $-4 < 20 - 3y \leq 31$. How many integer values of y are possible?

2.211 If $w + x + 9 \leq 13 - y - z$, what is the maximum possible average of w, x, y and z?

2.212 If $2p + q \geq 15$, what is the minimum integer value of p for which: (i) $q \leq 19$, (ii) $q < 19$?

2.213 $p^2 > 0 > q > r$. Verify if the following statements always hold true:

(i) $p - q < p - r$

(ii) $p > qr$

(iii) $\dfrac{q}{r} < 1$

(iv) $\dfrac{p}{q} < \dfrac{p}{r}$

(v) $\dfrac{p}{r} < \dfrac{q}{r}$

(vi) $pqr > 0$

ABSOLUTE VALUE INEQUALITIES

An **absolute value inequality** is one which has a modulus in it. For example, $|2x + 3| \leq 5$. It can be solved, and the *interval (or range) of values* of the variable (x) can be graphed on the number line. Similarly, given a graph of the interval of a variable, an absolute value inequality can be formulated from it. If the variable *cannot assume* the boundary values (or value) of its interval, then those values (or value) are denoted on the number line by a hollow circle. If the variable can assume such values, then a solid circle is used to denote them. The following graphs clarify this notation.

$4 < p < 31$ p cannot assume 4 or 31.

$-11 \leq c \leq 7$ c can assume -11 as well as 7.

$-9 \leq n < 20$ n can assume -9, but not 20.

GRAPHING ABSOLUTE VALUE INEQUALITIES

An absolute value inequality can be graphed using the method we applied while solving an absolute value equation. First, solve the inequality considering the *value of the modulus*. Then do it using the *negative of value of the modulus*. Finally, on the number line, show the range of values obtained. Let's solve a few examples to understand this method.

2.214 **Graph $|x - 5| \leq 12$.**

Consider the *value* and the *negative of value* of the modulus.

$x - 5 \leq 12$ $\Rightarrow$ $x \leq 17$

$-(x - 5) \leq 12$ $\Rightarrow$ $-x + 5 \leq 12$ $\Rightarrow$ $-x \leq 7$ $\Rightarrow$ $x \geq -7$

The graph on the right shows the interval of x.

x can assume -7 as well as 17. Therefore, solid circles.

2.215 **Graph $|5 + 3y| < 11 - y$.**

Consider the *value* and the *negative of value* of the modulus.

$$5 + 3y < 11 - y \quad \Rightarrow \quad 4y < 6 \quad \Rightarrow \quad y < 1.5$$

$$-(5 + 3y) < 11 - y \quad \Rightarrow \quad -5 - 3y < 11 - y \quad \Rightarrow \quad -2y < 16 \quad \Rightarrow \quad y > -8$$

The graph on the right shows the interval of y.

y cannot assume -8 or 1.5. Hence, hollow circles.

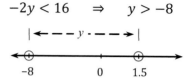

2.216 **Graph $|2c + 3| \geq 15$.**

Consider the *value* and the *negative of value* of the modulus.

$$2c + 3 \geq 15 \quad \Rightarrow \quad 2c \geq 12 \quad \Rightarrow \quad c \geq 6$$

$$-(2c + 3) \geq 15 \quad \Rightarrow \quad -2c - 3 \geq 15 \quad \Rightarrow \quad -2c \geq 18 \quad \Rightarrow \quad c \leq -9$$

The graph below shows the *discontinuous interval* of c.

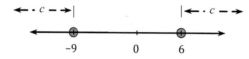

2.217 **Graph $|10 - 3d| > 7 + d$**

Consider the *value* and the *negative of value* of the modulus.

$$10 - 3d > 7 + d \quad \Rightarrow \quad -4d > -3 \quad \Rightarrow \quad d < 0.75$$

$$-(10 - 3d) > 7 + d \quad \Rightarrow \quad -10 + 3d > 7 + d \quad \Rightarrow \quad 2d > 17 \quad \Rightarrow \quad d > 8.5$$

The graph below shows the discontinuous interval of d.

PRACTICE EXAMPLES (Graph these inequalities)

2.218 $|a + 19| \geq 4$

2.219 $|9 - 5b| < 6$

2.220 $|4m + 7| > 19 - 2m$

2.221 $|13 - 5n| \leq 11 + 3n$

SPACE FOR CALCULATIONS

FORMULATING ABSOLUTE VALUE INEQUALITIES FROM GRAPHS

Given the graph of an interval of a variable, an absolute value inequality can be generated by using the steps outlined as follows:

1) Shift the interval to the left or right along the number line and align it with zero. Accordingly, add or subtract the 'number of paces during the shift' from the variable.
2) Shift the interval further and make it *symmetric* about zero. Modify the variable again.
3) Whatever you see, write it in the form of a three-sided inequality.
4) Split it into two inequalities, and express each in terms of the same *positive* constant.
5) Combine the two inequalities (*value* and *negative of value* of a modulus) into one absolute value inequality.

Let's try to systematically understand these steps by graphing the following inequalities.

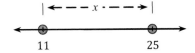

Notice that the interval spans 14 units (from 11 to 25). Move it 11 units to the left and align it with zero. The interval will now be $(x - 11)$, not just x.

Now, make it symmetric about zero by pushing it 7 more units leftwards. The interval will now be $(x - 18)$, and will span 7 units each to the left and right of zero.

What we see is the inequality: $-7 \le (x - 18) \le 7$ It can be rewritten as two separate inequalities:

First inequality: $x - 18 \le 7$

Second inequality: $x - 18 \ge -7$.   Notice the solid circles.

If we multiply both sides of the second inequality by -1, we can rewrite it in terms of a positive constant $(+7)$ as: $-(x - 18) \le 7$.   Notice that the sign reverses.

These two inequalities resemble the *value* and *negative of value* of a modulus. We can combine them into the absolute value inequality: $|x - 18| \le 7$

6-unit interval, moved 8 units to the right, and aligned with zero. p becomes $(p + 8)$.

Moved 3 more units to the right, and placed symmetrically about zero.

$-3 < (p + 11) < 3$ Notice the hollow circles.

First inequality: $p + 11 < 3$

Second inequality: $p + 11 > -3$ $\Rightarrow$ $-(p + 11) < 3$

We multiplied the second inequality (and not the first) by -1 because the inequalities are to be expressed in terms of a *positive* constant (here, $+3$).

Combining the two (*value* and *negative of value* of modulus): $|p + 11| < 3$

2.224
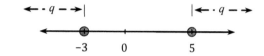

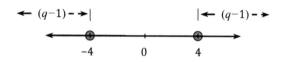

8-unit *discontinuous* interval, moved only 1 unit to the left, and placed symmetrically about zero. No need for two separate steps. q becomes $(q - 1)$.

First inequality: $q - 1 \le -4$ $\Rightarrow$ $-(q - 1) \ge 4$

Second inequality: $q - 1 \ge 4$

We multiplied the first inequality by -1 because the inequalities are to be expressed in terms of a *positive* constant (here, $+4$).

Combining the two (*value* and *negative of value* of modulus): $|q - 1| \ge 4$

2.225

11-unit *discontinuous* interval, moved 3 units to the left, and aligned with zero. w becomes $(w - 3)$.

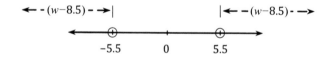

Moved 5.5 more units to the left, and placed symmetrically about zero.

First inequality: $w - 8.5 \le -5.5$ $\Rightarrow$ $-(w - 8.5) \ge 5.5$

Second inequality: $w - 8.5 \ge 5.5$

We multiplied the first inequality by -1 because the inequalities are to be expressed in terms of a *positive* constant (here, $+5.5$).

Combining the two (*value* and *negative of value* of modulus): $|w - 8.5| \ge 5.5$

If you don't like decimals, then multiply it throughout by 2 and rewrite it as: $|2w - 17| \ge 11$

Both the results are equivalent.

PRACTICE EXAMPLES (Formulate absolute value inequalities)

2.226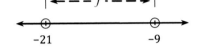

2.227

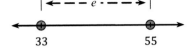

2.228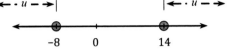

2.229

SPACE FOR CALCULATIONS

CHAPTER 3

WORD PROBLEMS

The last two chapters were the building blocks for what we will now be covering: the *application* of mathematical concepts to solve real-world problems. We will only be studying a few new topics. On the GRE/GMAT, word problems present a brief description of a situation, and then follow it up with a question. In chapters 1 and 2, equations were already set up for you. But in word problems, no equations will be given; you will be expected to correctly formulate them using the given information, and then solve them to figure out the unknown(s). It means double the work for you! That translates into more time required for such questions.

On the test, an efficient strategy would be to save time on questions on arithmetic and algebra, and use it on word problems. It is therefore very important that you be well versed with the concepts from the previous chapters, so that you can readily apply them here.

In order to correctly interpret a word problem, it is necessary to understand the mathematical equivalents of some key phrases outlined below.

PHRASE	WHAT IT MEANS
5 more than x	$x + 5$
x less than 3	$3 - x$
x less 3	$x - 3$
Up to 4 (4 or less)	≤ 4
At least 4 (4 or more)	≥ 4
a times b	$a \times b$
The quotient of a and b	$a \div b$
p increased by 7	$p + 7$
p decreased by 7	$p - 7$
One-third the difference of a and b	$\frac{1}{3}(a - b)$
¾ of q	$\frac{3}{4} \times q$
Is the same as (is equal to)	$=$

Let's apply what we've learned thus far to discuss word problems by topic.

NUMBERS

Questions pertaining to **numbers** involve one or more unknowns to be determined. Regardless of the number of unknowns, it is always possible to use *only one variable* to find *all* of the unknowns. More variables simply mean more equations, more calculations, and more time! Remember, you only have an average of 1½ minutes per question on the GRE, and 2 minutes per question on the GMAT.

3.001 **One-fourth a number less than $\frac{6}{8}$ the same number is 8. What is the number?**

Assume the number as x. One-fourth of it would be $\frac{x}{4}$ and $\frac{6}{8}$ of it would be $\frac{6}{8}x$, or in a reduced

form, $\frac{3}{4}x$. Now simply convert the given sentence into an algebraic expression:

$$\frac{3x}{4} - \frac{x}{4} = 8 \quad \Rightarrow \quad x\left(\frac{3}{4} - \frac{1}{4}\right) = 8 \quad \Rightarrow \quad x \cdot \frac{2}{4} = 8 \quad \Rightarrow \quad \frac{x}{2} = 8 \quad \Rightarrow \quad x = 16$$

3.002 **When twice the difference between a number and 7 is decreased by 3, the result is 4 greater than the number. What is the number?**

Assume the number as x.

The *difference* between x and 7 would be: $x - 7$.

Twice this *difference* would be: $2(x - 7)$

Decreased by 3 would give: $2(x - 7) - 3$

4 greater than the *number* would be: $x + 4$

It is given that *these two* are equal. So set them equal to each other, and solve for x.

$2(x - 7) - 3 = x + 4$

$2x - 14 - 3 = x + 4$

$2x - 17 = x + 4$

$x = 21$

3.003 **A number less 8 is equal to 3 less than the quotient of the number and 2. Find the number.**

Notice the use of the phrases 'less than' and 'less.' They both have different meanings.

The number *less* 8: $x - 8$

Quotient of the number and 2: $\frac{x}{2}$

3 *less than* this quotient: $\frac{x}{2} - 3$

Set them equal: $x - 8 = \frac{x}{2} - 3 \quad \Rightarrow \quad x - \frac{x}{2} = -3 + 8 \quad \Rightarrow \quad \frac{x}{2} = 5 \quad \Rightarrow \quad x = 10$

3.004 **The larger of two numbers is 13 more than 11 times the smaller. If their difference is 3, what is the larger number?**

There's no need to assume the two numbers as x and y. Consider *only one variable*, i.e. only x. Let's also develop the habit of assuming *the number to be determined* as the variable. Here, it is the *larger* of the two numbers that is to be determined.

If the difference between the two numbers is 3, then the *smaller* number must be $x - 3$.

11 times the smaller number: $\qquad\qquad\qquad\qquad$ $11(x - 3)$

13 more than that: $\qquad\qquad\qquad\qquad\qquad$ $11(x - 3) + 13$

Set this equal to the larger number, and solve.

$$11(x - 3) + 13 = x \quad \Rightarrow \quad 11x - 33 + 13 = x \quad \Rightarrow \quad 10x = 20 \quad \Rightarrow \quad x = 2$$

If we had assumed the *smaller* number as x, then we would've had to determine the *larger* number by adding 3 to it. Notice that we avoided that step by assuming *the number to be determined* as the variable. On the GRE/GMAT every second counts. Cut steps; save time!

3.005 **The tens digit of a 2-digit number is thrice the units digit. If the digits are reversed, the resulting number is 36 less than the original number. What is the original number?**

Before we solve this example, consider a concept: For some random 2-digit number, say 75, its *value* can be determined by considering its digits individually. Here's how:

$75 = (7 \times 10) + (5 \times 1)$ $\qquad\qquad$ Review page 28 for 'expanded notation.'

Now let's look at the problem at hand. *Both digits are to be determined.* Assume the units digit as x. Then, the tens digit would be thrice of that, or $3x$. On similar lines as the above number 75, the *value* of the 2-digit number would be $(3x \times 10) + (x \times 1)$ or simply $31x$.

If we reverse the digits, then x would become the new tens digit, whereas $3x$ would go into the units place. The *value* of the new 2-digit number would be $(x \times 10) + (3x \times 1)$ or simply $13x$. This new number is 36 less than the original number:

$$\text{New} = \text{Original} - 36 \quad \Rightarrow \quad 13x = 31x - 36 \quad \Rightarrow \quad -18x = -36 \quad \Rightarrow \quad x = 2$$

Therefore, the original number had 2 as its units digit. The tens digit was thrice that, or 6. Meaning, the original number was 62.

3.006 **In a class of 39 students, there are twice as many girls as boys. How many boys are in the class?**

Assume only one variable: x for boys (the number to be determined).

Twice of that number is the number of girls, or $2x$ girls.

Boys + Girls = 39 $\quad \Rightarrow \quad x + 2x = 39 \quad \Rightarrow \quad 3x = 39 \quad \Rightarrow \quad x = 13$

3.007 **A sandwich shop sells \$2 salads and \$5 sandwiches. With \$38, I bought a total of 13 items, some being sandwiches while others being salads. How many of them were sandwiches?**

Assume x sandwiches (the number to be determined). Now consider this:

If I had bought 10 sandwiches, then I'd have bought $13 - 10 = 3$ salads.

If I had bought 7 sandwiches, then I'd have bought $13 - 7 = 6$ salads.

If I had bought 4 sandwiches, then I'd have bought $13 - 4 = 9$ salads. $\qquad$ And so on......

See the point? If I had bought x sandwiches, then I'd have bought $(13 - x)$ salads.

Total cost of sandwiches $+$ Total cost of salads $=$ Total money spent

$$5x + 2(13 - x) = 38$$

$$5x + 26 - 2x = 38 \quad \Rightarrow \quad 3x = 12 \quad \Rightarrow \quad x = 4$$

3.008 **$4,000 were invested unequally in two stocks which paid 13% and 12% dividends at the end of the year. A total dividend of $490 was earned from the stocks. How much was invested in the stock that paid a 13% dividend?**

Assume x invested in the stock that paid 13% (to be determined).

If x were invested in the 13% stock, then the remaining $(4000 - x)$ must have been invested in the 12% stock.

Dividend from the 12% stock $+$ Dividend from the 13% stock $=$ Total dividend

12% of $(4000 - x) + 13\%$ of $x = 490$

$$\frac{12}{100}(4000 - x) + \frac{13}{100}x = 490$$

$$12(4000 - x) + 13x = 49000 \qquad \text{Make it easier. Multiply both sides by 100.}$$

$$48000 - 12x + 13x = 49000$$

$$x = 1000$$

3.009 **The annual membership fee at health club A is $125, and that at health club B is $35. Additionally, club A charges its members $3 per visit, whereas club B charges $5 per visit. In how many visits a year do members at club A as well as club B end up spending the same amount of money?**

Assume x number of visits (to be determined).

By the end of x visits, money spent at club A = Annual Fee + $3 per visit = $125 + 3x$

By the end of x visits, money spent at club B = Annual Fee + $5 per visit = $35 + 5x$

Both these amounts would be the same at the end of x visits. Therefore, equate the two.

$$125 + 3x = 35 + 5x \quad \Rightarrow \quad -2x = -90 \quad \Rightarrow \quad x = 45$$

PRACTICE EXAMPLES (*Only one variable*, please.)

3.010 When three times a number is increased by 8, the result is four times the same number decreased by 3. What is the number?

3.011 When twice a number is added to twice the difference between the number and 4, the result is 4. What is the number?

3.012 12 less a number is 4 times the quotient of the number and 12. What is the number?

3.013 A number is 6 times another number. Their sum is 28. What is the smaller of the two numbers?

3.014 The units digit of a 2-digit number is twice the tens digit. If the digits are reversed, the resulting number is 12 less than twice the original number. What is the original number?

3.015 In a graduating batch of 65 students, there are 15 more women than men. How many women are in the batch?

3.016 Twice the length of a cardboard is three times its width. If the length and width together span 15 inches, what is the width?

3.017 A beauty shop sold 200 shampoo bottles for a total of $1,250. One type of shampoo sells for $5 a bottle, whereas the other sells for twice as much. How many bottles of the higher priced shampoo did the shop sell?

SPACE FOR CALCULATIONS

CONSECUTIVE INTEGERS

Integers are consecutive if they occur immediately next to one another on the number line. For example, 3, 4, 5 are consecutive integers, as are −14, −13, −12, −11. In a problem on **consecutive integers**, if the smallest number is to be determined, then assume that as x, and accordingly, the next larger integer would be $x + 1$, the one larger than that would be $x + 2$ etc. On the other hand, if the largest number is to be determined, then assume that as x, the one smaller than that as $x - 1$, the one smaller than that as $x - 2$ etc. Let's continue our practice of assuming *the number to be determined* as x.

Consecutive odd integers are those *odd* integers that occur right after one another on the number line. Similarly, **consecutive even integers** are those *even* integers that occur right after one another on the number line. For example, 5, 7, 9 are consecutive odd integers, whereas 4, 6, 8 are consecutive even integers. Notice that consecutive odd integers are spaced out by 2. So are consecutive even integers. When setting up an equation involving such numbers, whether odd or even, assume *the number to be determined* as x, and the larger ones as $x + 2$, $x + 4$, $x + 6$ etc. if the smallest number is to be determined. If the largest number is to be determined, then assume that as x, and the smaller ones as $x - 2$, $x - 4$, $x - 6$ etc.

3.018 **What is the largest of three consecutive odd integers whose sum is 39?**

Assume the largest of them as x. The smaller ones would be $x-2$ and $x-4$.

Add them up, and equate that to 39:

$$x + (x-2) + (x-4) = 39 \quad \Rightarrow \quad 3x - 6 = 39 \quad \Rightarrow \quad 3x = 45 \quad \Rightarrow \quad x = 15$$

3.019 **If the sum of three consecutive integers is 10 more than twice the smallest, what is the largest of them?**

Assume the largest of them as x. Notice that now we are talking about consecutive integers, not consecutive *odd* or *even* integers. Therefore, the numbers would be x, $x-1$, and $x-2$; not x, $x-2$, and $x-4$.

Their sum would be: $\qquad\qquad\qquad\qquad\qquad x + (x-1) + (x-2) \quad$ or $\quad 3x - 3$

10 more than twice the smallest would be: $\quad 10 + 2(x-2) \quad$ or $\quad 10 + 2x - 4 \quad$ or $\quad 2x + 6$

Equate the two: $\qquad 3x - 3 = 2x + 6 \quad \Rightarrow \quad 3x - 2x = 6 + 3 \quad \Rightarrow \quad x = 9$

3.020 **Find the smallest of three consecutive even integers such that the sum of the first and third integers is three times the second integer less 4.**

Assume the smallest of them as x. the larger two would be $x+2$, and $x+4$.

Sum of the first and third integers would be: $\quad x + (x+4) \quad$ or $\quad 2x + 4$

Thrice the second integer less 4 would be: $\quad 3(x+2) - 4 \quad$ or $\quad 3x + 6 - 4 \quad$ or $\quad 3x + 2$

Equate the two: $\qquad 2x + 4 = 3x + 2 \quad \Rightarrow \quad -x = -2 \quad \Rightarrow \quad x = 2$

3.021 **What is the minimum sum of three consecutive even integers whose product is zero?**

Here, it doesn't matter whether we assume the smallest or largest as x. Let's assume the smallest as x. The larger two numbers would be $x+2$, and $x+4$.

Their product is zero: $\qquad x(x+2)(x+4) = 0$

It means, $\qquad$ either $\quad x = 0$

$\qquad\qquad$ or $\qquad x + 2 = 0 \quad \Rightarrow \quad x = -2$

$\qquad\qquad$ or $\qquad x + 4 = 0 \quad \Rightarrow \quad x = -4$

If $x = 0$, then the three numbers are $0, 2, 4$. $\qquad$ Their sum is 6.

If $x = -2$, then the three numbers are $-2, 0, 2$. $\qquad$ Their sum is 0.

If $x = -4$, then the three numbers are $-4, -2, 0$. $\qquad$ Their sum is –6. (The minimum sum)

Instead of all the above analysis, think logically: If the product of three consecutive even integers is zero, then one of them *must be zero*. For their minimum sum, the other two *must be less than zero*, i.e. -2 and -4. There's no other possible combination of numbers.

PRACTICE EXAMPLES

3.022 Find the sum of three consecutive odd integers such that 12 more than the product of the first and second is equal to the product of the second and third.

3.023 If the sum of three consecutive integers is 4 less than four times the smallest, what is the largest of them?

3.024 What is the smallest of five consecutive integers whose sum is zero?

FRACTIONS

A fraction, as we know, is the quotient of two quantities. What fraction of a year is a month? That would be 1 month out of 12 months, or $\frac{1}{12}$. Basically, it means *a part* out of *the whole*, or $\frac{part}{whole}$.

3.025 **What fraction of a day comprises 180 minutes?**

That would be 180 minutes out of the total number of minutes in a day.

The total minutes in a day are 24 hours times 60 minutes each, or 24 × 60, or 1440.

Therefore, the fraction is $\frac{180}{1440}$ or $\frac{18}{144}$ or $\frac{1}{8}$.   Be sure to *reduce* to lowest terms.

3.026 **A classroom has 4 red chairs and 6 green chairs. What fraction of the chairs is *not* green?**

The fraction that is *not* green, must be red.

That fraction would be $\frac{part}{whole}$ or $\frac{red\ chairs}{total\ chairs}$ or $\frac{4}{10}$ or $\frac{2}{5}$.

3.027 **Out of the 27 students in my class, 15 are women, 9 of whom are married. What fraction of the women in my class comprises married women?**

That fraction would be $\frac{part}{whole}$ or $\frac{married\ women}{total\ women}$ or $\frac{9}{15}$ or $\frac{3}{5}$.

Notice that the total *students* (=27) didn't matter. The question was "what fraction of the *women*," and not "what fraction of the *students*." On the GRE/GMAT please read the wording very carefully.

3.028 **After swimming 30 laps, Bradley has completed $\frac{5}{6}$ of the race. How many laps does the race comprise?**

Let's assume that the race has x laps (to be determined). It appears to me that 30 laps is the equivalent of $\frac{5}{6}$ of the race, or $\frac{5}{6}$ of x, or $\frac{5}{6}x$. Equate this to 30:

$$\frac{5}{6}x = 30 \quad \Rightarrow \quad \frac{5x}{6} = \frac{30}{1} \quad \Rightarrow \quad 5x \times 1 = 6 \times 30 \quad \Rightarrow \quad x = \frac{6 \times \cancel{30}^{6}}{\cancel{5}^{1}} \quad \Rightarrow \quad x = 36$$

3.029 **Four students lease an apartment for a semester for $1200 a month. They agree to split the rent equally. In the middle of the semester, one of them transfers to another school, and has to move out. By how much will the monthly rent *increase* for each of the remaining students?**

Initially, each of the 4 students is paying $\frac{\$1200}{4} = \$300$ a month.

After a student moves out, each of the remaining 3 will have to pay $\frac{\$1200}{3} = \$400$ a month.

Thus, the monthly rent for each of the remaining 3 students will increase from $300 to $400, or increase by $100.

3.030 **A certain car dealership greets about 500 visitors each month. Two-thirds of them test drive a car. Two-fifths of those who test drive don't buy. Nobody buys a car without test driving it. Around how many cars does the dealership sell each month?**

Number of visitors each month: 500

How many test drive a car? $\frac{2}{3} \times 500$

$\frac{2}{5}$ of them don't buy. So, $\frac{5}{5} - \frac{2}{5} = \frac{3}{5}$ do buy: $\frac{3}{5} \times \frac{2}{3} \times 500$ or $\frac{2}{5} \times 500$ or   200

Therefore, the car dealership sells around 200 cars each month.

3.031 **If I put anywhere from 15 to 25 gallons of gasoline in my truck, then depending upon the traffic conditions, my truck goes anywhere from 150 to 250 miles in that much fuel. What are the most and least number of miles my truck goes for every gallon of gasoline it uses?**

The most miles per gallon would be when the truck goes the greatest distance while consuming the least gasoline, or $\frac{\text{greatest distance}}{\text{least gasoline}}$ or $\frac{250 \text{ miles}}{15 \text{ gallons}}$ or $16.\overline{6}$ miles per gallon.

The least miles per gallon would be when the truck goes the least distance while gulping the most gasoline, or $\frac{\text{least distance}}{\text{most gasoline}}$ or $\frac{150 \text{ miles}}{25 \text{ gallons}}$ or just 6 miles per gallon!

3.032 **If $\frac{1}{5}$ of a number is 7, then what is $\frac{3}{5}$ of the same number?**

Assume the number as x. Then, $\frac{1}{5}x = 7$, given. What would $\frac{3}{5}x$ be?

Notice that $\frac{3}{5}x$ would simply be 3 times of $\frac{1}{5}x$, or 3 times of 7, or 21.   No need to find x!

3.033 **How many distinct fractions (in their lowest terms) can be formed by picking a number from 1, 3 and 8, and then dividing it by picking a number from 3 and 4?**

First set of numbers: 1, 3, 8   Second set of numbers: 3, 4

Picking 1 from the first set, and dividing it by the numbers from the second set, we get:

$\frac{1}{3}$, $\frac{1}{4}$

Picking 3 from the first set, and dividing it by the numbers from the second set, we get:

$\frac{3}{3}$ (or 1), $\frac{3}{4}$

Picking 8 from the first set, and dividing it by the numbers from the second set, we get:

$\frac{8}{3}$, $\frac{8}{4}$ (or 2)

In their *lowest terms*, the only *fractions* are: $\frac{1}{3}, \frac{1}{4}, \frac{3}{4}$, and $\frac{8}{3}$.   Four distinct fractions.

3.034 **List each integer value of x from 90 to 100 that will yield non-integer values when divided by 2, 3 and 4.**

Integer values of x would be 90, 91, 92, 93, 94, 95, 96, 97, 98, 99, 100.

Which of these values when divided by 2, 3 and 4 yield fractions? *Even numbers* will not yield fractions when divided by 2. So they can all be eliminated.

The shortened list is: 91, 93, 95, 97, 99.

These numbers when divided by 2 or 4 will yield fractions. But when divided by 3, some of them will still yield integers. Which ones? 93 and 99. So remove them too.

What remains is just 91, 95 and 97. Each of them will yield non-integer values (fractional values) when divided by 2, 3 and 4.

Notice that 94 when divided by 4, yields a fraction. But it is excluded because when divided by 2 also, yields an integer, not a fraction.

PRACTICE EXAMPLES

3.035 What fraction of a decade comprises 4½ years?

3.036 A passenger jet has 30 business class seats, and 80 economy class seats. What fraction of the seats on the jet comprises business class seats?

3.037 Comfy Airways owns 45 helicopters, and 60 jets, 35 of which are small jets. What fraction of Comfy Airways' fleet of aircraft is *not* made up of small jets?

3.038 I just started a road trip that spans 450 miles in all. After completing $\frac{4}{15}$ of the trip, how many more miles would I have to go to finish the trip?

3.039 Six students reserve a beach house for spring break for a total of P dollars. They decide on sharing the cost equally. Just before the spring break begins, one of them drops out because of an illness. How much *more* will the beach house now cost each of the rest?

3.040 In response to the 16 job openings a company advertised, it received several applications. One-fourth of the applicants didn't qualify for the jobs, and were rejected. The ones that qualified were interviewed, and one-third of them were selected to fill all the vacant positions. How many job applications had the company received?

3.041 My car goes anywhere from 35 to 40 miles per gallon of gasoline. If I refuel my car anywhere from 10 to 15 gallons, inclusive, what are the least and most miles my car could cover in that much fuel?

3.042 If 10 is $\frac{2}{3}$ of a number, then what is $\frac{1}{3}$ of the same number?

3.043 List the unique integers formed by picking a number from 2, 4 and 6, and then dividing it by picking a number from 1, 3 and 5.

3.044 How many integers between 50 and 80 will result in integers when divided by each of the numbers 5, 10, 15 and 20?

PERCENTAGES

A percent, as we know, is a part *per hundred*. The percent sign (%), or the word 'percent' is the same as $\div 100$, or $\times \frac{1}{100}$. Sentences involving percentages can directly be converted into equations, word by word. For example:

$$\underline{\text{What}} \quad \underline{\text{percent}} \quad \underline{\text{of}} \quad \underline{500} \quad \underline{\text{is}} \quad \underline{300}\,?$$

$$x \qquad \div 100 \qquad \times \qquad 500 \qquad = \qquad 300$$

Solving this equation, we'd get $x = 60$, NOT $x = 60\%$. Please do not add the % sign again! Remember, the value of x is 60, not 60%. If you put 60 in the original sentence, it will all become clear to you:

60 percent of 500 is 300   NOT   60% percent of 500 is 300

On the GRE/GMAT, such confusions run rampant. Some answer choices may have the % sign; others may not. You will have to be alert to avoid being tricked!

3.045 **What is 4% of 120?**

Simply interpret the sentence, word by word, to form an equation:

$$\underline{\text{What}} \quad \underline{\text{is}} \quad \underline{4\%} \quad \underline{\text{of}} \quad \underline{120}$$

$$x \quad = \quad \frac{4}{100} \quad \times \quad 120 \qquad \Rightarrow \qquad x = \frac{4 \times \cancel{120}^{\,6}}{\cancel{100}^{\,5}} \qquad \Rightarrow \qquad x = 4.8$$

3.046 **What percent of 60 is 18?**

$$\underline{\text{What}} \quad \underline{\text{percent}} \quad \underline{\text{of}} \; \underline{60} \; \underline{\text{is}} \; \underline{18}$$

$$x \quad \div 100 \quad \times \quad 60 \; = \; 18 \qquad \Rightarrow \qquad \frac{x}{100} \times 60 = 18 \qquad \Rightarrow \qquad \frac{x}{5} \times 3 = 18$$

$$x = \frac{\cancel{18}^{\,6} \times 5}{\cancel{3}^{\,1}} \qquad \Rightarrow \qquad x = 30$$

Please stay in the habit of first cancelling and reducing. Deal with smaller numbers.

3.047 **What is 25% of 80% of 15q?**

What	is	25	%	of	80	%	of	15q
x	$=$	25	$\times \dfrac{1}{100}$	$\times$	80	$\times \dfrac{1}{100}$	$\times$	15q

$$x = \frac{25^1 \times 80^4}{100^4 \times 100^5} \times 15q \quad \Rightarrow \quad x = \frac{4^1 \times 15^3 q}{4^1 \times 5^1} \quad \Rightarrow \quad x = 3q$$

3.048 **1 kilometer = 1,000 meters. 15,000 meters is what percent of 300 kilometers?**

Focus on the sentence that forms the *question*.

15000	meters	is	what	percent	of	300	kilometers
15000	meters	$=$	x	$\times \dfrac{1}{100}$	$\times$	300	$\times 1000$ meters

I substituted 1000 meters for 1 kilometer, because that's what's given. Now, divide both sides of the equation by 1000 meters. Treat 'meters' as if it is a variable.

$$15 = x \times \frac{1}{100} \times 300 \quad \Rightarrow \quad 15 = 3x \quad \Rightarrow \quad 5 = x$$

3.049 **The quality control department of a toothbrush maker noted that the batch of toothbrushes manufactured in July 2008 had $\frac{1}{6}$% defective pieces. If 3 million toothbrushes were manufactured in July 2008, then how many non-defective pieces did the batch have?**

It appears to me that $\frac{1}{6}$% of 3 million toothbrushes were defective. Let's calculate that:

$$\frac{1}{6} \times \frac{1}{100} \times 3{,}000{,}000 \quad \Rightarrow \quad \frac{30{,}000}{6} \quad \Rightarrow \quad 5{,}000 \text{ toothbrushes}$$

Clearly, the remaining toothbrushes were non-defective. That would be:

$$3{,}000{,}000 - 5{,}000 \quad \Rightarrow \quad 2{,}995{,}000 \text{ toothbrushes}$$

3.050 **A flat-screen TV is priced at $2,500 if you pay all cash. Instead, if you opt for the installment plan, you'd need to pay 40% of the price up front, and $75 a month for 24 months. Which payment option is more expensive?**

The installment plan would cost: 40% of $2,500   + $75/month $\times$ 24 months

$$40\%(2500) + 75 \times 24 = \frac{40}{100^1} \times 2500^{25} + 1800 = 1000 + 1800 = \$2{,}800$$

Compared with the cash price of $2,500, the installment price of $2,800 is more expensive.

3.051 **I paid $25 to a pizza delivery man. My bill was $20, while the rest was his tip. What percent of the bill was his tip?**

His tip = $25 − $20 = $5

What	percent	of	the bill	was	his tip
x	$\times \dfrac{1}{100}$	$\times$	20	$=$	5

$$\Rightarrow \quad x \times \frac{20^1}{100^5} = 5 \quad \Rightarrow \quad x = 25$$

3.052 **Lucy has twice the money as Peter, who has one-third the money as Samantha. What percent of the total money between the three of them does Samantha have?**

We can assume that Samantha has x dollars. But that will give us fractional values in terms of x for Lucy's money and Peter's money. Instead, an easier way is to assume that the person with the least amount of money has x dollars. That would be Peter. Then Lucy would have twice that, or $2x$ dollars. Since Peter has one-third the money as Samantha, it follows that Samantha has thrice the money as Peter, or $3x$ dollars.

What fraction of the total money is with Samantha? That would be:

$$\frac{\text{Money with Samantha}}{\text{Total money}} = \frac{3x}{x + 2x + 3x} = \frac{3x}{6x} = \frac{1}{2}$$

What *percent* of the total money is with Samantha? That would be:

$\frac{1}{2} \times 100\% = 50\%$ Remember converting fractions to percents?

3.053 **An airline company has 250 pilots to fly its airplanes and helicopters. Each pilot is licensed to fly only one type of aircraft. 40% of the pilots are women. Of the 50 helicopter pilots, 60% are women. What percent of the airplane pilots are men?**

The easiest way to solve such examples is to prepare a table with the following fields: Men and women along one side, and airplane and helicopter along the other. Then, fill up the cells using the given information. Finally, fill the remaining cells by calculating what's missing.

	Airplane	Helicopter	TOTAL
Men		40% of 50	60% of 250
Women		60% of 50	40% of 250
TOTAL		50	250

Total 250 pilots:
40% women (Therefore, 60% men)

50 helicopter pilots:
60% women (Therefore, 40% men)

Now calculate the *number of pilots* using the percentages given.

	Airplane	Helicopter	TOTAL
Men		20	150
Women		30	100
TOTAL		50	250

Subtract the helicopter pilots from the total to find the airplane pilots.

	Airplane	Helicopter	TOTAL
Men	130	20	150
Women	70	30	100
TOTAL	200	50	250

$$\text{Airplane pilots that are men} = \frac{\text{Male airplane pilots}}{\text{Total airplane pilots}} \times 100\% = \frac{130}{200} \times 100\% = 65\%$$

In the above final step, please be sure to first *reduce* the 100 with 200. Then divide.

3.054 **A $25,000 investment yields simple interest at the rate of 15% per year. How much interest accumulates every four months?**

Simple interest is the interest that accumulates only on the investment. In contrast, *compound interest* accumulates on the investment as well as on the accumulated interest.

In this example, the simple interest is 15% per year. You could break it down to 5% three times a year, i.e. 5% every four months. That would be:

$$5\% \text{ of } \$25,000 = \frac{5}{100} \times 25,000 = 5 \times 250 = \$1,250$$

PRACTICE EXAMPLES

3.055 12.1% of 400 is what number?

3.056 35 is what percent of 175?

3.057 $19p$ is 25% of what number?

3.058 A travel company generally ends up cancelling 5% of the number of airline tickets it sells. If last year it cancelled 95 tickets, then how many tickets did it originally sell?

3.059 I just bought a new car priced at $20,000. But since I didn't have enough money to pay for it in full, I made a down payment of 20% of the price. In addition, I'll be paying $300 a month for the next 60 months. How much more will I end up paying above the $20,000 price tag?

3.060 A corporate executive earns a $45,000 bonus. If this represents 20% of her annual salary, then how much is her annual salary?

3.061 My sister is $\frac{4}{5}$ my age, and my mom is twice my age. What percent of my mom's age is my sister's age?

3.062 An aquamarine zoo has sharks and dolphins, 30 altogether. 30% of them are sharks. Of the 12 males, 25% are sharks. What is the ratio of male sharks to female dolphins?

3.063 In a certain year, a soccer team plays 20 matches. If it wins 80% of the first 15 matches, then what percent of the remaining 5 matches does it need to win to have an overall 75% wins for the year?

3.064 An equal investment is made in two funds, *A* and *B*, which yield simple interest at the rate of 13% and 12% per year, respectively. What percent of the total interest at the end of the year would come from fund *A*?

PERCENTAGE INCREASE/DECREASE

When a quantity increases or decreases, the *actual change* can be measured by taking the difference between its final and initial values. Similarly, the *percentage change* can be calculated as:

$$\text{Percentage change} = \frac{\text{Change}}{\text{Original value}} \times 100\%$$

Here are some important facts about percentage increase/decrease:

> If a number is increased by x%, and the result is again increased by x%, then the final number obtained is *greater than* if the original number were increased directly by $2x$%. For example, 100 increased by 10%, results in $100 + 10 = 110$. Further, 110 increased by 10%, results in $110 + 11 = 121$. Instead, if 100 were increased only once by 20%, then the result would be 120. Thus, the two increases of 10% each produce a greater result than a one-time increase of 20%.

> If a number is decreased by x%, and the result is again decreased by x%, then the final number obtained is *greater than* if the original number were decreased directly by $2x$%. For example, 100 decreased by 10%, results in $100 - 10 = 90$. Further, 90 decreased by 10%, results in $90 - 9 = 81$. Instead, if 100 were decreased only once by 20%, then the result would be 80. Thus, the two decreases of 10% each produce a greater result than a one-time decrease of 20%.

> If a number is decreased by x%, and the result is increased by x%, then the final number is slightly less than the original number, i.e. the original value is never regained. For example, 100 decreased by 10%, results in $100 - 10 = 90$. Further, 90 increased by 10%, results in $90 + 9 = 99$, not 100. The original value is forever lost.

3.065 **After following a certain diet for a year, Humpty's weight came tumbling down from 315 pounds to 126 pounds. What was his percent weight loss?**

$$\text{Percentage change} = \frac{\text{Change}}{\text{Original value}} \times 100\%$$

$$\text{Percentage weight loss} = \frac{315 - 126}{315} \times 100\% = \frac{189}{315} \times 100\% = \frac{3}{5} \times 100\% = 60\%$$

3.066 **I recently got a 15% raise on my annual salary. I'm now paid \$69,000 a year. What was my annual salary prior to the raise?**

Assume my initial salary as x dollars.

$$\text{Percentage change} = \frac{\text{Change}}{\text{Original value}} \times 100\%$$

$$15\% = \frac{69,000 - x}{x} \times 100\% \quad \Rightarrow \quad 15 = \frac{69,000 - x}{x} \times 100 \quad \Rightarrow$$

$$15x = 6,900,000 - 100x \quad \Rightarrow \quad 115x = 6,900,000 \quad \Rightarrow \quad x = \frac{6,900,000}{115} = 60,000$$

In case you're wondering how I calculated the last part so easily, here's how:

$$\frac{6,900,000}{115} = \frac{690 \times 10^4}{115} = \frac{\cancel{690}^{\,6} \times 10^4}{\cancel{115}^{\,1}} = 6 \times 10^4 = 60,000$$

3.067 **A top-notch investment fund guarantees a 20% return, compounded annually. What is the value of a \$10,000 investment after 3 years?**

Compound interest accumulates on the investment as well as on the interest accumulated year after year. The money grows faster this way than with simple interest. Here's how:

After 1 year: $10,000 + (20\% \text{ of } 10,000) = 10,000 + 2,000 = \$12,000$

After 2 years: $12,000 + (20\% \text{ of } 12,000) = 12,000 + 2,400 = \$14,400$

After 3 years: $14,400 + (20\% \text{ of } 14,400) = 14,400 + 2,880 = \$17,280$

3.068 **In order to leave your present job, you're asking your potential new employer for a 10% increase over your present salary. That employer is actually willing to offer you 10% over your *asking salary*. Your present employer has made you a counter offer: a 20% increase in your current salary. Which employer is offering you a higher salary?**

Assume your current salary as x dollars.

Your *asking salary* is: $x + (10\% \text{ of } x) = x + 0.1x = 1.1x$

Your new employer is offering you: $1.1x + (10\% \text{ of } 1.1x) = 1.1x + 0.11x = 1.21x$

Your present employer's offer is: $x + (20\% \text{ of } x) = x + 0.2x = 1.2x$

It's clear that your potential new employer is offering you a higher salary.

3.069 **Since the year 2001, the price of milk has gone up 25%. How many gallons of milk can be bought today with the money that could buy 20 gallons in 2001?**

Let's say the price of a gallon of milk was x dollars back in 2001. Then, 20 gallons of milk in 2001 would have cost $20x$ dollars.

Today, the price of a gallon of milk is: $x + (25\% \text{ of } x) = x + 0.25x = 1.25x$ dollars

The amount of milk that can be bought with $20x$ dollars (from 2001) today would be:

$$\frac{\text{Amount of money}}{\text{Unit price of milk}} = \frac{20x}{1.25x} = \frac{20}{1.25} = \frac{20}{\frac{5}{4}} = 20 \times \frac{4}{5} = 16 \text{ gallons}$$

In the above step, I converted 1.25 into $\frac{5}{4}$ for ease of calculations. I suggest doing a quick review of equivalent fractions, decimals and percents from page 39.

3.070 **During the holiday season, a local bookstore sells greeting cards for 20% less than the sticker price. If the cards cost the store $4 each, then how much should their sticker price be if the store is to make a 25% profit on them?**

If you read the question carefully, you will realize that the store does not *sell* the cards at the advertised sticker price. It sells the cards at 20% *below* that price.

First, the cards cost the store $4 each, and the store needs to make a 25% profit on them. Meaning, the store needs to sell them at: $4 + (25\% \text{ of } 4) = 4 + 1 = \$5$ each.

The sticker price (assume x dollars) needs to be higher than $5, so that when it is discounted by 20%, results in $5. What could this price be?

(Sticker price) $-$ (20% of Sticker price) $= \$5$ $\Rightarrow$ $x - (20\% \text{ of } x) = 5$ $\Rightarrow$

$x - 0.2x = 5$ $\Rightarrow$ $0.8x = 5$ $\Rightarrow$ $x = \dfrac{5}{0.8} = \dfrac{5}{\frac{4}{5}} = 5 \times \dfrac{5}{4} = \dfrac{25}{4} = \$6.25$

3.071 **I make a fixed monthly payment on my home loan. If in a certain year, I won't be able to make 3 payments, then by what percent should I step up the remaining 9 monthly payments to make up for the lost amount?**

Let's say my monthly payment is x dollars. It means that I pay $12x$ dollars a year on my loan. Instead, if in a certain year, I make only 9 payments of y dollars each ($y > x$), then that total amount ($9y$) should be the same as the total yearly amount I've always paid ($12x$).

$9y = 12x$ $\Rightarrow$ $y = \dfrac{12x}{9}$ $\Rightarrow$ $y = \dfrac{4}{3}x$ $\Rightarrow$ $y = 1.\overline{3}x$

Percentage increase in the monthly amount $= \dfrac{\text{Change}}{\text{Original value}} \times 100\%$

$= \dfrac{y - x}{x} \times 100\% = \dfrac{1.\overline{3}x - x}{x} \times 100\% = \dfrac{0.\overline{3}x}{x} \times 100\% = 33.\overline{3}\%$

An alternate way to solve this example is:

Assume $100 monthly payments. Therefore, $1200 a year would need to be paid.

If only 9 payments are made, then each payment would have to be for $\dfrac{\$1200}{9}$, or $133.33.

A payment of $100 increased to $133.33 means it's a $33.\overline{3}\%$ increase.

3.072 **An airline pilot is advised of an approaching severe thunderstorm reaching a height of just 10% below his present flying altitude. So, as a safety precaution, he increases his flying altitude by an additional 10%. He's now flying the plane 6,000 feet above the thunderstorm. How high was he originally flying the plane?**

Assume x feet as the original altitude.

The thunderstorm reaches: $x - (10\% \text{ of } x) = x - 0.1x = 0.9x$

He raises the plane to: $x + (10\% \text{ of } x) = x + 0.1x = 1.1x$

The difference in the two altitudes is: $1.1x - 0.9x = 0.2x$

This difference is given as 6,000 feet. So, equate the two:

$0.2x = 6,000$ $\Rightarrow$ $x = \dfrac{6,000}{0.2}$ $\Rightarrow$ $x = \dfrac{6,000}{0.2} \times \dfrac{10}{10} = \dfrac{60,000}{2} = 30,000$

PRACTICE EXAMPLES

3.073 After hiring 12 new employees, my firm now has a total of 72 employees. What is the percent increase in the number of employees?

3.074 I sold my TV for $100 through a website, thus taking a $25 loss. What was my percent loss?

3.075 Due to the sustained downturn in the housing market, the $100,000 house I bought 3 years ago lost 10% of its value the first year, 10% of its remaining value the second year, and again 10% of its remaining value the third year. What is my house worth today?

3.076 An electronics store sells all of its items at a 10% discount on Saturdays. There's a 10% sales tax on every item sold. How much will a $200 TV cost on Saturdays?

3.077 Today's average car consumes 40% less gasoline per mile than the average car in 1970. If in 1970, 10 gallons of gasoline could take a car 120 miles, how far would it take a car today?

3.078 I want to sell my car on the internet. I bought it for $1,000 at an auction today. I'm interested in making a 40% profit on the sale. The website on which I'm listing my car sells cars for 30% below the listed price. At what price should I list my car on the website?

3.079 If a house loses 30% of its value during a recession, then by what percent would the resulting value need to increase for the house to be at its original worth again?

3.080 A furniture dealer has 137 equally priced yellow and blue chairs. The blue chairs are in great demand, so he raises their price by 10%. The yellow ones don't sell, so he discounts them by 20%. This results in a $12 difference between their prices. What was their original price?

SPACE FOR CALCULATIONS

RATIO AND PROPORTION

A ratio expresses the comparison between quantities. Any number of quantities can be compared by using ratios. Ratios can be reduced (or raised) by dividing (or multiplying) all of its terms by the same number. Ratios can also be rewritten as fractions when only two quantities are being compared. When an increase in one quantity results in an increase in another quantity, the two quantities are said to be **directly proportional** to each other. When an increase in one causes a decrease in another, the two are **inversely proportional** to each other.

3.081 **What is the ratio of the values of a nickel to a dime to a quarter?**

A nickel is worth 5 cents, a dime worth 10 cents, and a quarter worth 25 cents.

nickel : dime : quarter = 5 : 10 : 25

This ratio can be reduced by 5 to get 1 : 2 : 5.

3.082 **Township _A_ has 25 buildings, each of which has 12 apartments, each of which is home to 2 rats. Township _B_ has 100 houses, each of which is home to 3 cats. What is the ratio of number of cats in township _B_ to the number of rats in township _A_?**

Number of rats in township _A_ = 25 buildings × 12 apartments × 2 rats = 600 rats

Number of cats in township _B_ = 100 houses × 3 cats = 300 cats

(# of cats in township _B_) : (# of rats in township _A_) = 300 : 600 = 1 : 2 (when reduced by 300)

3.083 **If $\frac{3}{5}$ of the patrons in a cinema hall are women, what is the ratio of men to women in the cinema hall?**

$\frac{3}{5}$ are women. So, $\frac{2}{5}$ must be men, because $\frac{5}{5} - \frac{3}{5} = \frac{2}{5}$, where $\frac{5}{5}$ represents ALL those present.

Therefore,   men : women $= \frac{2}{5} : \frac{3}{5} = \frac{\frac{2}{5}}{\frac{3}{5}} = \frac{2}{5} \times \frac{5}{3} = \frac{2}{3} = 2 : 3$

The same result can be obtained by simply multiplying $\frac{2}{5} : \frac{3}{5}$ by 5 (raising the ratio by 5).

3.084 **A \$24 million lottery prize is to be divided among the winner, the first runner-up and the second runner-up in the ratio 3 : 2 : 1. How much does the first runner-up get?**

We're given that   (winner) : (1ˢᵗ runner-up) : (2ⁿᵈ runner-up) = 3 : 2 : 1

The 1ˢᵗ runner-up's share is 2 parts out of a total of 3 + 2 + 1 = 6 parts.

\$24 million can be divided into 6 parts, each part being worth $\frac{\$24M}{6}$, or \$4 million.

The 1ˢᵗ runner-up gets 2 such parts worth \$4 million each, or a total of \$8 million.

3.085 **Health clubs _A_ and _B_ had 300 and 200 members. During their promotional campaigns, club _B_ acquired twice as many new members as club _A_ did. As a result, the ratio of members of club _A_ to club _B_ became 7 : 6. How many new members joined club _B_?**

If we assume _x_ members joined club _B_, then we would end up with $\frac{x}{2}$ members having joined club _A_. Avoid fractions! So, assume _x_ members joined club _A_, and 2_x_ members joined club _B_.

After the promotional campaign, the new ratio of members of club A to members of club B would be $(300 + x) : (200 + 2x)$. This ratio is given as $7 : 6$. Therefore, equate the two.

$$(300 + x) : (200 + 2x) = 7 : 6 \quad \Rightarrow \quad \frac{300 + x}{200 + 2x} = \frac{7}{6} \quad \Rightarrow \quad 6(300 + x) = 7(200 + 2x)$$

$$1800 + 6x = 1400 + 14x \quad \Rightarrow \quad -8x = -400 \quad \Rightarrow \quad x = \frac{-400}{-8} \quad \Rightarrow \quad x = 50$$

It means 50 new members joined club A.   So, $2x = 2 \times 50 = 100$ new members joined club B.

3.086 **Columbia is 200 miles away from Atlanta. On a wall map of the United States, this distance is represented by 4 inches. On the same map, if Seattle is 5 feet away from Columbia, then what is the actual distance between Seattle and Columbia?**

On the map, 4 inches represents 200 miles. So, 1 inch must be worth $\frac{200}{4}$, or 50 miles.

On that map, Seattle is 5 feet (or $5 \times 12 = 60$ inches) away from Columbia. The actual distance between Seattle and Columbia must be 60 inches $\times$ 50 miles (for every inch) = 3,000 miles.

3.087 **If I can eat 3 cookies every 4 minutes, then how many minutes will it take me to finish 12 cookies, eating at the same rate?**

This is a problem on direct proportion. Simply match up what's given with what's asked.

3 cookies : 4 minutes = 12 cookies : x minutes

$$3 : 4 = 12 : x \quad \Rightarrow \quad \frac{3}{4} = \frac{12}{x} \quad \Rightarrow \quad 3x = 4 \times 12 \quad \Rightarrow \quad x = \frac{48}{3} \quad \Rightarrow \quad x = 16$$

3.088 **If I can walk y yards in h hours, then how many feet can I walk in m minutes, walking at the same rate? (Note: 1 yard = 3 feet)**

In such examples, first *standardize the units*. Convert yards into feet, because the question is in terms of feet, not yards. For the same reason, convert hours into minutes.

1 yard = 3 feet	2 yards = 3(2) = 6 feet	3 yards = 3(3) = 9 feet	y yards = $3y$ feet
1 hr = 60 min	2 hrs = 60(2) = 120 min	3 hrs = 60(3) = 180 min	h hrs = $60h$ min

y yards : h hrs = $3y$ feet : $60h$ min   Using the above two conversions.

$3y$ feet : $60h$ min = x feet : m min   Equating what's known to what's asked.

$$3y : 60h = x : m \quad \Rightarrow \quad \frac{3y}{60h} = \frac{x}{m} \quad \Rightarrow \quad \frac{3y \times m}{60h} = x \quad \Rightarrow \quad x = \frac{ym}{20h} \text{ feet}$$

PRACTICE EXAMPLES

3.089 What is the ratio of the value of a dollar to that of a quarter to that of two dimes?

3.090 On planet X, a calendar day comprises 60 hours, each hour having 24 minutes, each minute having 24 seconds. What is the ratio of the number of seconds in a calendar day on planet X to that on planet Earth?

3.091 In a certain nation, if the ratio of those who don't eat meat to those who do is $13 : 4$, then what fraction of the people in that nation eats meat?

3.092 A graduate school accepts a total of 60 students from economics, engineering and liberal arts backgrounds in the ratio $5 : 2 : 3$. How many students with an engineering background are accepted?

.093 A consulting firm employs 40 men and 50 women. Due to recession, it lays off the same number of men and women employees. As a result, the ratio of men to women becomes $5:7$. How many employees does the firm lay off?

094 The scale on a map says 1 cm = 75 miles. If the distance between two towns on the map is 2 inches, what is the actual distance between the two towns? (Note: 1 inch = 2.5 cm)

)95 If you can read at a constant rate of 300 words per minute, then how many hours will it take you to read 27,000 words?

3.096 A printing press can print P pages in S seconds. At that rate, how many pages would it print in ½ hour?

SPACE FOR CALCULATIONS

SIMPLE AVERAGE

An average, or mean, or arithmetic mean is the sum of elements divided by the number of elements. Average can also be calculated when a quantity changes from one value to another with respect to another variable. For example, if the atmospheric temperature increases from 50°F to 70°F over 5 hours, then the average *hourly* increase in temperature would be the change in temperature divided by the number of hours over which that change occurred, i.e. 20°F ÷ 5 hours, or 4°F/hour.

3.097 **A contestant on a quiz has scored 42, 50, 39 and 44 in the first four rounds. What must she score in the fifth round to average 45 across all five rounds?**

Assume she'd need to score x.

$$\frac{42 + 50 + 39 + 44 + x}{5} = 45 \quad \Rightarrow \quad \frac{175 + x}{5} = 45 \quad \Rightarrow \quad 175 + x = 5 \times 45 \quad \Rightarrow$$

$175 + x = 225 \quad \Rightarrow \quad x = 50$

An alternate (faster) way would be to see how far each value is from the average of 45:

42 is 3 short of 45	−3
50 is 5 more than 45	+5
39 is 6 short of 45	−6
44 is 1 short of 45	−1
Add up these points:	−5

Overall, it looks like her score is 5 short of 45. So x needs to be 5 more than 45, i.e. $x = 50$.

3.098 **The sum of 20 numbers is what percent of their average?**

$$\text{average} = \frac{\text{sum}}{\text{\# of elements}} \quad \Rightarrow \quad \text{average} = \frac{\text{sum}}{20} \quad \Rightarrow \quad a = \frac{s}{20} \quad \Rightarrow \quad s = 20a$$

Now, convert the given sentence into an equation:

The sum of 20 numbers	is	what	percent	of	their average
s	$=$	x	$\times \dfrac{1}{100}$	$\times$	a

Substitute $20a$ in the place of s.

$$20a = \frac{x}{100} \times a \quad \Rightarrow \quad 20 = \frac{x}{100} \quad \Rightarrow \quad x = 2000$$

The sum of 20 numbers is 2000 percent of their average.

3.099 **Freshly prepared cheesecakes at the room temperature of 75° were put in a freezer. Four hours later, they were at 5° below zero. What was the average hourly drop in temperature of the cheesecakes?**

Total drop in temperature = Drop from 75° to 0° + Drop from 0° to −5° = $75 + 5 = 80°$

$$\text{Average hourly drop in temperature} = \frac{\text{total drop}}{\text{time elapsed}} = \frac{80°}{4 \text{ hours}} = 20°/\text{hour}$$

3.100 **The average of four numbers is 26. If you increase two of those numbers by 9 each, reduce the third one by 10, and leave the fourth one unchanged, what is the new average of the four numbers?**

$$\frac{\text{sum}}{\text{\# of numbers}} = \text{average} \quad \Rightarrow \quad \frac{\text{sum}}{4} = 26 \quad \Rightarrow \quad \text{sum} = 26 \times 4 = 104$$

If two of those numbers increase by 9 each, and the third drops by 10, then the sum will increase by $9 + 9 - 10 = 8$.

The new sum will be $104 + 8 = 112$, and the new average will be $\dfrac{112}{4}$, or 28.

3.101 **A school's entrance exam has five parts, each of which is scored on a scale of 50. An overall 80% score is required for admission. I've scored 39, 44 and 41 on the first three parts of the exam. What's the least I could score on the fourth part, and still be able to get into the school?**

The total maximum possible score is $5 \times 50 = 250$.

To get into the school, I'd need to get to 80% of that, i.e. 80% of 250, i.e. 200.

On the first three parts, I've scored a total of $39 + 44 + 41 = 124$.

It means, I still need to score $200 - 124 = 76$ between parts four and five. For me to think "what's the least I can score on the fourth part," I'd also need to consider "what's the most I can possibly score on the fifth part," which is 50.

So, if I do score a perfect 50 on the fifth part, then I'd need to score a minimum of $76 - 50 = 26$ on the fourth part. Only then I'd have a total score of 200 needed to get into the school.

3.102 **A charity received 200 monetary donations, each worth either \$20 or \$50. If the average value of all donations was \$32, how many \$50 donations did the charity receive?**

Assume it received x number of \$50 donations. Then it must have received $(200 - x)$ number of \$20 donations.

$$\text{average value of donations} = \frac{\text{total value of donations}}{\text{number of donations}} \quad \Rightarrow \quad 32 = \frac{50x + 20(200 - x)}{200} \quad \Rightarrow$$

$$32 \times 200 = 50x + 4000 - 20x \quad \Rightarrow \quad 6400 = 30x + 4000 \quad \Rightarrow \quad 2400 = 30x \quad \Rightarrow \quad 80 = x$$

PRACTICE EXAMPLES

3.103 I remember in my second semester of college, I had grades of 85, 79, 80 and 92 in four of the five courses I took. My average grade for that semester was 85. What was my grade in the fifth course?

3.104 The average of 17 numbers is what fraction of their sum?

3.105 A bullet train starts from a station at 9:00 a.m. At 9:05 a.m. it speeds up to 160 mph. What is its average gain in speed per minute up to that time?

3.106 The average top speed of three specific cars in a race is 120 mph. If, in the next race, the top speed of one of them increases by 12 mph, that of another decreases by 3 mph, and that of the third decreases by 6 mph, what is the average top speed of the three cars in that race?

3.107 In three of the six tests scored out of 100 each, Abigail's grades are 79, 88 and 86. Just how low can she score on the sixth test, and still be able to average 90 over the six tests?

3.108 Fifteen families were relocated to new apartments, each home having an area of either 500 sq. ft. or 800 sq. ft. If the average area of the 15 apartments was 660 sq. ft., then how many families were relocated into 500 sq. ft. homes?

WEIGHTED AVERAGE

Consider an automobile dealership that has 10 new trucks priced at $20,000 each, and 50 used cars priced at $10,000 each. What is the average price of all 60 vehicles? Is it just the average of $20,000 and $10,000? No.

To answer that question, we'd need to calculate the total worth of the 60 vehicles, and then divide it by 60, the number of vehicles. A little intuition will tell us that the average won't be at the center point, i.e. at $15,000. It will be closer to $10,000 than to $20,000, because much more vehicles are priced at $10,000 than at $20,000. Meaning, the $10,000 cars collectively exert much more *weight* on the average than do the $20,000 trucks. Therefore, such an average is called **weighted average**.

3.109 **In a class of 30 students, the average weight of the 16 girls is 75 pounds, and that of the 14 boys is 90 pounds. What is the average weight of the 30 students?**

Total weight of the 16 girls = 75 pounds × 16 = 1200 pounds

Total weight of the14 boys = 90 pounds × 14 = 1260 pounds

Total weight of the 30 students = Total weight of boys + Total weight of girls

= 1200 + 1260 = 2460 pounds

$$\text{Average weight of the 30 students} = \frac{\text{Total weight of the 30 students}}{30} = \frac{2460}{30} = 82 \text{ pounds}$$

3.110 **A big law firm has 5 partners making $400,000 a year each, 8 attorneys making $250,000 a year each, and 3 paralegals making $80,000 a year each. What is the average annual income of all of them?**

Total income of the 5 partners = 400,000 × 5 = $2,000,000

Total income of the 8 attorneys = 250,000 × 8 = $2,000,000

Total income of the 3 paralegals = 80,000 × 3 = $240,000

Total income of all 16 of them = 2,000,000 + 2,000,000 + 240,000 = $4,240,000

$$\text{Their average income} = \frac{\text{Total income of all 16 of them}}{16} = \frac{4,240,000}{16} = \$265,000$$

3.111 **A certain graduate school has 40 male students, 30 female students and 10 professors. The average age of male students is 28, and that of female students is 26. If the average age of the students and professors combined is 30, then what is the average age of the professors?**

Total age of the 40 male students = 28 × 40 = 1120 years

Total age of the 30 female students = 26 × 30 = 780 years

Total age of the 10 professors $= x \times 10 = 10x$ years

$$\text{Average age of all} = \frac{\text{Total age of all}}{40 + 30 + 10} \quad \Rightarrow \quad 30 = \frac{1120 + 780 + 10x}{80} \quad \Rightarrow$$

$30 \times 80 = 1900 + 10x \quad \Rightarrow \quad 2400 = 1900 + 10x \quad \Rightarrow \quad 500 = 10x \quad \Rightarrow \quad x = 50$ years

PRACTIC EXAMPLES

3.112 In a science experiment, 12 electrons are accelerated to a speed of 200 km/sec, 3 protons are accelerated to 300 km/sec, and 5 neutrons are accelerated to 700 km/sec. What is the average speed of all the accelerated particles?

3.113 A rental car company has 30 cars valued at $18,000 each, 25 vans valued at $26,000 each, and 15 trucks valued at $14,000 each. What is the average value of all the vehicles the rental car company has?

3.114 In the last decade, on an average, there have been 200 UFO sightings in each of the 7 southern states, 150 in each of the 6 northeastern states, and some in the 5 western states. If the average number of sightings for all those 18 states is 300, then an average of how many UFO sightings have occurred in each of the 5 western states?

SPACE FOR CALCULATIONS

MIXTURE

When two or more consumables are mixed, the *monetary value* of the mixture is the sum of the monetary values of the individual consumables. For example, if 50 pounds of yellow corn worth $2 per pound is mixed with 70 pounds of white corn worth $3 per pound, then the monetary value of the mixture would be:

Value of yellow corn + Value of white corn $= (50 \times 2) + (70 \times 3) = 100 + 210 = \$310$

If this mixture is to be priced *per pound*, then simply divide the monetary value by the quantity:

$$\text{Price} = \frac{\$310}{120 \text{ pounds}} = \$2.58/\text{pound}$$

3.115 **Calculate the dollar value of a mixture consisting of 100 barrels of crude oil worth $50 per barrel and 250 barrels of refined oil worth $120 per barrel.**

Value of mixture = Value of crude oil + Value of refined oil $= (100 \times 50) + (250 \times 120) =$

$5,000 + 30,000 = \$35,000$

3.116 **I added some lima beans worth $3 per pound to 30 pounds of kidney beans worth $2.50 per pound to make a mixture worth $2.80 per pound. How many pounds of lima beans did I add?**

Assume I added x pounds of lima beans. Then the total mixture would be of $(x + 30)$ pounds.

Value of mixture = Value of lima beans + Value of kidney beans $\quad \Rightarrow$

$$2.80(x + 30) = 3x + 2.50(30) \quad \Rightarrow \quad 28(x + 30) = 30x + 25(30)$$

I multiplied both sides of the equation by 10 to eliminate the decimal. It's easier that way.

$$28x + 840 = 30x + 750 \quad \Rightarrow \quad -2x = -90 \quad \Rightarrow \quad x = 45$$

3.117 **A 200 pound mixture of wheat flakes and corn flakes is worth \$3.25 per pound. If corn flakes are worth \$4 per pound, and wheat flakes are worth \$1.50 per pound, then how many pounds of corn flakes are in the mixture?**

Assume x pounds of corn flakes. The mixture comprises 200 pounds total. It means there must be $(200 - x)$ pounds of wheat flakes in it.

Value of mixture = Value of corn flakes + Value of wheat flakes $\Rightarrow$

$$3.25(200) = 4x + 1.50(200 - x) \quad \Rightarrow \quad 325(200) = 400x + 150(200 - x) \quad \Rightarrow$$

I multiplied both sides of the equation by 100 to eliminate the decimals.

$$65,000 = 400x + 30,000 - 150x \quad \Rightarrow \quad 35,000 = 250x \quad \Rightarrow \quad \frac{35,000}{250} = x \quad \Rightarrow \quad x = 140$$

3.118 **A 5 gallon fuel mixture consists of 80% alcohol and 20% gasoline, by volume.**
(i) **How many gallons of gasoline should be added to it to change its composition to 30% gasoline and 70% alcohol?**
(ii) **How many gallons of alcohol should be added (to the original mixture) to change its composition to 90% alcohol and 10% gasoline?**

First find out just how much alcohol and gasoline are there in the mixture.

$$\text{Amount of alcohol} = 80\% \text{ of 5 gallons} = \frac{80}{100} \times 5 = 4 \text{ gallons}$$

$$\text{Amount of gasoline} = 20\% \text{ of 5 gallons} = \frac{20}{100} \times 5 = 1 \text{ gallon}$$

(i) Assume x gallons of gasoline added to the original 1 gallon of gasoline in the mixture. The new amount of gasoline would then be $(x + 1)$ gallons. The alcohol would remain unchanged at 4 gallons within the mixture.

The original ratio of   alcohol : gasoline   was   80% : 20%,   or   80 : 20,   or simply   8 : 2.

The new ratio is 70% : 30%,   or 70 : 30,   or just 7 : 3. This is the same as the new ratio of 4 gallons : $(x + 1)$ gallons.

The proportion can be written as $7 : 3 = 4 : (x + 1)$

$$\frac{7}{3} = \frac{4}{x + 1} \quad \Rightarrow \quad 7(x + 1) = 3 \times 4 \quad \Rightarrow \quad 7x + 7 = 12 \quad \Rightarrow \quad 7x = 5 \quad \Rightarrow \quad x = \frac{5}{7}$$

(ii) Assume y gallons of alcohol added to the original 4 gallons of alcohol in the mixture. The new amount of alcohol would then be $(y + 4)$ gallons. The gasoline would remain unchanged at 1 gallon within the mixture.

The original ratio of   alcohol : gasoline   was   8 : 2.

The new ratio is 90% : 10%,   or 90 : 10,   or just 9 : 1. This is the same as the new ratio of $(y + 4)$ gallons : 1 gallon.

The proportion can be written as $9 : 1 = (y + 4) : 1$

$$\frac{9}{1} = \frac{y+4}{1} \quad \Rightarrow \quad 9 = y + 4 \quad \Rightarrow \quad y = 5$$

PRACTICE EXAMPLES

3.119 Find the dollar value of a mixture consisting of 35 gallons of peanut oil worth $4 per gallon and 3 gallons of olive oil worth $45 per gallon.

3.120 Brown rice worth ¥150/kg was added to 60 kg of basmati rice worth ¥180/kg to make a rice mixture worth ¥168/kg. How many kg of brown rice was added?

3.121 A 15 pound mixture of cashews and pistachios is worth $3.20 per pound. If cashews are worth $2.50 per pound, and pistachios are worth $6 per pound, then how many pounds of pistachios are in the mixture?

3.122 A 20 gallon bio-fuel mixture contains 75% vegetable oil and 25% diesel by volume. (i) How many gallons of vegetable oil should be added to it to change its composition to 90% vegetable oil and 10% diesel? (ii) How many gallons of diesel should be added (to the original mixture) to change its composition to 75% diesel and 25% vegetable oil?

SPACE FOR CALCULATIONS

AGE

Age-related problems can be solved simply by using basic arithmetic and algebra. We will not be talking about age-related problems such as arthritis, knee replacement surgeries, cholesterol control, or colon cancer prevention! We will limit our discussion to math problems only.

The interpretation of the following two phrases is key to solving such questions correctly:

'x years from now'	means	x years added to the variable
'y years ago'	means	y years subtracted from the variable

3.123 **Sam is 4 years younger than Pam. Four years ago, Pam was twice Sam's age back then. How old is Pam now?**

Assume Pam is x years old. Use an easy chronological set-up, the following way:

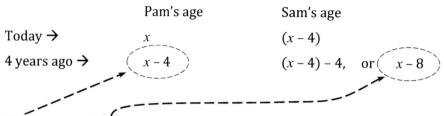

	Pam's age	Sam's age
Today →	x	$(x-4)$
4 years ago →	$x-4$	$(x-4)-4$, or $x-8$

This is twice of that, given.   Write the equation, and solve for x:

$$x-4 = 2(x-8) \quad\Rightarrow\quad x-4 = 2x-16 \quad\Rightarrow\quad -x = -12 \quad\Rightarrow\quad x = 12$$

3.124 **Veronica is 7 times as old as her daughter today. If 4 years from now, Veronica will be 4 times as old as her daughter will be then, how old is Veronica today?**

Assume Veronica's daughter (not Veronica) is x years old. Think convenience. Avoid fractions.

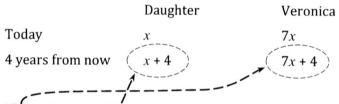

	Daughter	Veronica
Today	x	$7x$
4 years from now	$x+4$	$7x+4$

That is 4 times of this, given.   Write the equation, and solve for x:

$$7x+4 = 4(x+4) \quad\Rightarrow\quad 7x+4 = 4x+16 \quad\Rightarrow\quad 3x = 12 \quad\Rightarrow\quad x = 4 \quad\Rightarrow\quad 7x = 28$$

3.125 **Eric is 26 and Erica is 20. How many years ago was Eric twice Erica's age back then?**

Slightly different question, but use the same way to analyze it. Assume x years ago Eric was twice Erica's age.

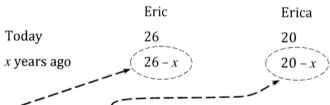

	Eric	Erica
Today	26	20
x years ago	$26-x$	$20-x$

This is twice of that, given.

$$26-x = 2(20-x) \quad\Rightarrow\quad 26-x = 40-2x \quad\Rightarrow\quad x = 14$$

3.126 **My wife is 3 years younger than I. If 7 years from today, I'll be twice as old as my wife was 10 years ago, how old am I today?**

Three different points in time: 10 years ago, today, 7 years from now. Assume I'm x years old.

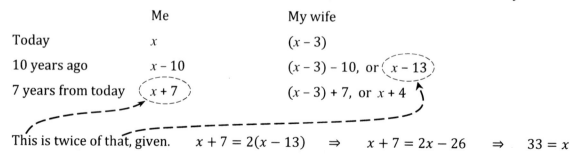

	Me	My wife
Today	x	$(x-3)$
10 years ago	$x-10$	$(x-3)-10$, or $x-13$
7 years from today	$x+7$	$(x-3)+7$, or $x+4$

This is twice of that, given. $x+7 = 2(x-13) \quad\Rightarrow\quad x+7 = 2x-26 \quad\Rightarrow\quad 33 = x$

PRACTICE EXAMPLES

3.127 Peter is 6 years older than Rebecca. Eighteen years ago, he was twice her age back then. How old is Rebecca today?

3.128 Coleen is twice as old as Liz today. Five years from now, twice Coleen's age would equal thrice Liz's age then. How old is Coleen now?

3.129 Ulrich is now twice as old as Ulster. If 9 years from now, the sum of their ages will be 30, how old is Ulster now?

3.130 I'm 3 years younger than my sister. If 4 times my age 6 years ago equals my sister's age 6 years from now, how old am I?

SPACE FOR CALCULATIONS

EXPONENTIAL GROWTH/DECAY

Consider a country C in which there were 100 car owners in the year 1900. Since then, every decade the number of car owners has doubled. How many car owners were there in country C in 2000?

In 1900:	100 cars	or	100×2^0
In 1910:	100×2	or	100×2^1
In 1920:	$100 \times 2 \times 2$	or	100×2^2
In 1930:	$100 \times 2 \times 2 \times 2$	or	100×2^3

...

...

| In 2000: | | | 100×2^{10} |
| In n decades: | | | 100×2^n |

Generally speaking, if a certain quantity starts at A_1 (here, number of cars), and increases to f times itself (doubles, triples etc.) every n units of time (years, hours, minutes etc.), then that quantity is said to be experiencing **exponential growth**. Instead, if it decreases to f times itself every n units of time, then it is said to be experiencing **exponential decay (decline)**. Whether increasing or decreasing, the value of that quantity (A_2) at any point in time is given by the formula: $A_2 = A_1 \cdot f^n$

What if we were to find the number of cars in 1926? Then we'd need to express the exponent n (decade) as a fraction: $\frac{10}{10} = 1$ decade, $\frac{20}{10} = 2$ decades, $\frac{26}{10} = 2.6$ decades etc.

The formula would then become: $A_2 = A_1 \cdot f^{\frac{t}{T}}$ where t = point in time under consideration, and T = the time it takes for the quantity A_1 to increase (or decrease) to f times itself.

The number of cars in 1926 would be $100 \times 2^{\frac{26}{10}}$, or $100 \times 2^{2.6}$.   On the GMAT, calculators are not allowed, so the exact value may be hard to determine. But such questions, if asked, will have answer choices in exponential form (as above). Only if the exponents are integers or easy roots, will you be expected to perform the actual calculations by hand.

3.131 **Every 12 years, a certain village has been losing 10% of its existing population to urban migration. If 1,000 people lived in that village in 1960, then what was its population in 1996?**

Losing 10% means decreasing to 90% of its existing size, or decreasing to 0.9 times itself.

At the start:	1960	1,000	
After 12 years:	1972	$1,000 \times 0.9$	900
After 12 more years:	1984	$1,000 \times 0.9 \times 0.9$	810
After 12 more years:	1996	$1,000 \times 0.9 \times 0.9 \times 0.9$	729

Instead, by the formula: $A_2 = A_1 \cdot f^{\frac{t}{T}} = 1,000 \times 0.9^{\frac{36}{12}} = 1,000 \times 0.9^3 = 1,000 \times 0.729 = 729$

3.132 **A certain bacteria triple in number every 20 minutes. How long will it take a million such bacteria to reach 27 million in number?**

At the start:	0 minutes	1 million	
After 20 minutes:	20 minutes	(1×3) million	3 million
After 20 more minutes:	40 minutes	$(1 \times 3 \times 3)$ million	9 million
After 20 more minutes:	60 minutes	$(1 \times 3 \times 3 \times 3)$ million	27 million

By the formula: $27 = 1 \times 3^{\frac{t}{20}}$ $\Rightarrow$ $3^3 = 3^{\frac{t}{20}}$ $\Rightarrow$ $3 = \frac{t}{20}$ $\Rightarrow$ $t = 60$ minutes

PRACTICE EXAMPLES

3.133 At a certain place, the atmospheric temperature drops by half with every mile rise. If the surface temperature at that place is 96°F, then what is the temperature 6 miles above it?

3.134 Rats quadruple in number every month. How long will it take 20 rats to grow to 1280 rats?

SPACE FOR CALCULATIONS

SPEED (RATE)

Speed is the rate of travel. It is measured in terms of the distance covered per unit time, for example miles per hour (mph), kilometers per hour (km/h), meters per second (m/s) etc. Simply looking at the unit of measurement of speed gives us an idea about the formula for speed:

Miles per hour, or $\frac{\text{miles}}{\text{hour}}$. Distance is measured in miles, whereas time is measured in hours. Clearly, speed must be distance divided by time:

$$\text{speed} = \frac{\text{distance}}{\text{time}} \quad \text{or} \quad s = \frac{d}{t} \qquad \text{Variations of this formula are often useful:} \quad d = st \qquad t = \frac{d}{s}$$

3.135 **If I left Atlanta at 11 a.m., and arrived 330 miles away at Charleston at 4:30 p.m., then what was my average speed for the trip?**

The time elapsed from 11 a.m. to 4:30 p.m. is 5.5 hours. The distance covered is 330 miles.

Please note that the term 'average speed' is often used because, during travel, the speed of a vehicle often changes depending upon traffic conditions. Therefore, the overall speed is expressed in terms of average speed.

$$\text{speed} = \frac{\text{distance}}{\text{time}} = \frac{330 \text{ miles}}{5.5 \text{ hours}} = \frac{330}{5.5} \times \frac{10}{10} = \frac{3300}{55} = \frac{300}{5} = 60 \text{ mph}$$

In the last step, I reduced by the numerator and the denominator by 11.

3.136 **Kurt can drive from Vänersborg to Oslo in 3 hours, driving at an average rate of 55 mph. In just 2 more hours, Susan can drive from Vänersborg to Oslo and back to Vänersborg, using the same road as Kurt. What is Susan's average speed of driving?**

To find Susan's average speed, we would need to know the distance and the time she takes.

The time it takes her is 2 hours more than the 3 hours it takes Kurt, i.e. a total of 5 hours.

Distance $= d = st = $ Kurt's speed $\times$ The time it takes Kurt $= 55 \times 3 = 165$ miles

Therefore,   Susan's speed $= \dfrac{\text{Twice that distance}}{\text{The time it takes her}} = \dfrac{2 \times 165 \text{ miles}}{5 \text{ hours}} = 66 \text{ mph}$

3.137 **A train travels 210 miles from Boston to NYC at an average speed of 70 mph. It travels another 90 miles at an average rate of 45 mph to reach Philadelphia. What is its average speed for the entire journey from Boston to Philadelphia?**

210 miles		90 miles	
Boston	70 mph	NYC	45 mph	Philadelphia

Drawing a diagram as above helps visualize such questions more quickly. To find the average speed for the entire journey, we would need to find the total distance and total time.

Total distance $= 210 + 90 = 300$ miles

Time from Boston to NYC $= t = \dfrac{d}{s} = \dfrac{210}{70} = 3$ hours

Time from NYC to Philadelphia $= t = \dfrac{d}{s} = \dfrac{90}{45} = 2$ hours

Total time $=$ Time from Boston to NYC $+$ Time from NYC to Philadelphia $= 3 + 2 = 5$ hours

Average speed for the entire journey $= \dfrac{\text{total distance}}{\text{total time}} = \dfrac{300 \text{ miles}}{5 \text{ hours}} = 60 \text{ mph}$

3.138 **Along a 1,680 mile route, an airplane flies the first 900 miles in 3 hours. At what speed must it fly the remaining miles to average 280 mph for the entire journey?**

Let's assume x mph as the speed during the second part of the journey.

$$\underset{\substack{\longleftarrow \\ \text{3 hours}}}{\overset{\text{900 miles}}{\vert \quad\longrightarrow\vert}} \quad \underset{\substack{\\ x \text{ mph}}}{\overset{\text{780 miles}}{\vert\longrightarrow\vert}}$$

We know the total distance is 1,680 miles, and that the average speed for that total distance needs to be 280 mph. Therefore, the total time available $= \dfrac{d}{s} = \dfrac{1{,}680 \text{ miles}}{280 \text{ mph}} = 6$ hours.

Out of those 6 hours, 3 hours are already used up in the first part of the journey.

Meaning, 3 more hours remain, in which 780 miles need to be covered.

$$s = \frac{d}{t} = \frac{780 \text{ miles}}{3 \text{ hours}} = 260 \text{ mph}$$

3.139 A freight truck travels from Seattle to Vancouver and back to Seattle in 4 hours, considering only the travel time. If its average speed along one direction is 60 mph, and along the other direction on the same route is 45 mph, then how far is Seattle from Vancouver?

$$\underset{\substack{\\ \text{Seattle} \qquad\quad 60 \text{ mph}}}{\overset{x \text{ miles}}{\vert \longleftrightarrow \vert}} \quad \underset{\substack{\\ \text{Vancouver} \qquad 45 \text{ mph} \qquad\quad \text{Seattle}}}{\overset{x \text{ miles}}{\vert \longleftrightarrow \vert}}$$

In such examples, it's best to unfold the routes in one direction, instead of superimposing them back and forth. Assume Seattle is x miles from Vancouver.

Time from Seattle to Vancouver $= t = \dfrac{d}{s} = \dfrac{x}{60}$

Time from Vancouver to Seattle $= t = \dfrac{d}{s} = \dfrac{x}{45}$

Total time taken $=$ Time from Seattle to Vancouver $+$ Time from Vancouver to Seattle

We're given that total time taken $= 4$ hours

Therefore, $4 = \dfrac{x}{60} + \dfrac{x}{45} \quad \Rightarrow \quad 60 = \dfrac{x}{4} + \dfrac{x}{3}$ \qquad Multiplying both sides by 15

$60 = x\left(\dfrac{1}{4} + \dfrac{1}{3}\right) \quad \Rightarrow \quad 60 = x\left(\dfrac{3+4}{12}\right) \quad \Rightarrow \quad 60 = x\left(\dfrac{7}{12}\right) \quad \Rightarrow \quad x = \dfrac{60 \times 12}{7} = \dfrac{720}{7} \approx 103$

PRACTICE EXAMPLES

3.140 It's 3:45 p.m., and you're leaving for a destination 225 miles away. If you must reach there by 7:30 p.m., then what needs to be your average speed (in mph) of travel?

3.141 Walking at an average rate of 2 mph, a student can walk from his house to his school in m minutes. In only 15 more minutes, he can jog from his house to his school and back to his house, using the same route. What is his average speed of jogging, in mph?

3.142 Vinay drives from his home to a flower shop, 5 miles away, at an average rate of 30 mph. From there, he drives to Vinaya's house, 3 more miles ahead, at an average rate of 36 mph. What is his average speed for the entire trip?

3.143 On a 550 mile road trip, Jessica drives the first 375 miles in 7½ hours. At what speed must she drive the remaining distance to average 55 mph for the entire trip?

3.144 I spend 1¾ hours each day in traffic going to my office and coming back home along the same route. My average speed while driving to my office is 64 mph, and while driving back home is 48 mph. How far is my office from my home?

SPACE FOR CALCULATIONS

MEETING/SEPARATION OF TWO OBJECTS

When two objects, say trucks, are approaching each other from opposite directions along the same route, the speed at which they are closing in on each other is *the sum of their individual speeds*. For example, if one of them is doing 40 mph, and the other is doing 50 mph, then the distance between them is *shrinking* at the rate of $40 + 50 = 90$ mph. This concept is depicted below:

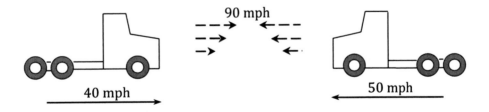

Similarly, when two objects are going away from each other in opposite directions along the same route, the speed at which they are running away from each other is *the sum of their individual speeds*. For example, if one of them is doing 60 mph, and the other is doing 70 mph, then the distance between them is *growing* at the rate of $60 + 70 = 130$ mph. The diagram below shows this:

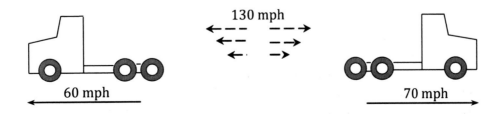

When a faster moving object is approaching a slower moving object from behind along the same route, the speed at which it is closing in on the slower moving object is the *difference between their individual speeds*. For example, if one truck is doing 55 mph, and the other 75 mph, then the gap between them is *shrinking* at the rate of 75 – 55 = 20 mph. Once the faster truck overtakes the slower truck, the gap between them will *grow* at the rate of 20 mph. The figure below depicts this:

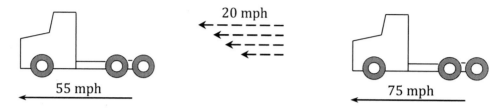

3.145 **At 2:30 p.m., two trains start toward each other along the same railroad from two different stations 230 miles apart. One train averages 55 mph, while the other averages 60 mph. At what time will they cross each other?**

The two trains are closing in on each other at the rate of 55 + 60 = 115 mph.

The distance between them is 230 miles.

$$\text{time} = \frac{\text{distance}}{\text{speed}} \quad \Rightarrow \quad t = \frac{d}{s} = \frac{230}{115} = 2 \text{ hours}$$

The two trains will cross each other 2 hours after 2:30 p.m., i.e. at 4:30 p.m.

3.146 **At 10 a.m., Jason leaves home and starts driving at an average speed of 50 mph. At 10:30 a.m., his roommate leaves the same home and starts driving in the opposite direction at an average rate of 45 mph. How far will the two of them be from each other at 1:30 p.m.?**

From 10 a.m. to 10:30 a.m., Jason will have already travelled 25 miles.

Thereafter, the two of them will further go away from each other at the rate of 50 + 45 = 95 mph for a full 3 hours, until 1:30 p.m. Meaning, the distance between them will increase by an additional 95 × 3 = 285 miles.

Add the first 25 miles, and you get 310 miles, the distance between them at 1:30 p.m.

Alternative method:

Jason drives from 10 a.m. to 1:30 p.m., a total of 3.5 hours. His roommate drives only 3 hours.

Distance covered by Jason in one direction = $d = st$ = 50 mph × 3.5 hours = 175 miles

Distance covered by his roommate in the opposite direction = 45 mph × 3 hours = 135 miles

Therefore, at 1:30 p.m., the two of them will be 175 + 135 = 310 miles away from each other.

3.147 **A fighter jet is covertly flying over enemy territory at 700 mph. A missile is fired at it from an airbase 300 miles away. If the missile chases the plane at 1900 mph, how long will it take the missile to strike the jet?**

The missile is closing in on the jet at the rate of 1900 – 700 = 1200 mph.

The instant the missile is fired, the jet is 300 miles away from it. Meaning, the missile has 300 miles to cover at the rate of 1200 mph in order to reach the jet.

$$t = \frac{d}{s} = \frac{300}{1200} = \frac{1}{4}\text{hour} = 15 \text{ minutes}$$

PRACTICE EXAMPLES

3.148 Two buses leave the same bus station in opposite directions. One averages 50 mph, while the other averages 45 mph. How long will it take the two buses to be 380 miles apart?

3.149 At noon, a freight train leaves from town *A* toward town *B* which is 429 miles away, traveling at an average speed of 57 mph. At 2 p.m., another freight train leaves from town *B* toward town *A*, along the same railroad, averaging 48 mph. At what time will the two trains cross each other?

3.150 At 1 p.m., a plane flying high above Heathrow airport at 300 mph radios a request for mid-air refueling. At 1:15 p.m., a tanker plane takes off from Heathrow airport and follows the first plane at an average speed of 450 mph. At what time will the second plane meet the first?

SPACE FOR CALCULATIONS

WORK (DIFFERENT INDIVIDUAL RATES)

If I can water a garden in 2 hours, and you can water the same garden in 3 hours, then how long do you think it will take the two of us to water the garden together? Well, you and I work at **different individual rates**. I take 2 hours to finish the job. Meaning, in one hour, I would finish $\frac{1}{2}$ the job; $\frac{1}{2}$ being a fraction of 1 (1 = the total work to be done). Similarly, you take 3 hours to do it. Or, in one hour, you would finish $\frac{1}{3}$ of it.

Together, in one hour, we would finish $\frac{1}{2} + \frac{1}{3} = \frac{3+2}{6} = \frac{5}{6}$ the job. (Remember to use the *right-arm product first* rule). Therefore, $1 - \frac{5}{6} = \frac{6}{6} - \frac{5}{6} = \frac{1}{6}$ the job would still remain to be done. Remember, '1' means 'whole,' or the 'complete job.'

In 1 hour, we would get $\frac{5}{6}$ the job done; in 2 hours, we would get $2 \times \frac{5}{6}$ the job done; in 3 hours, $3 \times \frac{5}{6}$; in 4 hours, $4 \times \frac{5}{6}$; in *x* hours, $x \times \frac{5}{6}$. Let's say we end up finishing the whole job (=1) in *x* hours.

Then, $x \times \frac{5}{6} = 1$ $\Rightarrow$ $x = 1 \times \frac{6}{5}$ $\Rightarrow$ $x = 1\frac{1}{5}$ hours $\Rightarrow$ $x = 1$ hour, 12 minutes

When solving such examples, *first always find the individual rates of work.*

3.151 **A set of furniture is to be painted. Sarah alone can paint it 3 hours, Kelly alone can paint it in 4 hours, and Amy alone can paint it in 6 hours. If the three of them work together, how long would it take them to paint the set of furniture?**

In one hour, Sarah gets $\frac{1}{3}$ the job done; Kelly gets $\frac{1}{4}$ the job done; Amy gets $\frac{1}{6}$ the job done.

Together, in 1 hour they do: $\frac{1}{3}+\frac{1}{4}+\frac{1}{6}=\left(\frac{1}{3}+\frac{1}{4}\right)+\frac{1}{6}=\frac{7}{12}+\frac{1}{6}=\frac{42+12}{72}=\frac{54}{72}=\frac{3}{4}$ the job.

If they work for x hours, and get the *whole* job done, then:

$$x \times \frac{3}{4} = 1 \quad \Rightarrow \quad x = 1 \times \frac{4}{3} \quad \Rightarrow \quad x = 1\frac{1}{3} \text{ hours} \quad \Rightarrow \quad x = 1 \text{ hour, 20 minutes}$$

3.152 **An ant can eat a sugar crystal in 15 minutes. A cockroach can eat the same sugar crystal thrice as fast. If they eat it together for 3 minutes, without fighting, then what fraction of the sugar crystal *remains to be eaten*?**

The cockroach eats the crystal thrice as fast, i.e. it can finish eating it in only 5 minutes.

In one minute, the ant eats $\frac{1}{15}$ the crystal, whereas the cockroach eats $\frac{1}{5}$ the crystal.

Together, in one minute, they finish eating: $\frac{1}{15}+\frac{1}{5}=\frac{5+15}{75}=\frac{20}{75}=\frac{4}{15}$ the crystal.

If they eat the crystal together for 3 minutes, the fraction they would finish eating would be:

$3 \times \frac{4}{15}=\frac{12}{15}=\frac{4}{5}$ The fraction that would *remain* would be: $1-\frac{4}{5}=\frac{5}{5}-\frac{4}{5}=\frac{1}{5}$

3.153 **Pump *X* alone can empty a pool in 9 hours. Pump *Y* alone can empty it in 6 hours. The two pumps together start emptying the pool that is initially full. After 3 hours, pump *Y* overheats and shuts down. How much longer will it take pump *X* to empty the remaining pool by itself?**

In one hour, pump X empties $\frac{1}{9}$ the pool; pump Y empties $\frac{1}{6}$ the pool.

Together, in one hour, they would empty: $\frac{1}{9}+\frac{1}{6}=\frac{6+9}{54}=\frac{15}{54}=\frac{5}{18}$ the pool.

In 3 hours, they would empty $3 \times \frac{5}{18}=\frac{5}{6}$ the pool.

So, after 3 hours, $1-\frac{5}{6}=\frac{6}{6}-\frac{5}{6}=\frac{1}{6}$ the pool would still remain to be emptied.

If pump X takes x more hours to empty this remaining portion, then:

$$x \times \frac{1}{9}=\frac{1}{6} \quad \Rightarrow \quad x=\frac{9}{6}=1.5 \text{ hours} = 1 \text{ hour 30 minutes.}$$

3.154 **Vikrant alone can assemble a sofa set in *h* hours. If Vikrant and Sundeep together can assemble the same sofa set in 2 hours, how long would it take Sundeep alone to assemble the sofa set?**

Assume it takes Sundeep x hours to assemble the sofa by himself.

In one hour, Vikrant can assemble $\frac{1}{h}$ of it; Sundeep can assemble $\frac{1}{x}$ of it.

Together, in one hour, they can assemble: $\dfrac{1}{h} + \dfrac{1}{x} = \dfrac{x+h}{hx}$ of it.

Working together for 2 hours, they get it all done.

Therefore, $2 \times \dfrac{x+h}{hx} = 1 \quad \Rightarrow \quad \dfrac{2(x+h)}{hx} = 1 \quad \Rightarrow \quad 2x + 2h = hx \quad \Rightarrow$

$2x - hx = -2h \quad \Rightarrow \quad x(2-h) = -2h \quad \Rightarrow \quad x = \dfrac{-2h}{(2-h)} \quad \Rightarrow \quad x = \dfrac{2h}{(h-2)}$ hours

Do not be afraid of variables (x, h etc.). Such questions are quite common on the GRE/GMAT.

PRACTICE EXAMPLES

3.155 Working alone, John can plaster an 8′ × 25′ brick wall in 8 hours; Fred can do it in 12 hours; and Chris can do it in 6 hours. Working together, how long would it take them to plaster the wall?

3.156 Aurash can mow a stretch of lawn in 20 minutes. His father can mow it twice as fast. If they work together for 5 minutes, what percent of the lawn *remains to be mowed*?

3.157 A large pipe can fill a tank in 3 hours. A small pipe can fill the same tank in 12 hours. The large pipe alone starts to fill the empty tank. After 2 hours, the small pipe joins in. How much longer will it take the two pipes working together to fill the rest of the tank?

3.158 Sophia can wash her car in h hours. If she and her husband together can wash the same car in m minutes, how many minutes would it take her husband alone to wash the car?

SPACE FOR CALCULATIONS

WORK (SAME INDIVIDUAL RATE)

Let's say there's a bridge being constructed somewhere. It's going to take 8 months for 5 workers to build it. Every worker is equally well-trained and skilled. Therefore, every worker works at the **same individual rate**. What if the bridge is to be built in only 4 months? Well, then there would need to be twice the number of workers, or 10 workers. What about finishing the bridge in only 2 months? In that case, 4 times the number of workers, or 20 workers. Clearly, the sooner the job needs to be completed, the more workers needed. It sounds like common sense. That's because the *amount of work* to be done is the same. It doesn't matter whether 10 workers do it, or 20 do it, or 50 do it.

The amount of work (or units of work) is the product of the number of workers doing it and the time it would take them to do it. If, instead of workers, machines are employed to do a certain type of work (for example, print cards) then the amount of work would be the product of the number of printing machines and the time it takes them to print.

Amount of work = (# of workers or machines) × (the time needed to do it)

In the above formula, the amount of work to be done is constant. Therefore, if the time increases, then the number of workers drops; if the time decreases, then the number of workers goes up. When solving such examples, simply equate this product given the initial and final conditions, and then solve for the unknown. Let's learn how with the following examples.

3.159 If 4 painters can paint my house in 5 days, then working at the same rate, how many more painters would be needed to paint my house in 3 days' less time?

Amount of work to be done = (# of painters) × (# of days) = 4 × 5 = 20

If this work is to be done by x more painters in 3 days' less time, i.e. in only 2 days, then:

Same amount of work = (new # of painters) × (new # of days) $\Rightarrow$ $20 = (4 + x) \times 2$

$\Rightarrow$ $20 = 8 + 2x$ $\Rightarrow$ $12 = 2x$ $\Rightarrow$ $6 = x$

3.160 If 2 masons can build a brick wall in 15 hours, then working at the same rate, how much less time would 5 masons take to build the same brick wall?

Amount of work = (# of masons) × (time needed) = 2 × 15 = 30

Assume it would take 5 masons x hours to build the same wall.

Same amount of work = (new # of masons) × (new amount of time) $\Rightarrow$

$30 = 5 \times x$ $\Rightarrow$ $x = 6$

The question is: How much *less* time? The answer is: 15 – 6 = 9 hours.

3.161 If w construction workers can repair a stretch of road in h hours, then at the same rate, how many more workers would it take to repair the same stretch of road in 12 hours' less time?

Amount of work = (# of workers) × (time needed) = $w \times h = wh$

Assume it would take x more workers to repair the road faster.

Then,   new # of workers = $(w + x)$ and   new time needed = $(h - 12)$ hours.

Same amount of work = (new # of workers) × (new time needed) $\Rightarrow$

$wh = (w + x)(h - 12)$ $\Rightarrow$ $wh = wh - 12w + xh - 12x$ $\Rightarrow$ $xh - 12x = 12w$ $\Rightarrow$

$x(h - 12) = 12w$ $\Rightarrow$ $x = \dfrac{12w}{h - 12}$

3.162 If 2,000 pints of glucose can last 50 patients for 30 days, then for how many weeks would 1,680 pints of glucose last 60 patients?

Remember: *Same individual rate* of work, *same individual rate* of consumption, etc. Here, we need to find how much glucose every patient uniformly consumes every day. That would be # of pints per patient per day, or # of pints divided by # of patients divided by # of days:

$$\text{Individual rate of consumption} = \dfrac{\dfrac{2{,}000}{50}}{30} = \dfrac{40}{30} = \dfrac{4}{3} \text{ pints/patient/day}$$

Assume 1,680 pints will last 60 patients x weeks, or $7x$ days (the question is in *days*).

$$\text{Individual rate of consumption} = \dfrac{\dfrac{1{,}680}{60}}{7x} = \dfrac{168}{6} \times \dfrac{1}{7x} = \dfrac{168}{42x} = \dfrac{4}{x} \text{ pints/patient/day}$$

Equate the two individual rates of consumption, and find x.

$$\frac{4}{3} = \frac{4}{x} \quad \Rightarrow \quad x = 3 \text{ days}$$

PRACTICE EXAMPLES

3.163 If 3 tractors can plow a field in 8 hours, how many more tractors would be needed to plow the same field in 2 hours' less time?

3.164 If 2 cows consume a sack of fodder in 3 days, then at the same rate, in how many fewer days would 6 cows consume the same sack of fodder?

3.165 If h hoses can fill a pond in m minutes, then at the same rate, how many more hoses would it take to fill the pond in 15 minutes' less time?

3.166 If 5 machines manufacture 2,100 wheels in a week, then how many days would it take 8 machines to manufacture 4,800 wheels?

SPACE FOR CALCULATIONS

BASIC STATISTICS

Given a set of numbers, there are five statistical terms important from the GRE/GMAT point of view: mean, median, mode, range and standard deviation.

Mean (μ) is the arithmetic mean or the average of the numbers.

Median (m) is literally the midpoint of the set of numbers, when arranged in either ascending or descending order. If two numbers occur at the middle of the set, then the median would be the average of those two numbers.

Mode (M) is the most frequently occurring number in the set. If several numbers in the set occur with the same frequency (the same number of times), then each of them is a mode. If every number in the set occurs with the same frequency, then every number is a mode. Thus, a set of numbers can have as many modes as there are numbers in it.

Range (R) is the difference between the largest number and the smallest number in the set.

Standard deviation (σ) is the measure of the dispersion of the numbers, calculated by the formula

$$\sigma = \sqrt{\frac{\Sigma(\mu - x)^2}{n}}$$

where n = the number of numbers in the set

x = each number in the set, considered one at a time

Σ stands for 'summation.' The value of $(\mu - x)^2$ is calculated for each value of x, and then added up to get $\Sigma(\mu - x)^2$.

Standard deviation is never negative. So, only the positive root should be considered.

3.167 **Find the mean, median, mode, range and standard deviation for: 2, –1, 5, 1, 3.**

First, arrange the numbers in increasing order: $-1, 1, 2, 3, 5$

Mean $(\mu) = \dfrac{-1 + 1 + 2 + 3 + 5}{5} = \dfrac{10}{5} = 2$

Median (m) = midpoint of the set of numbers = 2

Mode (M) = most commonly occurring number = all numbers = $-1, 1, 2, 3, 5$

Range (R) = largest number $-$ smallest number = $5 - (-1) = 5 + 1 = 6$

Standard deviation $(\sigma) = \sqrt{\dfrac{\Sigma(\mu - x)^2}{n}}$

To find σ, let's first find $\Sigma(\mu - x)^2$ by building the following table from left to right.

x (given numbers)	$(\mu - x)$	$(\mu - x)^2$
-1	$2-(-1) = 2 + 1 = 3$	$3^2 = 9$
1	$2 - 1 = 1$	$1^2 = 1$
2	$2 - 2 = 0$	$0^2 = 0$
3	$2 - 3 = -1$	$(-1)^2 = 1$
5	$2 - 5 = -3$	$(-3)^2 = 9$
		$\Sigma(\mu - x)^2 = 20$

Standard deviation $(\sigma) = \sqrt{\dfrac{\Sigma(\mu-x)^2}{n}} = \sqrt{\dfrac{20}{5}} = \sqrt{4} = 2$ (positive root only)

This seems like a lot of work to do in 1½ to 2 minutes. On the GRE/GMAT, it is very unlikely that you will come across a question asking you to *find* the standard deviation. Instead, questions are typically asked on *one of the steps* of finding standard deviation. Therefore, it was necessary to learn the method, step by step, using an example.

3.168 **Find the mean of the median, mode and range of the numbers 12, 4, –5, 8, 12 and –1.**

The question may sound confusing, but the wording is typical of any competitive exam. Basically, find the median, mode and range, and then find *their* mean.

Arrange the numbers in increasing order: $-5, -1, 4, 8, 12, 12$

Median (m) = midpoint of the set of numbers = average of 4 and 8 = $\dfrac{4 + 8}{2} = 6$

Mode (M) = most commonly occurring number = 12

Range (R) = largest number $-$ smallest number = $12 - (-5) = 12 + 5 = 17$

Mean of the median, mode and range = $\dfrac{m + M + R}{3} = \dfrac{6 + 12 + 17}{3} = \dfrac{35}{3} = 11\dfrac{2}{3} = 11.\overline{6}$

COUNTING (MENU)

If you've ever been to a restaurant that boasts of having over 100,000 different meals on its menu, you may have wondered how big their kitchen might be. Well, actually it's not a question about their kitchen; it's a question about their menu. Let's see how it all works.

Consider a restaurant menu that offers 3 different entrées (E_1, E_2, E_3) and 4 different drinks (D_1, D_2, D_3, D_4). Let's say a meal comprises an entrée and a drink. How many different choices of meals would you have? You would have:

E_1D_1	E_1D_2	E_1D_3	E_1D_4
E_2D_1	E_2D_2	E_2D_3	E_2D_4
E_3D_1	E_3D_2	E_3D_3	E_3D_4

That's a total of 12 different meals. How did we come up with that? We simply combined entrée E_1 with drinks D_1 through D_4, one at a time. Then we took up entrée E_2, and combined it with drinks D_1 through D_4, and finally E_3 with D_1 through D_4. In short, we considered *each entrée* with *each drink*. Mathematically speaking, we multiplied the number of entrées with the number of drinks.

What if a restaurant offered 10 appetizers, 5 salads, 10 entrées, 5 soups, 10 drinks and 10 desserts? If a full course meal comprises an appetizer, a salad, an entrée, a soup, a drink and a dessert, then how many different meals could one choose from? Well, in that case, we would combine each appetizer with each salad with each entrée with each soup with each drink with each dessert. In other words, we would multiply the number of appetizers with the number of salads with the number of entrées with.............. The end result would be $10 \times 5 \times 10 \times 5 \times 10 \times 10 = 250,000$!

Now you see how a modest menu can be advertised as a grand menu?

3.169 **A pizzeria offers a special meal consisting of one of 5 types of pizza, one of 12 flavors of wings, one of 6 types of beverages, and either salted or unsalted breadsticks. How many different variations of the special meal does the pizzeria offer?**

5 types of pizza, 12 flavors of wings, 6 types of beverages, 2 types of breadsticks

Total different variations of the special meal = $5 \times 12 \times 6 \times 2 = 720$

3.170 **A contest team is to be assembled by choosing one of 12 engineering students, one of 15 liberal arts students, and one of several business students. If it's possible to assemble 900 unique teams with those students, then how many business students are there to choose from?**

This may not be a restaurant menu question, but the concept is the same.

12 engineering students, 15 liberal arts students, x business students (assume). Pick one from each. Total possible combinations are 900.

$$12 \times 15 \times x = 900 \quad \Rightarrow \quad x = \frac{900}{12 \times 15} \quad \Rightarrow \quad x = 5$$

PRACTICE EXAMPLES
3.171 An automaker offers 2-door and 4-door versions of a certain car in 5 choices of exterior color, 3 choices of interior color, 3 choices of engine, and 2 choices of transmission. How many different variations of that car does the automaker offer?

3.172 A smoothie can be prepared using one of several types of fruit, one of 3 types of yogurt, and one of 4 types of sweeteners. If 84 different smoothies with unique combination of ingredients can be prepared using the above ingredients, then how many types of fruit can be used?

COUNTING (SLOT MACHINE)

Consider a phone company that is issuing 7-digit phone numbers. How many phone numbers can it issue at most? To find that, let's imagine seven slots in a casino type slot machine. In all, how many different numbers can show up in each of the slots?

⊠ ⊠ ⊠ ⊠ ⊠ ⊠ ⊠

The first digit from the left can be one of 10 different numbers: 0, 1, 2, 3, ..., 9; same with the second digit from the left; same with the third digit; same with *all seven digits*. To find the total number of phone numbers, we would need to combine each number in the first digit with each number in the second digit with each number in the third digit, and so on. Does that sound familiar? Correct, just like the restaurant menu type questions.

10 10 10 10 10 10 10
⊠   ⊠   ⊠   ⊠   ⊠   ⊠   ⊠

The total number of phone numbers would be $10 \times 10 \times 10 \times 10 \times 10 \times 10 \times 10 = 10^7 = 10$ million. Keep in mind that there are no restrictions on the phone numbers starting or ending with a certain number, for example 1 or 0 or 5 etc.

What if the only restriction is that phone numbers can't begin with a 0, 1 or 9? Then we'd have the following slot machine type configuration:

7 10 10 10 10 10 10
⊠ ⊠   ⊠   ⊠   ⊠   ⊠   ⊠

Barring 0, 1 and 9, the first digit can be one of seven numbers; the remaining digits can each be all 10 numbers. Total number of phone numbers = $7 \times 10 \times 10 \times 10 \times 10 \times 10 \times 10 = 7,000,000$

What if the only restriction is that phone numbers can only have odd digits? Then each digit can be one of 1, 3, 5, 7, 9. (Same logic applies with only even digits.)

5 5 5 5 5 5 5
⊠ ⊠ ⊠ ⊠ ⊠ ⊠ ⊠   Total phone numbers = $5 \times 5 \times 5 \times 5 \times 5 \times 5 \times 5 = 5^7$

What if the only restriction is that phone numbers can only be even *numbers*? First, here's the difference between *even digits only* and *even numbers only*.

Even digits only: Every digit must be even   Examples: 642   8,460   202

Even numbers only: *Units* digit must be even   Examples: 71<u>2</u>   62<u>8</u>   4,28<u>0</u>   39<u>4</u>

In an even *number*, the units digit must be even. The other digits may be odd or even. Similarly, in an odd *number*, the units digit must be odd; the other digits may be odd or even.

With the above restriction, the units digit can only be 0, 2, 4, 6 or 8. The others can be all 10 numbers.

10 10 10 10 10 10 5
☒ ☒ ☒ ☒ ☒ ☒ ☒   Total phone numbers = $10 \times 10 \times 10 \times 10 \times 10 \times 10 \times 5 = 5$ million

What if the only restriction is that each digit is to be unique in any phone number? Meaning, numbers can't repeat themselves in any phone number. Here's how to analyze this:

The first digit from the left can be one of 10 numbers: 0, 1, 2, 3........., 9. But once the first digit has assumed a number, then only nine more numbers are available for the second digit. Once the second digit has assumed a number, only eight more numbers are available for the third digit, and so on.

10 9 8 7 6 5 4
☒ ☒ ☒ ☒ ☒ ☒ ☒   Total phone numbers = $10 \times 9 \times 8 \times 7 \times 6 \times 5 \times 4 = 604,800$

What if there are two restrictions: (i) Numbers can't repeat themselves in any phone number, and (ii) phone numbers can't start with 0, 1 or 2? Now, this is the hard part. Ready?

If phone numbers can't start with 0, 1 or 2, then the first digit can only be one of the remaining seven numbers: 3, 4, 5,......, 9.

7
☒ ☒ ☒ ☒ ☒ ☒ ☒

Once the first digit assumes one of those seven numbers, say 5, the second digit can only be one of the *nine* remaining numbers, because it can be one of 10 numbers, but the number 5 is used up by the first digit, and therefore is not available, but others are. (Remember that the 0, 1, 2 restriction is *only on the first digit*.) Likewise, the third digit can be one of eight remaining numbers, fourth digit can be one of seven, and so on.

7 9 8 7 6 5 4
☒ ☒ ☒ ☒ ☒ ☒ ☒   Total phone numbers = $7 \times 9 \times 8 \times 7 \times 6 \times 5 \times 4 = 423,360$

The above calculations may seem daunting, but I've used 7-digit slot machine type examples only to demonstrate the mechanics of the concept. On the GRE/GMAT, calculations will be relatively simpler because of the use of several restrictions that will diminish the amount of work to be done.

3.173 **How many five-digit numbers can be formed using only even digits?**

'Only even digits' means each of the five digits in the number should be even: 0, 2, 4, 6, or 8. But for the number to be a five-digit number, the first digit can't be 0; it can only be 2, 4, 6, or 8. For example, 04426 is just 4426. It's not a five digit number.

4 5 5 5 5
☒ ☒ ☒ ☒ ☒   Total numbers = $4 \times 5 \times 5 \times 5 \times 5 = 2,500$

3.174 Using the digits 0, 1, 2, 3, 7, and 9, how many integers can be formed that are between 300 and 3000 if: (i) digits may repeat themselves within any integer, (ii) no digit may repeat itself within any integer?

(i) Let's first consider the 3-digit numbers from 300 to 999. The first digit from the left can be 3, 7, or 9, a total of three numbers. The second digit can be any of 0, 1, 2, 3, 7, or 9, a total of six numbers. Same with the third digit.

3   6   6
⊠   ⊠   ⊠   Total numbers = $3 \times 6 \times 6 = 108$ But these also include the number 300, which isn't right, because the question is "between 300 and 3000." Meaning, we should only consider the integers from 301 to 2999. So, total 3-digit numbers = $108 - 1 = 107$.

Now, let's find the 4-digit numbers from 1000 to 2999. The first digit from the left can only be 1 or 2, a total of two numbers. The second digit can be 0, 1, 2, 3, 7, or 9, a total of six numbers. Same with the third and fourth digits.

2   6   6   6
⊠   ⊠   ⊠   ⊠   Total numbers = $2 \times 6 \times 6 \times 6 = 432$

Total number of 3-digit and 4-digit integers = $107 + 432 = 539$.

(ii) If repetition of digits is not allowed, then in the 3-digit numbers from 300 to 999, the first digit from the left can only be 3, 7, or 9, or a total of three numbers. Once the first digit assumes one of them, say 7, the second digit will only have five numbers available. Likewise, the third digit will only have four numbers available.

3   5   4
⊠   ⊠   ⊠   Total numbers = $3 \times 5 \times 4 = 60$ Keep in mind that the number 300 doesn't exist in this case because repeated digits aren't a part of this calculation.

Considering the 4-digit numbers from 1000 to 2999, the first digit from the left can only be 1 or 2. Once the first digit assumes one of them, say 1, the second digit will only have five numbers available, the third will have four numbers available, and the fourth will have three.

2   5   4   3
⊠   ⊠   ⊠   ⊠   Total numbers = $2 \times 5 \times 4 \times 3 = 120$

Total number of 3-digit and 4-digit integers = $60 + 120 = 180$

3.175 How many 5-digit odd numbers exist that have their first three digits even?

'Odd numbers' means the units digit must be odd: 1, 3, 5, 7, or 9, a total of five numbers. The other digits can be odd or even. But the question also says "first three digits even." Meaning, the first three digits can only be 0, 2, 4, 6, or 8, a total of five numbers. But for it to be a 5-digit number, the first digit from the left can only be 2, 4, 6, or 8 (not 0), a total of four numbers.

4   5   5   10   5
⊠   ⊠   ⊠   ⊠   ⊠   Total numbers = $4 \times 5 \times 5 \times 10 \times 5 = 5,000$

3.176 Plural Wireless is issuing 7-digit phone numbers subject to the following restrictions: The phone numbers must start with 397. No digit may be repeated in any number. No number may end with a zero. How many 7-digit phone numbers can Plural Wireless issue?

If the 7-digit phone numbers must start with 397, then only the remaining four digits are flexible. So, this is not a 7-digit question; it's a 4-digit question. Since digits may not repeat themselves in any phone number, the numbers 3, 9 and 7 are not available anymore, because every single phone number will already have 397 in it.

In the 4-digit number under consideration, the last digit can't be 0. It can only be one of the remaining six numbers: 1, 2, 4, 5, 6, or 8.

$$\boxed{\times}\ \boxed{\times}\ \boxed{\times}\ \overset{6}{\boxed{\times}}$$

Once it has assumed a number, say 2, the first digit *from the left* can be one of *six* available numbers. Six, not five, because the number 0 is *allowed* in the first digit; it simply replaces the number no longer available (here, 2). The second digit from the left can be one of five remaining numbers; the third digit can be one of four.

$$\overset{6}{\boxed{\times}}\ \overset{5}{\boxed{\times}}\ \overset{4}{\boxed{\times}}\ \overset{6}{\boxed{\times}}\qquad \text{Total numbers} = 6 \times 5 \times 4 \times 6 = 720$$

3.177 **In how many different ways can the letters in the name *THATTE* be rearranged such that all the *T*s appear consecutively?**

If all the *T*s are to appear consecutively, then let's first group them together, and call them only *T*, not *TTT*. Wherever they go, they will always take up *one* space. So, this is not a 6-slot question; it's a 4-slot question: *T H A E*

While rearranging these four letters, the first slot from the left can assume one of the four letters: *T, H, A,* or *E*. Once it assumes one letter, say *H*, the second slot from the left can only hold one of the remaining three letters. Likewise, the third slot can only assume one of the remaining two letters; the fourth slot can only hold the last remaining letter.

$$\overset{4}{\boxed{\times}}\ \overset{3}{\boxed{\times}}\ \overset{2}{\boxed{\times}}\ \overset{1}{\boxed{\times}}\qquad \text{Total arrangements} = 4 \times 3 \times 2 \times 1 = 24$$

3.178 **Using the letters A, C, E, G and H, how many 4-letter arrangements can be made if the middle two letters must be the same, and the other two letters must be different from each other as well as from the middle two letters?**

4-letter arrangements = 4 slots. But the middle two slots are to hold the *same* letter. So, let's group the middle two slots into one slot. Meaning, this is a 3-slot question.

Please interpret the second part of the question correctly; it simply says that all three letters must be different from one another. Such confusing phrases are likely on the GRE/GMAT.

The first slot from the left can assume one of five letters. Once it assumes one letter, the second slot can hold one of the remaining four letters. Likewise, the third slot can hold one of three.

$$\overset{5}{\boxed{\times}}\ \overset{4}{\boxed{\times}}\ \overset{3}{\boxed{\times}}\qquad \text{Total arrangements} = 5 \times 4 \times 3 = 60$$

PRACTICE EXAMPLES

3.179 How many four-digit numbers can be formed by using only odd integers?

3.180 Using the digits 2, 3, 4, 6 and 9, how many 4-digit odd numbers can be formed if: (i) digits may be repeated in any number, (ii) no digit may be repeated in any number?

3.181 How many 6-digit odd numbers exist that have their odd-numbered digits even?

3.182 In a certain country, license plates for vehicles are issued with the following guidelines: Every plate must have 5 numerical digits. The 5-digit number must begin with a 9, and cannot end with a 4. No digit may be repeated on a plate. How many such license plates can be issued?

3.183 In how many different ways can the letters in the word *COTTON* be rearranged such that the *T*s appear consecutively <u>and</u> the *O*s also appear consecutively?

3.184 How many 3-letter displays can be formed from the name *KATARINA* such that the middle letter is always different from the other two, and the first letter is always *K*?

SPACE FOR CALCULATIONS

COUNTING (COMBINATIONS AND PERMUTATIONS)

In my years of instructing math, the topic I've enjoyed teaching the most is **combinatorics**, the branch of mathematics covering **combinations and permutations**. But before we discuss it, let's first define the **factorial** of a number n. The factorial of a positive integer n is the product of all the positive integers less than or equal to n itself. Factorials are so exciting that the exclamation mark (!) is used to denote the factorial of a number! n factorial is written as $n!$

Mathematically, $n! = n \times (n-1) \times (n-2) \times (n-3) \times \ldots\ldots\ldots \times 4 \times 3 \times 2 \times 1$

For example: $27! = 27 \times 26 \times 25 \times 24 \times \ldots\ldots\ldots\ldots \times 3 \times 2 \times 1$

$463! = 463 \times 462 \times 461 \times 460 \times \ldots\ldots\ldots\ldots \times 4 \times 3 \times 2 \times 1$

As you may notice, a factorial can be used to express very large numbers in a condensed notation. Factorials of negative numbers and factorials of fractions don't exist. Meaning, there's no such thing as $(-19)!$ or $\frac{3}{4}!$ or $(-\frac{1}{2})!$ Also, zero factorial and one factorial are both defined as equal to one.

$0! = 1$ and $1! = 1$

Now, let's discuss **combinations**. Suppose you have a list of n equally distinguished scientists. You've been asked to select r of them to form a team. In how many different ways could you do that? The answer is: In $_nC_r$ different ways, calculated by using the combination formula:

$$_nC_r = \frac{n!}{r!(n-r)!}$$ $_nC_r$ is also written as ^nC_r or $\binom{n}{r}$ or $C(n, r)$.

For example, imagine there are 5 scientists: John, Paula, Martha, Robert and Derrick. You've been asked to select 3 of them to form a team. You could select John, Paula and Martha, or John, Paula and Robert, or John, Paula and Derrick etc. In all, the number of different teams of 3 scientists you could come up with would be:

$$_nC_r = {_5}C_3 = \frac{n!}{r!(n-r)!} = \frac{5!}{3!(5-3)!} = \frac{5!}{3! \times 2!} = \frac{5 \times 4 \times 3 \times 2 \times 1}{(3 \times 2 \times 1) \times (2 \times 1)} = 10$$

Consider one random team that you could make: John, Paula and Robert. As a team, the three of them may only be counted once. It wouldn't be correct if you counted them again as Paula, Robert and John, or yet again as Robert, John and Paula etc. because a team is a team is a team.

Another example in which the combination formula ($_nC_r$) would be used is when selecting 4 fruits out of a box containing 7 non-identical fruits: an apple, a banana, a pear, a peach, a pineapple, an orange and a nectarine. If you randomly select an apple, a banana, a pineapple and a nectarine, then you have essentially formed a team. It wouldn't be correct to juggle the same 4 fruits in your hands, and count them again as a new team. $_7C_4$ would be the total number of unique teams you could come up with.

Generally speaking, the combination formula should be used when recounting of the same elements within the team (or group) is *not* allowed.

Let's explore a fun fact about $_nC_r$ using an example, say $_5C_3$:

$$_nC_r = {_5}C_3 = \frac{5!}{3!(5-3)!} = \frac{5!}{3! \times 2!} = \frac{5!}{2! \times 3!} = \frac{5!}{2!(5-2)!} = {_5}C_2 = {_5}C_{(5-3)} = {_n}C_{(n-r)}$$

Therefore, $_nC_r = {_n}C_{(n-r)}$

Next, it's **permutations**. Let's say a classroom has n number of chairs, and r number of students. Some chairs are empty, meaning $n > r$. In how many distinct configurations can the r students occupy the n chairs? Well, if they're all occupying certain r chairs out of n chairs, and only two occupants exchange places, then it's a new configuration. If more occupants exchange places, then it's again a new configuration. Meaning, while the chairs that are occupied are still the same, recounting the configuration by simply moving the occupants around in those very chairs is *allowed*. Also, if some occupants get up and occupy the empty chairs, then it's yet a new configuration. The total number of such configurations can be calculated by the permutation formula:

$$_nP_r = \frac{n!}{(n-r)!}$$ $_nP_r$ is also written as ^nP_r or $\binom{n}{r}$ or $P(n, r)$.

As an example, 3 people can occupy 11 chairs in $_{11}P_3$ unique ways, or:

$$_nP_r = {_{11}}P_3 = \frac{n!}{(n-r)!} = \frac{11!}{(11-3)!} = \frac{11!}{8!} = \frac{11 \times 10 \times 9 \times 8!}{8!} = 11 \times 10 \times 9 = 990$$

Notice that there's no need to write out the entire factorial: 11! can be conveniently written as $11 \times 10 \times 9 \times 8!$, so that the 8! in the numerator and denominator cancel out.

In my opinion, on the GRE/GMAT, questions on combinations and permutations require considerable amount of thinking, and are therefore the hardest ones. The key lies in being able to tell whether to use the combination formula ($_nC_r$), or the permutation formula ($_nP_r$). But solving a sufficient number of examples can make it much simpler.

3.185 In how many different configurations can 3 cars occupy 7 parking spaces? Assume every car enters the parking space nose-first.

Sounds similar to 3 people occupying 7 chairs, doesn't it? If two of them exchange places, then we've got a new configuration. Meaning, we're talking recounting the configuration by simply shuffling the same cars in the same parking spaces. This is permutation.

$$_nP_r = {_7}P_3 = \frac{n!}{(n-r)!} = \frac{7!}{(7-3)!} = \frac{7!}{4!} = \frac{7\times6\times5\times4!}{4!} = 7 \times 6 \times 5 = 210$$

3.186 If you had to assemble a team of 6 basketball players from 10 equally ranked players, in how many different ways could you do it?

When assembling a team, recounting simply by shuffling members within a team is not allowed. This is combination. 10 players; 6 to be chosen.

$$_nC_r = {_{10}}C_6 = \frac{10!}{6!(10-6)!} = \frac{10!}{6!\times4!} = \frac{10\times9\times8\times7\times6!}{6!\times4\times3\times2\times1} = \frac{10\times9\times8\times7}{4\times3\times2} = 210$$

In the final step above, please be sure to cancel and reduce first. Do not simply multiply out.

3.187 17 cowboys throw their hats in a horse shed. In how many different ways can they walk away with one hat each (not necessarily their own hats)?

This is similar to placing 17 hats on 17 cowboys' heads, or 17 cowboys occupying 17 chairs. They can be shuffled within the 17 chairs, and each configuration would be unique. This is permutation.

$$_nP_r = {_{17}}P_{17} = \frac{n!}{(n-r)!} = \frac{17!}{(17-17)!} = \frac{17!}{0!} = \frac{17!}{1} = 17!$$

Alternative method:

The first cowboy walks in, and walks out with hat #1, or hat #2, or hat #3, ... ,or hat #17. Therefore, there are 17 possibilities for him.

The second cowboy walks in, and walks out with one of the remaining 16 hats. Therefore, 16 possibilities for him.

15 possibilities for the third cowboy, 14 for the fourth, and so on. The 17th cowboy will have one last hat left to pick up.

In essence, this is a 17-slot question.

17 16 15 14   3   2   1
[×] [×] [×] [×] ················ [×] [×] [×]

Total possibilities = 17 × 16 × 15 × ·············· × 3 × 2 × 1 = 17!

3.188 A picnic set contains 6 forks, 3 spoons and 4 knives. In how many different ways can 6 silverware items be drawn at random such that 2 of each type are selected?

Let's call the forks F_1, F_2, F_3, F_4, F_5 and F_6; the spoons S_1, S_2 and S_3; and the knives K_1, K_2, K_3 and K_4. They are all unique, just like individual persons with names.

6 items can be drawn at random in many ways, but what is important is that during the course of drawing them, only some draws will result in 2 of each type of silverware.

2 forks can be drawn out of 6 forks in $_6C_2$ ways; 2 spoons out of 3 spoons in $_3C_2$ ways, and 2 knives out of 4 knives in $_4C_2$ ways. Remember, when drawing 2 of any item, we're essentially selecting two unique entities to form a team. Therefore, use combination, not permutation.

We need to consider the combination of each of the two 2 forks with each of the 2 spoons with each of the 2 knives. Similar to a restaurant menu, multiply out the numbers.

$$_6C_2 \times {}_3C_2 \times {}_4C_2 = \frac{6!}{2!(6-2)!} \times \frac{3!}{2!(3-2)!} \times \frac{4!}{2!(4-2)!} = \frac{6!}{2!4!} \times \frac{3!}{2!1!} \times \frac{4!}{2!2!} =$$

$$\frac{6 \times 5 \times 4!}{2 \times 4!} \times \frac{3 \times 2}{2} \times \frac{4 \times 3 \times 2}{2 \times 2} = \frac{30}{2} \times 3 \times 6 = 270 \text{ different ways}$$

In the above steps, save time by writing 2 for 2!, because 2! is 2 × 1, or just 2.

3.189

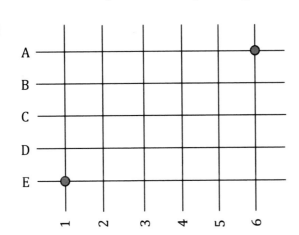

The figure on the left shows a grid of avenues and streets. In how many different ways can a car travel from the corner of 1st Street and E Avenue to the corner of 6th Street and A Avenue, while driving the least possible distance?

To analyze this problem, let's first downsize the grid as shown on the right. To go from the bottom left corner to the top right corner, the car would need to go a total of 5 blocks, with 2 blocks northwards (N_1 and N_2) and 3 blocks eastwards (E_1, E_2 and E_3).

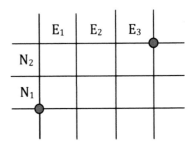

In any sequence of travel, block N_1 would always come before N_2. Similarly, block E_1 would always come before E_2, which would always come before E_3. Imagine the blocks are laid out in a straight line representing the car's travel.

| 1st | 2nd | 3rd | 4th | 5th |

So, the question essentially is: In how many ways can the blocks N_1 and N_2 be placed in these 5 places? This is the same as 2 persons occupying 5 chairs. The only difference being, the persons can't exchange their places, i.e. N_1 and N_2 can't swap places. N_1 is always to the left of N_2 because it always comes before N_2. So use combination $(_5C_2)$, not permutation $(_5P_2)$.

You can do the exact same analysis for the eastward blocks: In how many different ways can E_1, E_2 and E_3 occupy the 5 places? You would conclude that it's $_5C_3$.

They're both the same, i.e. $_5C_2 = {}_5C_3$ because $_nC_r = {}_nC_{(n-r)}$.

In general, such problems can be solved by substituting n = total # of blocks to be travelled, and r = # of blocks in one direction (either northward or eastward), in the combination formula.

Now, going back to the original question, we see that there are 4 northward blocks, and 5 eastward blocks. Total blocks = $n = 4 + 5 = 9$.

$$\text{Total number of ways to travel} = {}_9C_4 = \frac{9!}{4!(9-4)!} = \frac{9!}{4!5!} = \frac{9\times8\times7\times6\times5!}{4\times3\times2\times5!} = 126$$

3.190 There are 4 routes from Seattle to Denver, and 6 routes from Denver to Atlanta. I travel frequently between Seattle and Atlanta via Denver. If, on each trip, I decide to take different routes from Seattle to Denver and Denver to Seattle, and also different routes from Denver to Atlanta and Atlanta to Denver, then how many such trips could I make before repeating a route previously taken? (A route is considered different if retaken in the opposite direction.)

Seattle ◄———— 4 routes ————► Denver ◄———— 6 routes ————► Atlanta

Each of the 4 routes between Seattle and Denver can be combined with each of the 6 routes between Denver and Atlanta. Sounds like a restaurant menu type question. Simply multiply them out: $6 \times 4 = 24$ total onward routes. Similarly, 24 total return routes.

Each of the onward routes can be combined with each of the return routes. Therefore, the total number of unique routes for the entire trip would be $24 \times 24 = 576$.

3.191 My wallet has a $1 bill, a $5 bill, a $10 bill and a $20 bill. I pull out a bill, put it back in, and then I pull out a bill again (not necessarily the same one). In how many different ways can I pull out two bills such that the sum of their values is at least $20?

At least $20 means $20 or more. That can be obtained by pulling out 2 bills in these sequences:

$20	followed by	$1		$1	followed by	$20
$20	followed by	$5		$5	followed by	$20
$20	followed by	$10		$10	followed by	$20
$20	followed by	$20		$10	followed by	$10

Total number of ways = 8

Notice that there was no need to use combinatorics; simple analysis was enough.

3.192 A school library has 8 different newspapers. A student is allowed to borrow up to 2 newspapers at a time. In how many different combinations can the student borrow up to 2 newspapers?

Up to 2 newspapers means 1 or 2 newspapers, but not more than 2.

One newspaper can be borrowed out of 8 newspapers in 8 different ways.

Borrowing 2 newspapers out of 8 is like selecting a team of 2 persons out of 8 persons. Use combination ($_nC_r$), or $_8C_2 = \dfrac{8!}{2!(8-2)!} = \dfrac{8!}{2!6!} = \dfrac{8\times7\times6!}{2\times6!} = 28$

Therefore, total number of ways to borrow up to 2 newspapers out of 8 is: $8 + 28 = 36$

PRACTICE EXAMPLES

3.193 In how many different ways can 6 students be seated in 4 chairs?

3.194 At a certain college, a student can choose 5 out of 11 available courses in a semester. In how many different ways can the 5 courses be selected?

3.195 At the end of a theatrical performance, the 12 artists are to be introduced, one by one, to the audience. In how many unique sequences can this be done?

3.196 A kitchen drawer contains 5 butter knives, 8 steak knives and 4 carving knives. In how many different ways can 9 knives be drawn at random such that 3 of each type are selected?

3.197

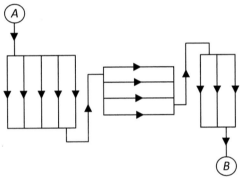

How many distinct paths can a person take while walking from point *A* to point *B*?

3.198 In a chess championship, there are 13 entrants. How many games should be arranged such that each entrant gets a chance to play against each of the other entrants?

3.199 *List A*: C, M, G, L, S, U, O
List B: X, Y, Z, P, T
How many different 4-letter arrangements can be made by selecting 2 letters each from *List A* and *List B* above?

3.200 Emily's purse has a $5 bill, a $10 bill, a $20 bill and a $50 bill. She takes out a bill, puts it back in, and then takes out a bill again (not necessarily the same one). In how many different ways can she take out two bills such that the sum of their values is $30 at the most?

3.201 A box has 5 distinct pieces of candy. In how many different ways can I pick up at least 4 candies?

COUNTING (VENN DIAGRAMS)

Let's say I'm hosting my birthday party. 40 guests show up. 25 of them bring along a greeting card each, whereas 20 of them bring along a food item each. 15 people bring along both a greeting card and a food item each. How many people bring along nothing?

Analyzing this question by using regular algebra can be a headache. Instead, using Venn diagrams is a much simpler way to do it. A Venn diagram, in its simplest form, uses a circle to represent each set (or group) of entities. The greater the number of sets, the greater the number of circles. If two groups overlap, then that would reflect in the intersection of the circles. The birthday party example is depicted below. You may notice that Venn diagrams are quite intuitive.

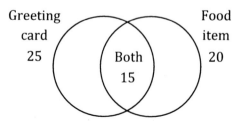

The whole circle on the left represents the number of people that bring along a greeting card. The whole circle on the right represents those who bring along a food item. The intersection of the two circles represents those who bring along both a card *and* a food item.

Assume x number of people bring along nothing. Then they would be included in neither of the two circles; they would lie outside.

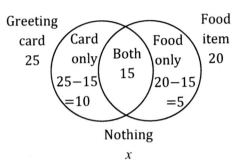

Those who bring along a card but not food would be represented by the *crescent* on the left. Their number is: Those who bring a card, minus those who bring both.   Card only = 25 – 15 = 10

Those who bring along food but not a card would be represented by the *crescent* on the right. Their number is: Those who bring food, minus those who bring both.

Food only = 20 – 15 = 5

Accounting for all the guests (40 of them) is only a matter of adding the two crescent regions, the intersecting region, and the quantity (x) that lies outside the figure. That should total 40.

$$10 + 15 + 5 + x = 40 \quad \Rightarrow \quad 30 + x = 40 \quad \Rightarrow \quad x = 10$$

3.202 At a certain library, 45% of the members borrow non-fiction books, 70% borrow fiction books, and 19% borrow both fiction and non-fiction books. What percent of the members borrow neither fiction nor non-fiction books?

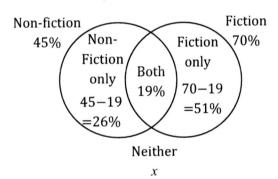

Non-fiction = 45%   Fiction = 70%

Non-fiction only = Non-fiction – Both = 45% – 19% = 26%

Fiction only = Fiction – Both = 70% – 19% = 51%

Summing up the two crescent regions, the intersection region and x should give us a total of 100%.

$$26 + 51 + 19 + x = 100 \quad \Rightarrow \quad 96 + x = 100 \quad \Rightarrow \quad x = 4\%$$

3.203 A school of music has 355 students. 56% of them are learning to play the viola, 38% are learning to play the guitar, and 16% are learning to play both the instruments. What *fraction* of the students is learning to play the viola, but not the guitar?

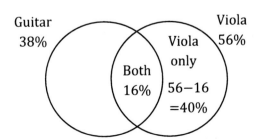

Construct the Venn diagram, and fill in the relevant values. Do not waste your time on calculating what you don't need.

Viola only = 40%   Convert this into fraction:

$$40\% = \frac{40}{100} = \frac{4}{10} = \frac{2}{5}$$

Notice that the total number of students, 355, didn't matter. Such questions do occur on the GRE/GMAT. Recognize them, and save time!

3.204 Among 200 of my health conscious friends, 135 exercise regularly, and 90 follow a healthy diet. *At least* how many of those 200 exercise daily <u>and</u> follow a healthy diet?

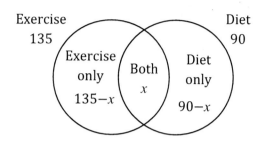

Assume x people do both: diet and exercise.

The easiest way to find the minimum value of x is to imagine the two circles pulling away from each other. x will start shrinking. When the two circles have no intersection region at all, $x = 0$.

Can x really become zero? Well, if it does, then the two circles should equal the total number of people, i.e. $135 + 90 = 200$, which isn't true.

$135 + 90 = 225$, which is 25 greater than 200. Meaning, x should be the overlapping 25. The overlap can always be greater than 25, but certainly not less. $x \geq 25$

3.205 Among the residents of an upscale neighborhood, 45% have RVs, 49% have classic cars, and 46% have electric cars. Also, 12% have only RVs and classic cars, 10% have only classic cars and electric cars, and 15% have only electric cars and RVs. 1% of those residents have none of the three types of vehicles. What percent have all three types of vehicles?

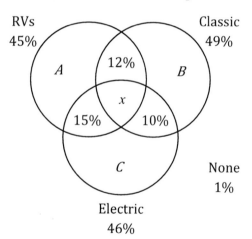

This example is on the same lines as what we've learned in the previous ones, except that there are three sets instead of two.

The given information leads to the adjacent Venn diagram. We need to find x, the common region to all three circles. To do that, let's first find the regions A, B and C.

$A = $ RVs only $= 45 - 12 - 15 - x = 18 - x$

$B = $ Classic only $= 49 - 12 - 10 - x = 27 - x$

$C = $ Electric only $= 46 - 15 - 10 - x = 21 - x$

Now let's add all the regions, plus a 1% that lies outside. That should give us 100%.

$A + B + C + 12 + 15 + 10 + x + 1 = 100$

$(18 - x) + (27 - x) + (21 - x) + 37 + x + 1 = 100$

$104 - 2x = 100 \quad \Rightarrow \quad 4 = 2x \quad \Rightarrow \quad x = 2\%$

I've added the % sign only at the end to keep the calculation steps looking neat.

PRACTICE EXAMPLES

3.206 In a class of 90 students, 53 can play football, 58 can play baseball, and 41 can play both games. How many students can play neither football nor baseball?

3.207   In a class of 55 students, 42 are majoring in English, 26 are majoring in Math, and 15 are majoring in both English and Math. What *percent* of the students are majoring in Math, but not in English?

3.208   In a township of 70 homes, 39 are equipped with solar panels, and 39 have water wells. *At least* how many homes in the township have both solar panels <u>and</u> water wells?

3.209   Let X be the set of integers from 50 to 60. Let Y be the set of even numbers. Let Z be the set of integers divisible by 3. If p is the number of integers common to sets X and Y, and q is the number of integers common to sets X and Z, find $(p + q)(p - q)$.

SPACE FOR CALCULATIONS

COUNTING (MISCELLANEOUS TYPES)

Up to this point in this chapter, every example on *counting methods* was systematically classified into one category or another. But on the GRE/GMAT, some questions on counting methods can't be classified; they're analytical in nature, and sometimes require mostly common sense. Let's solve a few such examples.

3.210   **If you read pages 136 through 244 of a book, how many pages would that be?**

Consider all 244 pages. From that number, simply take off the first 135 pages that you wouldn't read. The number of pages you'd end up reading would be: $244 - 135 = 109$

3.211 In a car race, Zeyad's car is in the 54th place. How many cars does he need to overtake to be in the 21st place?

Let's imagine the cars in their order:

1 2 3 4 ... ... 20 [21] 22 ... ... 53 [54]

For Zeyad to go from the 54th place to the 21st place, he would need to overtake car numbers 53 through 22. He would also need to overtake the 21st car, so that his car would then be the 21st car.

Meaning, consider all 53 cars ahead of Zeyad's car. From them, take off the first 20 cars.

$53 - 20 = 33$ That's the number of cars Zeyad would need to overtake.

Note: Simply subtracting 21 from 54 also gives the same answer: $54 - 21 = 33$. But it's important to analyze the problem before answering it, for every problem may not be solved by simply taking the difference. The analysis won't take much of your time.

3.212 *List A*: 1, 3, 5
List B: 2, 4, 6
How many different sums are possible by selecting a number each from *List A* and *List B* above?

First, find all the possible sums by selecting each number from *List A*, and adding it to each number from *List B*.

$1 + 2 = 3$	$1 + 4 = 5$	$1 + 6 = 7$
$3 + 2 = 5$	$3 + 4 = 7$	$3 + 6 = 9$
$5 + 2 = 7$	$5 + 4 = 9$	$5 + 6 = 11$

As you can see, some numbers get repeated among the sums. The only unique numbers are 3, 5, 7, 9 and 11. Therefore, there are only five different sums possible.

3.213 John Doe was born on August 21, 1921 and died on June 3, 1997. At what age did he die?

First, take the difference of the years to tell his age:

Year of death – Year of birth = $1997 - 1921 = 76$ years.

But did he make it to his 76th birthday? No, he died on June 3 which is before August 21.

Therefore, he was 75 and not 76 when he died.

3.214 A toolbox contains 7 wrenches, 4 pliers and 10 screw drivers. If you randomly remove tools from the box one by one, what's the minimum number you'd need to remove to be sure that you have at least 3 of each type of tool?

If you're lucky, you might end up removing 3 wrenches, followed by 3 pliers, followed by 3 screw drivers. But life isn't that simple, is it?

If you remove 10 tools, then maybe you'll have at least 3 of each. Or maybe you'll have all 10 screw drivers. So, no luck there.

If you remove 7 more tools, then you're very likely to have 3 of each type of tool among the 17 tools. Right? Maybe not; maybe you'll have all 10 screwdrivers and all 7 wrenches. No luck with the pliers! Don't give up.

If you remove 3 more tools, then you'll certainly have at least 3 of each type of tool. That's because you will have run out of the 10 screw drivers, followed by the 7 wrenches. The remaining 3 would have to be pliers.

Therefore, the minimum number of tools you'd need to remove is $10 + 7 + 3 = 20$.

Here's the method for such type of problems: Start with the largest number (here, 10) assuming the worst luck. Then move on to the next smaller number (7), followed by the next smaller number (4), and so on.

PROBABILITY

Probability is the likelihood (or the chance) of an event happening. It may sound vague, but it is a very specific branch of mathematics. Probability can be expressed as a fraction, or as a percent. Keep in mind that fractions and percents are interconvertible. When you see on the weather report that there's a 70% chance that it will rain tomorrow, it also means there's a 70 in 100 chance, or $\frac{70}{100}$ chance, or $\frac{7}{10}$ chance that it will rain tomorrow. The greatest possible value of probability is 100%, or $\frac{100}{100}$, or 1. Probability of an event can be determined by using several methods depending upon what information is available about the event. From the GRE/GMAT point of view, you're not expected to know any advanced methods of calculating probability. Let's study some basic methods:

Imagine yourself on a visit to an automobile factory that manufactures only cars and vans. You're standing at the end of the assembly line, recording the number of vehicles that are manufactured. It's been several hours since you've been there, and up to this point, you've counted 58 cars and 42 vans (a total of 100 vehicles) that have rolled out before you. You're standing at a location where you can see the vehicle only when it has exited the assembly line. Meaning, you can't see the next vehicle that is about to roll out.

Suddenly, a thought occurs to you: What's the probability that the next vehicle that rolls out will be a van? You look into your notes, and you see that out of the 100 vehicles recorded so far, 42 were vans. Therefore, there's a 42 in 100 chance, or a 42% likelihood that it will be a van. Next thought: What's the chance that it will be a car? That would be 58%. Next: What's the chance that it will NOT be a car? That would be the chance that it would be a van, which is 42%. See the point? Probability is mathematics, not sorcery! In the instances in which historical information is available (in the form of the vehicles you recorded), probability of an event happening, written as P(E), is:

$$P(E) = \frac{\text{number of expected outcomes}}{\text{total number of outcomes}}$$

The expected outcome was a van. You recorded 42 vans; so 42. The total outcomes were the total vehicles you recorded; so 100.

$$P(\text{van}) = \frac{\text{\# of vans}}{\text{total \# of vehicles}} = \frac{42}{100} = 42\%$$

What if historical information is not available? For example, you're about to roll a die. What's the chance that the outcome will be 2? In such an instance, you have to consider the *possible* outcomes, which are 1, 2, 3, 4, 5 or 6. The expected outcome is the number 2. There's only one such outcome.

Number of expected outcomes = 1   Total number of possible outcomes = 6

$$P(6) = \frac{\text{\# of expected outcomes}}{\text{total \# of possible outcomes}} = \frac{1}{6}$$

In fact, each of the numbers 1 through 6 has a $\frac{1}{6}$ chance of showing up.

What's the probability that the numbers 4 or 5 will show up?

Number of expected outcomes = 2   Total number of possible outcomes = 6

$$P(4 \text{ or } 5) = \frac{\text{\# of expected outcomes}}{\text{total \# of possible outcomes}} = \frac{2}{6} = \frac{1}{3}$$

Another context in which you might be asked to calculate probability is one in which there are several elements grouped together. One of them is randomly selected. What is the likelihood that a certain element X is selected? That would be the *proportion* in which element X is present in that group. For example, I have a list of 20 names. Out of them, 13 are engineers, and 7 are lawyers. If I randomly call out a name from that list, what's the chance that it is of an engineer? That would be 13 in 20, or $\frac{13}{20}$. Likewise, there's a $\frac{7}{20}$ chance that it is of a lawyer. Clearly, the probability of randomly picking a certain type of person from a group is the proportion in which such persons are present in the group.

3.215 A kitchen drawer contains 5 spoons, 4 butter knives and 7 forks. Without looking, one item is taken out at random. Find the probability that it will: (i) be a fork; (ii) NOT be a butter knife; (iii) be either a butter knife or a fork; (iv) be neither a spoon nor a fork?

Total number of items = $5 + 4 + 7 = 16$

(i)   Out of the 16 items, 7 are forks.

$$P(\text{fork}) = \frac{7}{16}$$

(ii) If it's not a butter knife, it must be a spoon or a fork, i.e. one of the $5 + 7 = 12$ items out of 16.

$$P(\text{not butter knife}) = \frac{12}{16} = \frac{3}{4}$$

(iii) Either a butter knife or a fork means one of $4 + 7 = 11$ items out of 16.   P(knife or fork)$= \frac{11}{16}$.

(iv) If it's neither a spoon nor a fork, it must be a butter knife, i.e. one of 4 out of 16 items.

$$P(\text{butter knife}) = \frac{4}{16} = \frac{1}{4}$$

3.216 **A flower pot contains one of each of these: rose, berlinia, sunflower, lily and jasmine. If three flowers are randomly picked, what's the chance that berlinia gets picked?**

Correct analysis:

If 1 flower is picked, there's a 1 in 5 chance that berlinia gets picked. $P(\text{berlinia}) = \frac{1}{5} = 20\%$

If 2 flowers are picked, there's a 2 in 5 chance that berlinia is picked. $P(\text{berlinia}) = \frac{2}{5} = 40\%$

If 3 flowers are picked, there's a 3 in 5 chance that berlinia is picked. $P(\text{berlinia}) = \frac{3}{5} = 60\%$

If 4 flowers are picked, there's a 4 in 5 chance that berlinia is picked. $P(\text{berlinia}) = \frac{4}{5} = 80\%$

If 5 flowers are picked, there's a 5 in 5 chance that berlinia is picked. $P(\text{berlinia}) = \frac{5}{5} = 100\%$

I've shown the percentage equivalents to put it in perspective. Meaning, if you pick all 5 flowers, then you can be 100% sure that berlinia is picked. The answer is the highlighted one.

Wrong analysis:

When the 1st flower is picked, it is 1 of 5 flowers. $P(\text{berlinia}) = \frac{1}{5} = 20\%$

When the 2nd flower is picked, it is 1 of 4 remaining flowers. $P(\text{berlinia}) = \frac{1}{4} = 25\%$

When the 3rd flower is picked, it is 1 of 3 remaining flowers. $P(\text{berlinia}) = \frac{1}{3} = 33.\overline{3}\%$

Add up the probabilities to get the probability that berlinia gets picked after 3 attempts:

$$P(\text{berlinia}) = 20\% + 25\% + 33.\overline{3}\% = 78.\overline{3}\%$$

The analysis seems reasonable. What's wrong with it? To find that out, let's continue with it and see it for ourselves:

When the 4th flower is picked, it is 1 of 2 remaining flowers. $P(\text{berlinia}) = \frac{1}{2} = 50\%$

When the 5th flower is picked, it is 1 of 1 remaining flowers. $P(\text{berlinia}) = \frac{1}{1} = 100\%$

Adding up the above five probabilities, we should get 100%.

$$P(\text{berlinia}) = 20\% + 25\% + 33.\overline{3}\% + 50\% + 100\% = 228.\overline{3}\%$$

That's impossible (228.$\overline{3}$%) because probability can never exceed 100%.

Similarly, if you multiply the probabilities, you'd get an erroneous result too:

After 3 attempts, $P(\text{berlinia}) = \frac{1}{5} \times \frac{1}{4} \times \frac{1}{3} = \frac{1}{60} = 1.\overline{6}\%$

The more flowers you pick, the greater the chance of picking berlinia. But after 3 attempts, it looks like the chance is actually less than that after 1 or 2 attempts. So it is incorrect too.

3.217 Amanda and Wes have 3 kids. What's the likelihood that they have at least 2 sons? Assume they do not have twins or triplets.

To analyze this problem, consider the chronological sequence in which Amanda and Wes could have had 3 kids. Underneath here: S = son, and D = daughter.

D D D					Zero sons (All daughters)
S D D	or	D S D	or	D D S	One son (Two daughters)
S S D	or	S D S	or	D S S	Two sons (One daughter)
S S S					Three sons (Zero daughters)

At least 2 sons means 2 or more sons. In this case, that would be 2 or 3 sons.

Two sons can be had in 3 different sequences, and three sons can be had in only one sequence. That's a total of 3 + 1 = 4 sequences out of a total of 8 possible sequences.

$$P(S \geq 2) = \frac{4}{8} = \frac{1}{2} = 50\%$$ On the GRE/GMAT, express the answer in percent if asked to.

3.218 A box has 4 scraps of paper, with the numbers 1, 2, 3 and 4 written on them. Pick one scrap, note its number, and put it back. Then pick a scrap again, and note its number. What's the chance that the sum of the two numbers you noted will be: (i) an odd number; (ii) at least 3; (iii) less than 6; (iv) 5?

It makes sense in quickly writing down the possible outcomes in the form of x, y where x is the first scrap you pick, and y is the second scrap you pick.

1, 1	1, 2	1, 3	1, 4	
2, 1	2, 2	2, 3	2, 4	
3, 1	3, 2	3, 3	3, 4	
4, 1	4, 2	4, 3	4, 4	Total possibilities = 16.

(i)

1, 1	1, 2	1, 3	1, 4
2, 1	2, 2	2, 3	2, 4
3, 1	3, 2	3, 3	3, 4
4, 1	4, 2	4, 3	4, 4

Recall number properties (page 6):
Odd + Even = Odd

The highlighted 8 pairs (odd + even) will result in odd sums.

$$P(\text{sum} = \text{odd}) = \frac{8}{16} = \frac{1}{2}$$

(ii)

1, 1	1, 2	1, 3	1, 4
2, 1	2, 2	2, 3	2, 4
3, 1	3, 2	3, 3	3, 4
4, 1	4, 2	4, 3	4, 4

The highlighted 15 pairs will have their sums equal to at least 3.

$$P(\text{sum} \geq 3) = \frac{15}{16}$$

(iii)

1, 1	1, 2	1, 3	1, 4
2, 1	2, 2	2, 3	2, 4
3, 1	3, 2	3, 3	3, 4
4, 1	4, 2	4, 3	4, 4

Less than 6 means 5 or less.

The highlighted 10 pairs will have their sums less than 6.

$$P(\text{sum} < 6) = \frac{10}{16} = \frac{5}{8}$$

(iv)

1, 1	1, 2	1, 3	1, 4
2, 1	2, 2	2, 3	2, 4
3, 1	3, 2	3, 3	3, 4
4, 1	4, 2	4, 3	4, 4

The highlighted 4 pairs will have their sums equal to 5.

$$P(\text{sum} = 5) = \frac{4}{16} = \frac{1}{4}$$

3.219 **If the individual letters of the word *GATES* are randomly reshuffled, what's the probability that the resulting word will be *STAGE*?**

First, let's find the number of ways in which *GATES* can be reshuffled. Five letters (*G, A, T, E* and *S*) are to be reshuffled in five slots.

Once the 1st slot assumes one of the 5 letters, the 2nd slot can have one of the 4 remaining letters. Once that happens, the 3rd slot can hold one of the 3 remaining letters, and so on.

5 4 3 2 1
⊠ ⊠ ⊠ ⊠ ⊠   Total possible arrangements = $5 \times 4 \times 3 \times 2 \times 1 = 120$

Among all these 120 possibilities, the word *STAGE* will result in only *one* arrangement.

Therefore, $P(STAGE) = \dfrac{1}{120}$

3.220 **Six differently priced cars are to be parked side by side for sale (nose first). What's the chance that the most expensive and the least expensive cars are parked at the extreme ends?**

First, parking 6 cars in 6 spaces is the equivalent of 6 people occupying 6 chairs. The number of ways in which it can be done is:

$$_6P_6 = \frac{6!}{(6-6)!} = \frac{6!}{0!} = \frac{6!}{1} = 6 \times 5 \times 4 \times 3 \times 2 = 720$$

Next, imagine the 6 parking spaces as 6 slots, with the two extreme slots occupied by the same two cars: C_1 = least expensive,   and C_6 = most expensive

C_1 ⊠ ⊠ ⊠ ⊠ C_6

This is now essentially a 4-slot problem, with 4 cars being reshuffled only within the middle 4 slots. That can be done with C_1 at extreme left, and C_6 at extreme right.

       4 3 2 1
C_1 ⊠ ⊠ ⊠ ⊠ C_6 Total arrangements = $4 \times 3 \times 2 \times 1 = 24$

Now exchange C_1 and C_6, and count again.

       4 3 2 1
C_6 ⊠ ⊠ ⊠ ⊠ C_1 Total arrangements = $4 \times 3 \times 2 \times 1 = 24$

Therefore, there are a total of 24 + 24 = 48 arrangements in which the least and most expensive cars would be at the extreme ends. We saw that the total number of ways of arranging 6 cars in 6 spaces is $_6P_6$, or 720.

The probability that the most and least expensive cars are parked at extreme ends is:

$$P = \frac{48}{720} = \frac{4}{60} = \frac{1}{15}$$

3.221 **A class has 80 students majoring in either English, or Chemistry, or Math. The probability of randomly selecting a student with English major is 0.25, and that with Math major is 0.4. How many students are majoring in Chemistry?**

The probability of randomly picking a certain type of person from a group is the proportion in which such persons are present in the group. The reverse is also true.

The probability of selecting a student majoring in English is 0.25, or 25%. It means that 25% of the students in the class are majoring in English. Likewise, 40% are majoring in Math. The remaining must be majoring in Chemistry. That is 100% − (25% + 40%) = 35%.

The number of students majoring in Chemistry would be: 35% of 80 $= \dfrac{35}{100} \times 80 = 28$

3.222 **A school has mostly female students. There's only a 32% chance of randomly selecting a male student from the school. *At least* how many female students could there be in the school?**

A 32% chance of randomly selecting a male student means that 32% of the students are male. The remaining students must be female: 100% – 32% = 68%

To find the minimum number of female students, think of the ratio in which male and female students are present: male : female = 32 : 68

Reduce the ratio (by 4) to its lowest terms: male : female = 8 : 17

So there must be a minimum of 8 male and 17 female students.

3.223 **I have a list of 153 countries ranked by population. Country *M* is ranked 37th, and country *G* is ranked 89th. If I randomly select a country from the list, what's the chance that it is ranked between countries *M* and *G*?**

$$\begin{array}{ccccccccccccccccc} & & & & & & & M & & & & & & G & & & \\ 1 & 2 & 3 & 4 & \dots & \dots & \dots & 36 & \boxed{37} & 38 & \dots & \dots & \dots & 88 & \boxed{89} & 90 & \dots & \dots & \dots & 153 \end{array}$$

Total number of countries = 153

The countries that lie between *M* (37th) and *G* (89th) are the ones from 38 through 88. To find that number, consider all 88 countries before *G*, and subtract the first 37:   88 – 37 = 51

The probability of randomly selecting a country from between ranks 37 and 89 is:

$$P(37 < rank < 89) = \frac{\text{\# of countries between 37 and 89}}{\text{total \# of countries}} = \frac{51}{153} = \frac{1}{3}$$

PRACTICE EXAMPLES

3.224 (i) What's the probability that it is between 7 a.m. and 10 p.m. right now?

(ii) What's the probability that it is NOT between 7 o'clock and 10 o'clock right now?

3.225 In the Principal's office, the Principal is talking to a parent and a child. What's the probability that at least one of the three of them is female?

3.226 If you roll two dice, what's the probability that the sum of the two numbers you get will be: (i) an even number; (ii) at least 7; (iii) 6; (iv) at most 5?

3.227 If the individual letters of the word *BATS* are randomly reshuffled, what's the probability that the resulting word will be either *TABS* or *STAB*?

3.228 In my office is a family photo, with 5 family members standing side by side. What's the likelihood that the oldest person is standing to the far left and the youngest person is standing to the far right?

3.229 A bag has black, red, and green marker pens. If you randomly select a pen from the box, there's a $\frac{5}{8}$ chance that it will be black, and a 20% chance that it will be green. *At least* how many red marker pens are in the box?

3.230 My brother, my cousin and I were in the same class of 70 students. My brother graduated at the top of the class, whereas I ranked 27th at graduation. What's the probability that my cousin ranked between me and my brother?

3.231 *List A*: 1, 3, 5

List B: 2, 5, 6

If you select a number each from *List A* and *List B* above, what's the chance that the product of the two numbers selected will be an odd number?

SPACE FOR CALCULATIONS

PROBABILITY OF DEPENDENT, INDEPENDENT AND MUTUALLY EXCLUSIVE EVENTS

The general formula to calculate the probability of one or both of two events A and B is:

$$P(A \text{ or } B) = P(A) + P(B) - P(A \text{ and } B)$$

where $P(A)$ = probability of event A happening
                 $P(B)$ = probability of event B happening
                 $P(A \text{ and } B)$ = probability of events A and B happening simultaneously
                 $P(A \text{ or } B)$ = probability of one or both (at least one) of events A and B happening

In the formula, the term $P(A \text{ and } B)$ depends upon what type of events A and B are. So let's discuss the three possible types of events.

Two events are said to be **dependent** if the outcome of one event is likely to influence the outcome of the other event. For example, if a box of candies contains 12 English toffees and 15 Swiss chocolates, and two candies are randomly drawn from it, then what's the probability that they are both English toffees?

Well, the probability that the first candy that is drawn is an English toffee is the proportion in which English toffees are present in the box, i.e. 12 English toffees out of a total of 27 candies, or $\frac{12}{27}$. *Presuming* the first candy drawn is an English toffee, the probability that the second one is an English toffee is: 11 *remaining* English toffees out of a total of 26 *remaining* candies.

Clearly, presuming the first event has occurred (presuming the first English toffee is drawn), the second event has slightly less probability. That's because $\frac{11}{26} < \frac{12}{27}$. Meaning, the outcome of the first event has *influenced the outcome of the second event*. Therefore, the two events are dependent.

The probability that both events will occur, or both candies will be English toffees, is the product of their individual probabilities: $\frac{12}{27} \times \frac{11}{26}$.

For dependent events A and B, in the general formula, substitute $P(A \text{ and } B) = P(A) \cdot P(B \text{ presuming } A)$

Note: Mathematically, there's no difference between drawing 2 candies one by one, and drawing 2 candies at once. The analysis is always to be done considering the items are drawn one by one.

On some GRE/GMAT questions, you will be told that one event is likely to cause another event, and you will also be given the probability of both the events occurring.

Two events are said to be **independent** if the outcome of one event does not influence the outcome of the other event, although they may happen at the same time. The probability of both the events happening at the same time is the product of their individual probabilities. For example, suppose there's a 40% chance it will rain in New York tonight, and a 70% chance it will rain in London tonight. The probability that it will rain in New York as well as London tonight is 40% × 70% = 28%.

For independent events A and B, in the general formula, substitute $P(A \text{ and } B) = P(A) \cdot P(B)$

Two events are said to be **mutually exclusive** if the outcome of one event does not influence the outcome of the other event, and they absolutely cannot happen at the same time. Therefore, the probability of both the events happening at the same time is zero. For example, I have a list of 100 vehicles. 26 of them are cars; 74 are trucks. If I randomly pick out one vehicle from the list, what's the probability that it is a car? 26%. What's the probability that it is a truck? 74%. What's the probability that it is a car as well as a truck? 0%, because it can't be a car and a truck at the same time.

For mutually exclusive events A and B, in the general formula, substitute $P(A \text{ and } B) = 0$

3.232   (i) What's the chance that I was born in April?

(ii) What's the probability that I was born in February, and my brother was born on a Monday?

(iii) What's the likelihood that I was born on August 15, and you were born on January 26?
      (In the above three questions, assume regular years, not leap years).

(i)   There are 30 April days in a regular year of 365 days. The probability that I was born in April is 30 in 365, or $\dfrac{30}{365}$ or $\dfrac{6}{73}$.

(ii) There's a $\dfrac{28}{365}$ chance that I was born in February. There's a $\dfrac{1}{7}$ chance that my brother was born on a Monday. My being born in February and my brother being born on a Monday are two independent events. Their joint probability is simply their product:

$$P(\text{Feb. and Mon.}) = P(\text{Feb.}) \cdot P(\text{Mon.}) = \frac{28}{365} \cdot \frac{1}{7} = \frac{4}{365}$$

(iii) The probability that I was born on August 15 (or any day for that sake) is 1 in 365, or $\dfrac{1}{365}$. Same with your being born on January 26. Once again, these are independent events. Their joint probability is their product:

$$P(\text{Aug. 15 and Jan. 26}) = P(\text{Aug. 15}) \cdot P(\text{Jan. 26}) = \frac{1}{365} \cdot \frac{1}{365} = \frac{1}{365^2} = \frac{1}{133,225}$$

3.233   **If a coin is tossed six times in a row, what's the chance that the outcomes will have:**
(i) all tails; (ii) at least 1 tail; (iii) up to 5 tails?

The first time a coin is tossed there could be 2 outcomes: heads (H) or tails (T).

The second time a coin is tossed, there could be 2 more outcomes: H or T. Each of the second outcomes could be considered in combination with each of the first outcomes. So the total number of outcomes could be 2×2, or 2^2.

The third time, there could be H or T again. Each of them could be combined with each of the second outcomes, which in turn could be combined with each of the first outcomes. The total number of outcomes could be $2 \times 2 \times 2$, or 2^3.

In general, if a coin is tossed n times, then there could be 2^n possible outcomes.

(i)   Total possible outcomes $= 2^6 = 64$ Out of all 64 outcomes, only one outcome will have all tails. Therefore, $P(T = 6) = \dfrac{1}{64}$

(ii) Out of the 64 outcomes, only one outcome will have all heads. The remaining 63 outcomes will each have at least one tail (one or more tails). Therefore, $P(T \geq 1) = \dfrac{63}{64}$

(iii) *Up to 5 tails* means 5 or fewer tails. Out of the 64 outcomes, only one outcome will have all 6 tails. The remaining 63 will have 5 or fewer tails. Therefore, $P(T \leq 5) = \dfrac{63}{64}$

3.234 **Seven of my students scored an *A* on the exam, and three scored a *B*. If I randomly call out the names of two students, what's the likelihood that both of them have an *A* on the exam?**

Number of students who scored an $A = 7$ Total number of students $= 10$

The probability that the first name I call out scored an A is $\dfrac{7}{10}$.

Presuming the first student scored an A, there would only be 6 more students with an A, out of a total of only 9 more students.

Therefore, the probability that the second name I call out scored an A is $\dfrac{6}{9}$, or $\dfrac{2}{3}$.

The probability that both the students scored an A is their joint probability, i.e. their product:
$$\frac{7}{10} \times \frac{2}{3} = \frac{14}{30} = \frac{7}{15}$$

3.235 **A circular pizza is cut into six equal slices. Four of them have only bacon, and two have only chicken. If two slices are selected at random, what's the chance that they both have bacon?**

This example is on similar lines as of the previous example.

Number of slices with bacon $= 4$ Total number of slices $= 6$

The probability that the first slice I select has bacon is 4 out of 6, or $\dfrac{4}{6}$, or $\dfrac{2}{3}$.

Presuming the first slice selected has bacon, the probability that the second slice has bacon is:
$$\frac{\text{\# of remaining slices with bacon}}{\text{total \# of remaining slices}} = \frac{3}{5}$$

The probability that both slices have bacon is their joint probability, i.e. their product:
$$\frac{2}{3} \times \frac{3}{5} = \frac{2}{5}$$

3.236 **Using a well-shuffled deck of playing cards, what's the chance of drawing two spades in a row: (i) if the first card is put back after drawing it; (ii) if it's not put back?**

(i)   The probability that the first card drawn is a spade is:
$$\frac{\text{\# of spades}}{\text{total \# of cards}} = \frac{13}{52} = \frac{1}{4}$$

The probability that the second card drawn is a spade is also ¼, because the card that was drawn was put back. So the # of spades and total # of cards remained unchanged.

The probability that both cards are spades is their joint probability, or their product:

$$\frac{1}{4} \times \frac{1}{4} = \frac{1}{16}$$

(ii) The probability that the first card drawn is a spade is ¼.

Presuming the first card drawn is a spade, the probability that the second card drawn is a spade is:

$$\frac{\text{\# of remaining spades}}{\text{total \# of remaining cards}} = \frac{12}{51}$$

The probability that both cards are spades is their joint probability, or their product:

$$\frac{1}{4} \times \frac{12}{51} = \frac{3}{51} = \frac{1}{17}$$

3.237 **What's the probability of drawing a face card or a spade from a deck of playing cards?**

There are 12 face cards (*FC*) in a pack of cards: Jack, Queen and King of each type.

There are 13 spades (*S*) in a pack of cards.

There are 3 cards that are both face cards and spades (*FC* and *S*).

$$P(FC \text{ or } S) = P(FC) + P(S) - P(FC \text{ and } S) = \frac{12}{52} + \frac{13}{52} - \frac{3}{52} = \frac{12 + 13 - 3}{52} = \frac{22}{52} = \frac{11}{26}$$

3.238 **A schoolbag contains 4 pens, 3 pencils and 2 markers. If 3 items are selected at random from the bag, what's the probability that one of each type of items will be selected?**

Total number of items = 9

3 items can be selected out of 9 items in $_9C_3$ ways.

$$_9C_3 = \frac{9!}{3!(9-3)!} = \frac{9!}{3!6!} = \frac{9 \times 8 \times 7 \times 6!}{3 \times 2 \times 6!} = \frac{9 \times 8 \times 7}{3 \times 2} = 84$$

Out of the 84 different combinations, only some will have one of each type of items.

One pen can be selected out of 4 pens in 4 different ways; one pencil out of 3 pencils in 3 ways, one marker out of 2 in 2 ways. To find the number of ways in which one of each type of items will be selected, we would need to consider the combination of each pen with each pencil with each marker, i.e. $4 \times 3 \times 2 = 24$.

Therefore, the probability that one of each type will be selected is $\frac{24}{84}$, or $\frac{2}{7}$.

3.239

	Beige	Black	White	TOTAL
Cats	-	8	-	8
Dogs	3	4	1	8
Mice	-	4	4	8
TOTAL	3	16	5	24

An animal shelter has cats, dogs and mice as listed in the adjacent table. If an animal is randomly selected, what's the probability of it being either white or a cat?

$$P(\text{white}) = \frac{\text{\# of white animals}}{\text{total \# of animals}} = \frac{5}{24}$$

$$P(\text{cat}) = \frac{\text{\# of cats}}{\text{total \# of animals}} = \frac{8}{24}$$

$$P(\text{white or cat}) = P(\text{white}) + P(\text{cat}) - P(\text{white and cat}) = \frac{5}{24} + \frac{8}{24} + \frac{0}{24} = \frac{13}{24}$$

3.240 When doing a road trip with my car, the probability of my car's alternator failing is 27%. The probability of my car's battery failing is 15%. The alternator keeps the battery charged. The probability of the alternator and the battery failing simultaneously is 4%. What's the probability that: (i) at least one of them will fail; (ii) exactly one of them will fail?

Let $P(A)$ = probability of alternator failing = 27%

$P(B)$ = probability of battery failing = 15%

$P(A \text{ and } B)$ = probability of alternator and battery failing = 4%

$P(A \text{ or } B)$ = probability of one or both (at least one) of them failing

(i) $P(\text{one or both}) = P(A \text{ or } B) = P(A) + P(B) - P(A \text{ and } B) = 27\% + 15\% - 4\% = 38\%$

(ii) $P(\text{exactly one}) = P(\text{one or both}) - P(\text{both}) = P(A \text{ or } B) - P(A \text{ and } B) = 38\% - 4\% = 34\%$

Please do not misunderstand the meaning of '*either A or B*.' It means '*A or B or both*'. In such a case, you are expected to find P(one or both), i.e. P(*A* or *B*).

On the other hand, a question asking you to find the probability of '*either A or B, but not both*' will clearly say so. In that case you are expected to find P(exactly one), i.e. P(*A* or *B*) – P(*A* and *B*). First, find P(*A* or *B*). From it, subtract P(*A* and *B*).

PRACTICE EXAMPLES

3.241 A tutor is teaching math to two students. (i) What's the likelihood that they were all born on a Thursday? (ii) What's the chance that the tutor was born on a Monday, and both the students were born in October? Assume a regular year, not a leap year.

3.242 An animal shelter only has 13 dogs and 12 cats. If 3 animals are adopted at random, what's the probability that they will all be cats?

3.243 Using a well-shuffled deck of playing cards, what's the chance of drawing three face cards in a row: (i) if each card is put back after drawing it; (ii) if it's not put back?

3.244 At a community college, 45% of the students attend part-time, and 40% of the students are majoring in liberal arts. What's the probability of selecting at random a part-time student or a student majoring in liberal arts?

3.245 An instrument box contains 5 ball pens, 3 pencils and 6 sketch pens. If 6 items are selected at random from the box, what's the likelihood that 2 of each type of items will be selected?

3.246 What's the probability of drawing a diamond or a black card from a deck of playing cards?

3.247 Not being interested in the coursework can cause Devin to fail on the exam. There's a 40% chance that Devin is not interested in the coursework. There's a 25% chance that he will fail on the exam. There's an 8% chance that both the things about Devin are true, i.e. he's not interested, and will fail. What's the chance that: (i) at least one of the things about him is true; (ii) exactly one of the things about him is true?

SPACE FOR CALCULATIONS

CHAPTER 4

GEOMETRY

Geometry requires considerable amount of imagination in terms of being able to visualize figures and objects. On the GRE/GMAT, some questions on geometry may be accompanied by figures alongside. If a figure is drawn to scale, then there will be a mention of that fact. But if there's no such note, then please do not assume that it is drawn to scale.

If a question is purely in terms of words, then it might help if you quickly draw a diagram to correctly visualize the problem. Some questions may be quite easy and may not need figures to solve them. As we discuss topics in this chapter, we will come across both types of questions.

Knowledge of trigonometry or calculus is not required to solve any geometry questions on the GRE/GMAT.

BASIC GEOMETRIC ENTITIES

A **point** has no shape, no length, no width and therefore no area. It is a dimensionless entity. It can be thought of as a dot that simply marks a position. It can be drawn as a small dot, or a cross or a dot/cross within a little circle. The following figure shows points A, B and C in various notations.

. A

$\cdot$ B

C
$\otimes$

An important property of points is that between any two points, there exists a third point. That is because points have no length or width. Therefore, however close two points may get, there will always be some room to accommodate a third point. As a matter of fact, it does not matter how many points are inserted between two points; there will still be room to accommodate unlimited number of points between those very two points.

A **line** can be defined in many ways depending upon which textbook you read. Here's a non-technical, common sense definition: A line is the straight path between two points that are infinitely far away from each other. It clearly means that a line has infinite length, and that it extends infinitely along two opposite directions. It is drawn as two arrows back to back. As an example, it can be written as $\overleftrightarrow{l}$, or as $\overleftrightarrow{AB}$, where A and B are not endpoints, but simply two reference points or anchors.

A **line segment** is a part of a line, and has a definite starting point and a definite endpoint. Therefore, it has finite length. As an example, it can be written as $\overline{AB}$, where A and B are its endpoints. Figure below shows $\overline{AB}$.

A line as well as a line segment has infinite points, because between any two points there exists a third point. Therefore, the number of points between any two points on a line as well as a line segment is countless. Please note that *a line segment has finite length, but infinite number of points.*

A **ray** is a half-line. Meaning, it has a definite starting point, but no endpoint; it extends infinitely in only one direction. As an example, it can be written as $\vec{r}$ or $\overrightarrow{AB}$, where A is the starting point, and B is not the endpoint, but just a point in the direction in which the ray infinitely extends. Figure below shows $\overrightarrow{AB}$.

An **angle** is formed by two rays sharing a common endpoint. The common endpoint is called the **vertex**. An angle is measured in degrees or radians. Figure on the right shows an angle formed by $\overrightarrow{BA}$ and $\overrightarrow{BC}$ at their vertex B. Angle ABC can be written as $\angle ABC$, or simply as $\angle B$.

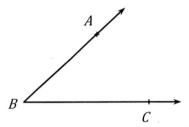

As you may imagine, an angle can also be formed by the intersection of two lines or two line segments, or a line with a line segment.

TYPES OF ANGLES

An **acute angle** measures less than 90°; a **right angle** measures 90°; an **obtuse angle** measures between 90° and 180°; a **straight angle** measures 180°. These angles are depicted below:

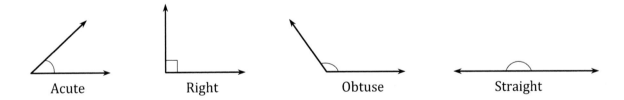

| Acute | Right | Obtuse | Straight |

Notice the unique notation for a right angle: A small block at the vertex.

When the measures of two angles add up to 90°, they are called **complementary angles**. When the measures of two angles add up to 180°, they are called **supplementary angles**.

If a straight angle is split into several smaller angles, then the measures of all those angles add up to the measure of the straight angle, i.e. 180°.

In the figure on the right, $a + b + c + d + e = 180°$

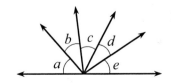

Similarly, one complete rotation of a ray about its vertex is the equivalent of 360°. A 360° angle can be split into smaller angles as shown on the right.

In the figure, $a + b + c + d + e + f = 360°$

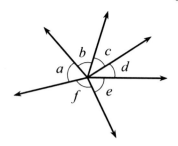

VERTICAL ANGLES (VERTICALLY OPPOSITE ANGLES)

When two lines or line segments intersect each other, they form two pairs of angles joined back to back at their vertex. Such angles are called **vertical angles**, or **vertically opposite angles**. Each of such pairs comprises two angles that are equal in measure. Figure on the right shows $\overleftrightarrow{m}$ and $\overleftrightarrow{n}$ intersecting each other. Notice that the vertical angles of equal measure are identified with similar marks.

When two lines intersect at a right angle (90° angle), they are said to be **perpendicular** to each other. In such an instance, all four resulting vertical angles would measure 90°. In mathematical notation, $\overleftrightarrow{a}$ perpendicular to $\overleftrightarrow{b}$ is written as: $\overleftrightarrow{a} \perp \overleftrightarrow{b}$.

LINE SEGMENT BISECTED

Consider a line segment PQ as shown on the right. Let point M be the midpoint of $\overline{PQ}$. Then, point M cuts $\overline{PQ}$ into two equal halves. Mathematically speaking, point M **bisects** $\overline{PQ}$. $PM = MQ$ In the figure, these line segments of equal length are identified with similar marks.

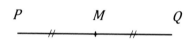

What if there were a line, say $\overleftrightarrow{m}$, that intersected $\overline{PQ}$ at its midpoint M? Then $\overleftrightarrow{m}$ would be the **bisector** of $\overline{PQ}$. If the angle of intersection of $\overleftrightarrow{m}$ and $\overline{PQ}$ were a right angle (90° angle), then $\overleftrightarrow{m}$ would be the **perpendicular bisector** of $\overline{PQ}$.

ANGLE BISECTED

Consider an angle ABC as shown on the right. Let $\overrightarrow{BM}$ make an angle with $\overrightarrow{BC}$ such that $\angle ABM$ and $\angle MBC$ have equal measure. Then, $\overrightarrow{BM}$ **bisects** $\angle ABC$. In other words, $\overrightarrow{BM}$ is the **angle bisector** of $\angle ABC$.

Mathematically, $\angle ABM = \angle MBC$. Notice the similar markers for angles with identical measures.

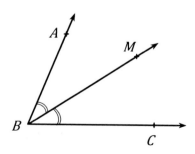

PARALLEL LINES

Two or more lines are said to be **parallel** if they lie in the same plane, and do not intersect each other no matter how far they are extended. The distance between parallel lines remains constant throughout their run. The concept can be extended to line segments as well: the distance between parallel line segments remains constant throughout their run.

Figure on the right shows $\overleftrightarrow{k}$, $\overline{AB}$ and $\overline{PQ}$ running parallel. Notice the arrow notation showing that $\overline{AB}$ and $\overline{PQ}$ are parallel.

The mathematical notation for writing the three of them being parallel is: $\overleftrightarrow{k} \parallel \overline{AB} \parallel \overline{PQ}$

PARALLEL LINES INTERSECTED BY A TRANSVERSAL

When two parallel lines are intersected by a third line (known as a **transversal**), several types of angles are formed. Let's understand them using the diagram on the right in which, $\overleftrightarrow{l}$ and $\overleftrightarrow{m}$ are intersected by the transversal $\overleftrightarrow{k}$.

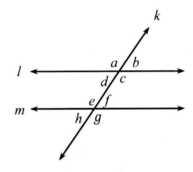

Interior angles are the pairs of angles on the inside of the two parallel lines. They add up to 180°. In the adjacent figure, c and f are interior angles. So are d and e. $c + f = 180°$ $d + e = 180°$

Alternate interior angles are the pairs of angles that look like the letter Z. They are equal in measure. In the above figure, d and f are alternate interior angles. So are e and c. $d = f$ $e = c$

Alternate exterior angles are the pairs of angles on opposite sides of the transversal, but on the outside of the two parallel lines. They are equal in measure. In the figure, b and h are alternate exterior angles. So are a and g. $b = h$ $a = g$

Corresponding angles are the pairs of angles that look like the letter F. They are equal in measure. In the figure, the corresponding angles are: $g = c$ $h = d$ $b = f$ $a = e$

If the transversal $\overleftrightarrow{k}$ were to cut the two parallel lines $\overleftrightarrow{l}$ and $\overleftrightarrow{m}$ at 90°, then $\overleftrightarrow{k} \perp \overleftrightarrow{l}$ and $\overleftrightarrow{k} \perp \overleftrightarrow{m}$. In that case, all eight angles a through h would measure 90° each.

ANGLE BETWEEN THE HANDS OF A CLOCK

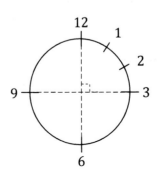

Here's an application of what we've learned about angles up to this point. Consider the dial of a clock. Every hour, the **minute hand** rotates through a full 360°. Meaning, every minute, the minute hand advances by 360° ÷ 60 minutes = 6°.

Looking at the diagram on the previous page, we can say that from 12 o'clock to 3 o'clock is 90°. From 12 o'clock to 1 o'clock would only be 30°, i.e. every hour (every 60 minutes), the **hour hand** travels by 30°. Therefore, every minute, the hour hand advances by 30° ÷ 60 minutes = ½° = 0.5°. For quick reference, we can summarize our findings as:

Hour hand	in 1 minute	advances ½°
Minute hand	in 1 minute	advances 6°

When solving examples on clocks, always draw figures. Things will be much simpler that way.

4.001 $\overleftrightarrow{a}$ and $\overleftrightarrow{b}$ **intersect as shown in the figure on the right. Find** $x : y$.

The portion above $\overleftrightarrow{a}$ can be considered a straight angle.

Meaning, $x + 25° = 180°$ $\Rightarrow$ $x = 155°$

Also, y is vertically opposite to the 25° angle.

Therefore, $y = 25°$

$x : y = 155 : 25$ $\Rightarrow$ $x : y = 31 : 5$ (after reducing the ratio by 5)

4.002 **In the adjoining figure, rays** $\overrightarrow{p}$ **and** $\overrightarrow{q}$ **originate at the point of intersection of** $\overleftrightarrow{l}$ **and** $\overleftrightarrow{m}$. **Find** $w : x : z$

Notice that z is vertically opposite to the top three x's grouped together.

Therefore, $z = 3x$ $\Rightarrow$ $3x = z$

$\Rightarrow$ $x = \dfrac{z}{3}$ $\Rightarrow$ $\dfrac{x}{z} = \dfrac{1}{3}$

This proportion can be written as $x : z = 1 : 3$

Also, w is vertically opposite to x.

Therefore, $w = x$ $\Rightarrow$ $\dfrac{w}{x} = \dfrac{1}{1}$

This proportion can be written as $w : x = 1 : 1$

In both the proportions above, x is represented by the number 1.

We can therefore combine them into one proportion: $w : x : z = 1 : 1 : 3$

4.003 **Using the figure on the right, calculate the arithmetic mean of** p **and** c.

The six contiguous angles that we see should add up to a total of 360°.

Therefore, $3p + 3c = 360°$

$\Rightarrow$ $3(p + c) = 360°$ $\Rightarrow$ $p + c = \dfrac{360°}{3}$

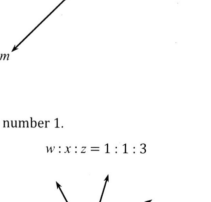

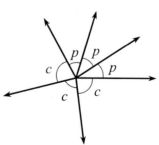

$$\Rightarrow \quad p + c = 120° \quad \Rightarrow \quad \frac{p + c}{2} = \frac{120°}{2} = 60°$$

Notice that the individual measures of p and c didn't matter. Simply dividing 360° by the number of angles (six) would yield the same result, i.e. 60°.

4.004 In the figure on the right, $\overleftrightarrow{l}$ bisects $\overline{PQ}$ at O. Point R is the midpoint of $\overline{PO}$, and point S is the midpoint of $\overline{RO}$. Find $PS:RQ$.

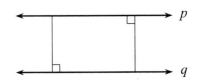

Since point S bisects $\overline{RO}$, $\quad RS = SO$

Let $RS = SO = x$

Then, since point R bisects $\overline{PO}$, $\quad PR = RO = 2x$

Also, $\overleftrightarrow{l}$ bisects $\overline{PQ}$ at O. Therefore, $PO = OQ = 4x$

Next, $\quad PS = PR + RS = 2x + x = 3x \quad$ and $\quad RQ = RS + SO + OQ = x + x + 4x = 6x$

Therefore, $PS:RQ = 3x:6x = 3:6 = 1:2$

4.005 In the figure on the right, is $\overleftrightarrow{p}$ necessarily parallel to $\overleftrightarrow{q}$?

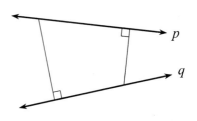

For $\overleftrightarrow{p}$ to be parallel to $\overleftrightarrow{q}$, at least one of the line segments between $\overleftrightarrow{p}$ and $\overleftrightarrow{q}$ would have to be perpendicular to both $\overleftrightarrow{p}$ and $\overleftrightarrow{q}$.

In the given figure, each of the line segments is perpendicular to either $\overleftrightarrow{p}$ or $\overleftrightarrow{q}$, but not to both. It is not correct to assume the value of an angle as 90°, unless the figure specifically shows it that way.

Therefore, $\overleftrightarrow{p}$ is not necessarily parallel to $\overleftrightarrow{q}$.

Alternative method:

Stretch your imagination. Grip the figure with your hands at the two right angles. Now twist your hands in opposite directions. The figure would look like the one on the right. Do $\overleftrightarrow{p}$ and $\overleftrightarrow{q}$ now look parallel? We can distort the figure because only two of the angles are given as 90°. The others could be anything.

4.006 What is the angle between the hour hand and the minute hand at 7:47?

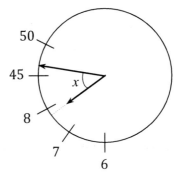

Assume x as the angle between the hour hand and the minute hand at 7:47, as shown in the figure on the right.

The minute hand travels 360° in one hour, or 6° in one minute. It has traveled for 2 minutes beyond the 45 minute marker, i.e. a total of $2 \times 6° = 12°$ beyond the 45 min. marker.

The hour hand travels 30° in one hour, or ½° in one minute. It has traveled for 47 minutes after 7 o'clock, i.e. a total of 47 × ½° = 23.5° beyond the 7 o'clock marker. It is only slightly away from the 8 o'clock marker, the exact angle being 30° − 23.5° = 6.5°.

Therefore, $x = 12° + 30° + 6.5° = 48.5°$

4.007 In the figure on the right, what is the average of the measures of all the angles a through x?

When two lines intersect, the sum of the four angles around their point of intersection is 360°.

In the figure, there are six such intersections. Therefore, the sum of the angles around those six intersections is 360° × 6.

The total number of angles a through x is 24.

Average of their measures $= \dfrac{\text{sum of their measures}}{\text{\# of angles}} = \dfrac{360° \times 6}{24} = 90°$

Alternative method:

The average of the four angles around a point of intersection of two lines is 360° ÷ 4 = 90°. This average will remain the same no matter how many two-line intersections we consider. End of story!

4.008 In the adjoining figure, $\overleftrightarrow{l} \parallel \overleftrightarrow{m}$. Find the value of b.

At the intersection of $\overleftrightarrow{l}$ and $\overleftrightarrow{n}$, the two angles labeled in the figure are vertically opposite to each other. Vertical angles are equal in measure.

$3a = 2a + 40 \quad \Rightarrow \quad a = 40°$

The two angles labeled below $\overleftrightarrow{n}$ look like the letter F (corresponding angles). Such angles are equal in measure too.

$2a + 40 = b + 13 \quad \Rightarrow \quad 2(40) + 40 = b + 13 \quad \Rightarrow \quad 120 = b + 13 \quad \Rightarrow \quad b = 107°$

PRACTICE EXAMPLES

4.009 In the figure on the right, $\overleftrightarrow{l}$ bisects $\overline{AB}$. $\overleftrightarrow{m}$ bisects $\overline{CD}$ and $\overline{OB}$. Find $DE : AP$.

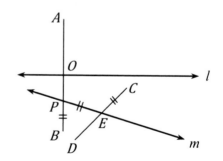

In the figure on the left, what is the value of x?

4.010

4.011 In the figure on the right, if a is greater than b by 80°, find $a^2 - b^2$.

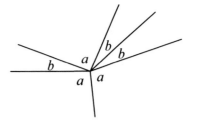

4.012

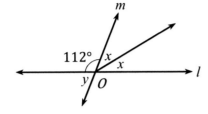

In the figure on the left, $\overrightarrow{m}$ and $\overleftrightarrow{l}$ intersect at O. Find $y - x$.

4.013 In relation to the figure on the right, which of the following statements is (are) true?

(i) $\overrightarrow{l} \parallel \overleftarrow{m}$
(ii) $\overrightarrow{m} \parallel \overleftarrow{n}$
(iii) $\overrightarrow{l} \parallel \overleftarrow{n}$

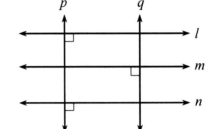

4.014 What is the angle between the hour hand and the minute hand of a clock at 3:39?

4.015

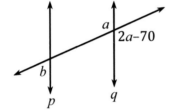

In the figure on the left, $\overrightarrow{p} \parallel \overleftarrow{q}$. What is the value of $b - a$?

4.016 In the figure on the right, a horizontal segment is intersected by two segments. If $p = s$, then is $q = r$?

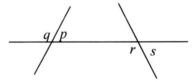

SPACE FOR CALCULATIONS

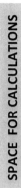

TRIANGLES

A **triangle** is a closed figure with three sides. Within a triangle, the three sides form three **interior angles**, the sum of which is 180°. Just the way there are several types of angles, there are several types of triangles as well.

TYPES OF TRIANGLES

An **acute angled triangle** (or acute triangle) is one in which all three interior angles measure less than 90°. A **right angled triangle** (or right triangle) is one in which *one* of the three interior angles measures 90°. The side opposite to the 90° angle is termed **hypotenuse**. The other two sides are called **legs**. An **obtuse angled triangle** (or obtuse triangle) is one in which *one* of the three interior angles measures more than 90°. Figure below shows these types. Triangle *ABC* is written as △*ABC*.

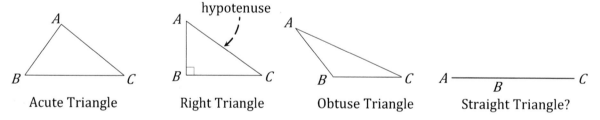

Acute Triangle	Right Triangle	Obtuse Triangle	Straight Triangle?

There's no such thing as a **straight angled triangle** (straight triangle), because you may notice that if one of the interior angles (up here, $\angle B$) increases all the way to 180°, then the largest side ($\overline{AC}$) will collapse onto the smaller two sides ($\overline{AB}$ and $\overline{BC}$), and there will be no triangle at all. The triangle will have collapsed onto itself! Meaning, there will only be a line segment ($\overline{ABC}$).

In the above figures, observe that in a triangle, *the smallest side faces (is opposite to) the smallest interior angle, the next larger side faces the next larger interior angle, and the largest side faces the largest interior angle.*

An **isosceles triangle** is one in which two sides are of equal length. Meaning, the two interior angles facing those two equal sides will also have equal measures. An **equilateral triangle** is one in which all three sides are of equal length. Consequently, the three interior angles facing the sides will also have the same measure (60°). Equilateral triangles are also referred to as **equiangular triangles**. A **scalene triangle** is one in which the three sides are of different lengths. Meaning, the three interior angles are also of different measures. Figure below shows these types of triangles. Notice the markers for equal sides and equal angles.

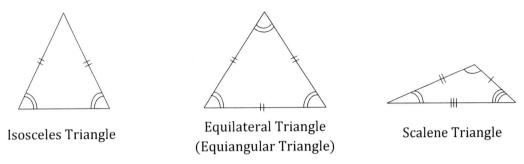

| Isosceles Triangle | Equilateral Triangle (Equiangular Triangle) | Scalene Triangle |

SPECIAL TYPES OF RIGHT TRIANGLES

A **30–60–90 triangle**, as the name suggests, is a right triangle in which the measures of the three interior angles are 30°, 60° and 90°. The ratio of these angles would be 30 : 60 : 90, or simply 1 : 2 : 3. The special property of a 30–60–90 triangle is that the smallest side (facing the 30° angle) is half as long as the hypotenuse (facing the 90° angle). A **3–4–5 triangle**, on the other hand, is a right triangle in which the lengths of the three sides are in the ratio 3 : 4 : 5. The sides could have any lengths, as long as all three lengths are the 3rd, 4th and 5th multiples of the same number. For example, 30–40–50; 300–400–500; 12–16–20; 15–20–25 etc. are all 3-4-5 triangles.

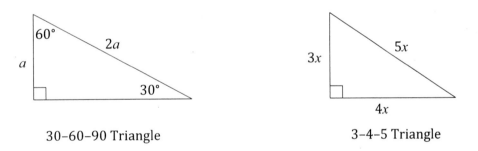

| 30–60–90 Triangle | 3–4–5 Triangle |

Please note that a 30–60–90 triangle is not the same as a 3–4–5 triangle. The 30, 60 and 90 pertain to the measures of angles. The 3, 4 and 5 pertain to the lengths of sides. **3–4–5 and 30–60–90 are the most commonly tested triangles on the GRE/GMAT.**

EXTERIOR ANGLES OF A TRIANGLE

Consider a triangle as shown on the right. Going counter-clockwise, extend its three sides (dotted) in only one direction. You will see three angles formed on the outside of the triangle, at the three vertices. Each of these angles is supplementary to the interior angle at that vertex, and is called an **exterior angle**. In the figure, $(180 - x)$, $(180 - y)$ and $(180 - z)$ are exterior angles. Their measures add up to:

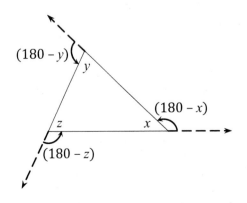

$$(180 - x) + (180 - y) + (180 - z) = 540 - x - y - z = 540 - (x + y + z) = 540 - 180 = \mathbf{360°}$$

The sum of measures of exterior angles, when considered in counter-clockwise direction, is 360°.

Similarly, you can form three other exterior angles by going clockwise and extending the three sides in the direction opposite to the prior one. Your analysis will conclude that the sum of measures of exterior angles, when considered in clockwise direction, is 360°.

Generally speaking, *the sum of measures of exterior angles is 360° when considered consistently in one direction: either clockwise or counter-clockwise.*

EQUILATERAL TRIANGLE BISECTED

A line segment perpendicular to a side and joining the vertex opposite to that side is called an **altitude**. Every triangle has three altitudes. Specifically, in the case of an equilateral triangle, an altitude bisects it into two 30–60–90 triangles. Figure on the right clarifies it.

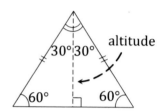

LENGTHS OF SIDES OF A TRIANGLE

In a triangle, the sum of the lengths of two sides is always greater than the length of the third side. This idea originates in the fact that the shortest distance between two points is a straight line. Any other route will always be longer. In $\triangle ABC$, the route B–A–C is longer than the straight distance BC. $x + y > z$

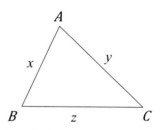

Similarly, $x + z > y$ and $z + y > x$

Perimeter is the length of the boundary. In $\triangle ABC$, it is the sum of the lengths of all sides, i.e. $x + y + z$.

AREA OF A TRIANGLE

$$A = \frac{b \times h}{2}$$ where A = area of the triangle b = length of base h = height

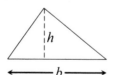   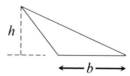

The figure above shows how to consider the base (b) and height (h) in different types of triangles.

PYTHAGOREAN THEOREM

In a **right triangle**, $(\text{hypotenuse})^2 = (\text{leg one})^2 + (\text{leg two})^2$

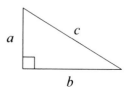

In the figure on the right, $c^2 = a^2 + b^2$

PYTHAGOREAN THEOREM EXTENDED

A square, as you probably know, is a four sided figure with all four sides being equal in length, and all four angles being equal in measure. Think of it as a tile. We will discuss a square in greater detail later in this chapter. The real life significance of Pythagorean Theorem is that the area of a square whose side is the hypotenuse of a right triangle is equal to the sum of the areas of the two squares whose sides are the respective legs of that same right triangle.

Instead of a right triangle, if we consider an acute triangle or an obtuse triangle, then the Pythagorean equality changes into two specific inequalities. Let's stretch our imagination, and try to understand them graphically as shown below. The dotted areas are squares generated by the sides of the triangles. Accordingly, the areas of the squares are $a \times a = a^2$, $b \times b = b^2$ and $c \times c = c^2$.

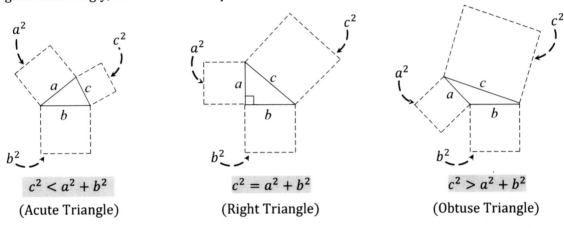

$$c^2 < a^2 + b^2$$
(Acute Triangle)

$$c^2 = a^2 + b^2$$
(Right Triangle)

$$c^2 > a^2 + b^2$$
(Obtuse Triangle)

From left to right, notice that the side with length a rotates counter-clockwise, increasing its angle with the side with length b, changing the triangle from acute to right to obtuse. Accordingly, the area c^2 rapidly increases from being less than area $(a^2 + b^2)$, to equal to it, to finally greater than it. This visual understanding is very important from mathematical reasoning point of view.

SIMILAR TRIANGLES

Two or more triangles are said to be **similar** if the same set of three angles appear in them. Please note that 'similar' does not mean 'identical.' Two or more similar triangles may or may not be identical. But they will certainly be scaled up or scaled down versions of one another. Similar triangles have the lengths of their three sides in the same proportion.

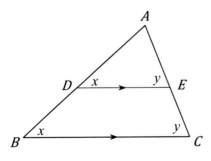

Consider $\triangle ABC$ as shown on the left. Draw $\overline{DE} \parallel \overline{BC}$.

Then, $\angle A$ is common to $\triangle ABC$ and $\triangle ADE$.

$\angle ADE = \angle ABC$ because they are corresponding angles of parallel line segments. (They look like the letter F).

Similarly, $\angle AED = \angle ACB$.   Clearly, $\triangle ABC$ and $\triangle ADE$ have the same set of three angles. Therefore, $\triangle ABC$ is similar to $\triangle ADE$. It is written as: $\triangle ABC \sim \triangle ADE$

In these similar triangles, the lengths of their sides are in proportion, i.e. $AD : DE : EA = AB : BC : CA$

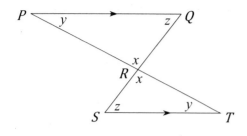

Here is another configuration of triangles. $\overline{PT}$ and $\overline{SQ}$ intersect at R. Also, $\overline{PQ} \parallel \overline{ST}$. Let's see how $\triangle PQR$ and $\triangle TSR$ are similar:

$\angle PRQ = \angle TRS$ because they are vertical angles.

$\angle PQR = \angle TSR$ because they are alternate interior angles of two parallel line segments. (They look like the letter Z). Similarly, $\angle QPR = \angle STR$.

Therefore, $\triangle PQR$ and $\triangle TSR$ have the same set of three angles, i.e. $\triangle PQR \sim \triangle TSR$. Their three corresponding sides are in proportion.　　　$PQ : QR : RP = TS : SR : RT$

4.017　**In the figure on the right, what is the degree measure of x?**

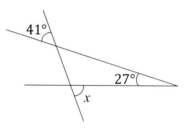

The three line segments form a triangle, one of the interior angles of which is shown as $27°$. The interior angle, vertically opposite to the $41°$ angle shown, also measures $41°$. The third interior angle, adjacent to x, measures $(180 - x)$.

The sum of the three interior angles in a triangle is $180°$.

Therefore, $27 + 41 + (180 - x) = 180$　　$\Rightarrow$　　$68 + 180 - x = 180$　　$\Rightarrow$　　$x = 68°$

4.018

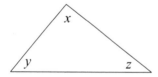

In the triangle on the left, if $x : y : z = 2 : 3 : 4$, then what is the value of z?

In the given proportion, z is represented by 4 parts. There are a total of $2 + 3 + 4 = 9$ parts. These 9 parts together mean a total of $180°$ in the triangle. So, 4 parts out of 9 parts would mean $z°$ out of $180°$.

$$\frac{4}{9} = \frac{z}{180} \quad \Rightarrow \quad z = \frac{4}{9} \times 180 = 80°$$

4.019　**In the figure on the right, area of $\triangle ABC = 18$. Find the area and the perimeter of the shaded $\triangle ADE$.**

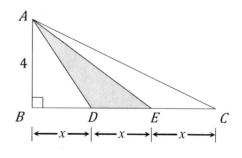

Area of $\triangle ABC = \dfrac{BC \times BA}{2}$　$\Rightarrow$　$18 = \dfrac{3x \cdot 4}{2}$

$\Rightarrow$　$18 = 6x$　$\Rightarrow$　$x = 3$

Area of $\triangle ADE = \dfrac{DE \times BA}{2} = \dfrac{x \cdot 4}{2} = 2x = 2(3) = 6$

Looks like $\triangle ABD$ is a 3–4–5 triangle in which, $AD = 5$.

Using Pythagorean Theorem in $\triangle ABE$, $(AE)^2 = (AB)^2 + (BE)^2 = 4^2 + 6^2 = 16 + 36 = 52$

Therefore, $AE = \sqrt{52} = \sqrt{13 \times 4} = 2\sqrt{13}$

Perimeter of $\triangle ADE =$ length of its boundary $= AD + DE + EA = 5 + 3 + 2\sqrt{13} = 8 + 2\sqrt{13}$

4.020

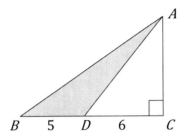

B   5   D   6   C

In the figure on the left, if area of $\triangle ADC$ is 24, then what is the area of the shaded triangle?

$$\text{Area of } \triangle ADC = \frac{DC \times CA}{2} \quad \Rightarrow \quad 24 = \frac{6 \cdot CA}{2}$$

$$\Rightarrow \quad 24 = 3 \cdot CA \quad \Rightarrow \quad CA = 8$$

$$\text{Area of } \triangle BDA = \frac{BD \cdot CA}{2} = \frac{5 \times 8}{2} = 20$$

4.021   **In the figure on the right, find x.**

Among the two adjacent triangles, do you notice something special about the lower triangle? It's a 3–4–5 right triangle.

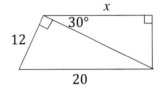

Meaning, the 3rd multiple and the 5th multiple of the same number (here, 4) are 12 and 20, respectively, i.e. $3 \times 4 = 12$ and $5 \times 4 = 20$. The other side must be the 4th multiple of 4, i.e. $4 \times 4 = 16$. This is the side common to both the triangles. You will reach the same result by using the Pythagorean Theorem, but it will take more calculations, and therefore more time. On the GRE/GMAT, use such shortcuts to save time.

Anything noticeable about the upper triangle? It's a 30–60–90 right triangle in which, *the side facing the 30° angle is half as long as its hypotenuse*. Its hypotenuse is the side common to both triangles. We just calculated that it measures 16. So the side facing the 30° angle must have a length of 8.

Now use the Pythagorean Theorem for the upper right triangle:

$$16^2 = x^2 + 8^2 \quad \Rightarrow \quad 256 = x^2 + 64 \quad \Rightarrow \quad x^2 = 192 \quad \Rightarrow \quad x = \sqrt{192} = \sqrt{64 \times 3} = 8\sqrt{3}$$

4.022   **The area of an equilateral triangle is $49\sqrt{3}$. What is the length of each of its sides?**

Figure on the right shows an equilateral triangle with an altitude (h) bisecting it into two 30–60–90 triangles.

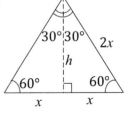

Area of the equilateral triangle $= \dfrac{(2x) \cdot h}{2} \quad \Rightarrow \quad 49\sqrt{3} = xh$

$$\Rightarrow \quad h = \frac{49\sqrt{3}}{x}$$

Using Pythagorean Theorem, $h^2 + x^2 = (2x)^2 \quad \Rightarrow \quad h^2 + x^2 = 4x^2 \quad \Rightarrow \quad h^2 = 3x^2$

$$\Rightarrow \quad h = x\sqrt{3}$$

These two have the same right hand side. Equate them:

$$\frac{49\sqrt{3}}{x} = x\sqrt{3} \quad \Rightarrow \quad \frac{49}{x} = x \quad \Rightarrow \quad x^2 = 49 \quad \Rightarrow \quad x = 7$$

The reason $x = 7$ and not $x = \pm 7$ is that length can never be negative. So, x is only $+7$.

The length of each side of the equilateral triangle is $2x = 2(7) = 14$

4.023 A building that is 90 feet tall casts a shadow that is 120 feet long. A man walking alongside the building casts a shadow 8 feet long. How tall is he?

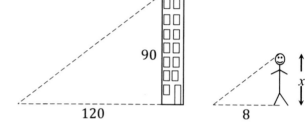

Assume the man is x feet tall. The two right triangles are similar because they are both formed by sunlight falling at the same angle.

Similar triangles have their sides in proportion. While writing the equation, we don't need to consider all three sides; just two sides of each triangle are enough to serve the purpose.

$$90 : 120 = x : 8 \quad \Rightarrow \quad \frac{90}{120} = \frac{x}{8} \quad \Rightarrow \quad \frac{3}{4} = \frac{x}{8} \quad \Rightarrow \quad x = \frac{3 \times 8}{4} = 6$$

4.024

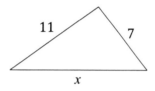

In the triangle on the left, which of the following is true about all the possible values of x?

(i) $4 < x < 18$
(ii) $7 < x < 11$
(iii) $0 < x < 18$

Stretch your imagination. If the sides with fixed lengths 7 and 11 start collapsing onto the third side, then they will stretch the third side, i.e. the value of x will go on increasing. The triangle will get flatter and flatter. If the two sides completely collapse onto the horizontal side, then the triangle will disappear, and there will only be a line segment of length $7 + 11 = 18$.

Therefore, in order to keep the triangle from collapsing onto itself, x would have to be slightly less than 18. Meaning, $x < 18$.

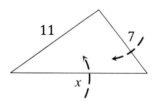

Now imagine that the sides with lengths 7 and x start collapsing onto the side with length 11. As they approach it, x will get smaller and smaller. If the two sides completely collapse onto the side with length 11, then the triangle will disappear, and there will only be a tilted line segment of length 11 out of which, $x = 4$.

Therefore, in order for there to be a triangle, x would have to be slightly greater than 4, i.e. $x > 4$ (or $4 < x$).

Putting the two inequalities together, $x < 18$ and $4 < x$, we can write $4 < x < 18$, which is statement (i). Clearly, statements (ii) and (iii) do not hold true.

PRACTICE EXAMPLES

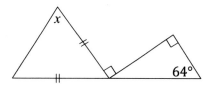

4.025 In the figure on the right, what is the
degree measure of *x*?

4.026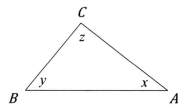

In $\triangle ABC$, if $x : y : z = 1 : 2 : 3$, then which of the
following statements is (are) true?

(i)   *ABC* is a right triangle.
(ii)   *ABC* is an isosceles triangle.
(iii) $\dfrac{1}{BC} = \dfrac{2}{AB}$

4.027 In the figure on the right, if $a : b : c = 2 : 3 : 5$,
and area of $\triangle PQR$ is 70, what is the area of the
shaded region?

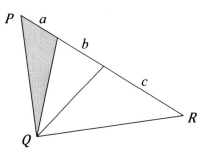

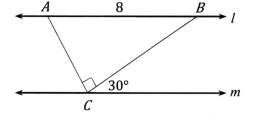

4.028

In the figure on the left, $\overleftrightarrow{l}$ and $\overleftrightarrow{m}$ run parallel.
What is the distance between them?

4.029 In the figure on the right, what is the
ratio of lengths $a : b$?

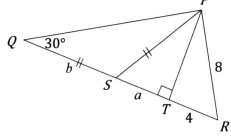

4.030 What is the measure of the smallest angle in a right triangle if it is 24° less than the other
non-right angle?

4.031 Two buses leave a bus station at 2:00 p.m. One travels straight north at 45 mph; the other
travels straight west at 60 mph. How far will the two buses be from each other at 6:00 pm?

4.032 In the figure on the right, what is the
area of $\triangle AED$?

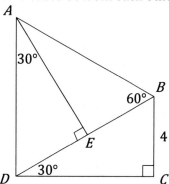

SPACE FOR CALCULATIONS

POLYGONS

In simple English, a **polygon** is a closed figure formed by three or more line segments joined end to end. It is important that all the segments be **coplanar** (in the same plane). A triangle is a three sided polygon; a quadrilateral is a four sided polygon; a pentagon is five sided; a hexagon is six sided; a heptagon is seven sided; and so on.

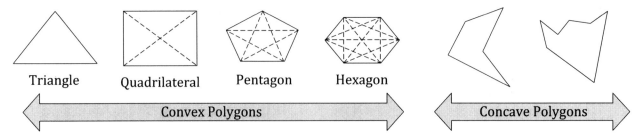

A **diagonal** of a polygon is a line segment joining two non-adjacent vertices. In the above figures, the dotted line segments are diagonals. Notice that a triangle has no diagonals. A **regular polygon** is one that has all of its sides of the same length, and all of its interior angles of the same measure.

A **convex polygon** is one that has all of its diagonals within its boundary, i.e. on the inside of the figure. In the figures, observe that every interior angle of a convex polygon is less than 180°. A **concave polygon** is one that has at least one diagonal outside of its boundary. Also, at least one of its interior angles exceeds 180°. For the GRE/GMAT, we will limit our discussion to convex polygons only.

HOW MANY DIAGONALS IN A POLYGON?

Consider a polygon with n number of sides (or n number of vertices). A diagonal is formed when two non-adjacent vertices in a polygon are joined by a line segment. Let's first find out in just how many different ways can two vertices be selected (to be joined) out of a total of n vertices. This is the same as selecting two candidates out of a total of n applicants. We're forming teams; therefore we should use combination, i.e. $_nC_2$ (recall theory from page 140). $_nC_2$ is the number of **all** the possible 2-vertex teams that can be formed out of n vertices. But this number would also include all the *sides* of the polygon, and we know that sides are not diagonals. Therefore, we would subtract the number of sides (n) from the total number of 2-vertex teams ($_nC_2$) to get:

The number of diagonals in an n-sided polygon $= {_nC_2} - n$

INTERIOR AND EXTERIOR ANGLES OF A POLYGON

The **sum of all interior angles** (measured in degrees) of an n-sided polygon is $180(n-2)$. The average of all interior angles is the sum divided by the number of vertices (n), i.e. $\dfrac{180(n-2)}{n}$.

The **sum of all exterior angles**, drawn in one direction, either clockwise or counter-clockwise, is 360°. This is simply the extension of the concept we learned on page 173. *The sum is 360° regardless of the number of sides.* As a simple proof of this, consider any polygon, say a hexagon ($n = 6$).

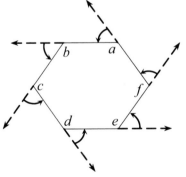

The sum of its interior angles is:

$a + b + c + d + e + f = 180(n - 2) =$
$180(6 - 2) = 180(4) = 720°$

The sum of its exterior angles (arrow-marked) is:
$(180 - a) + (180 - b) + (180 - c) + (180 - d) +$
$(180 - e) + (180 - f) = 1080 - a - b - c - d - e - f =$
$1080 - (a + b + c + d + e + f) = 1080 - 720 = $**360°**

The average of all exterior angles is the sum divided by the number of vertices (n), i.e. $\dfrac{360}{n}$.

QUADRILATERALS

A **quadrilateral** is a four sided polygon. It is also referred to as a **tetragon**. The interior angles of a quadrilateral add up to 360°. Depending upon the length and orientation of its sides, a quadrilateral can have several configurations, each with specific properties. Let's study them in order.

RECTANGLE

A **rectangle** is a quadrilateral in which all four interior angles are right angles. Also, the opposite sides are of equal length.

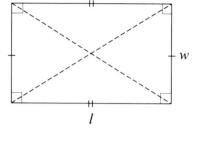

Area of rectangle $= l \times w$

Perimeter $= 2l + 2w = 2(l + w)$

In a rectangle:

- ✓ The diagonals (shown dotted) bisect each other
- ✓ The diagonals are not necessarily perpendicular to each other
- ✓ Each diagonal bisects the rectangle into two right triangles

SQUARE

A **square** is a special type of rectangle in which, all four sides are of equal length, and all four interior angles are right angles.

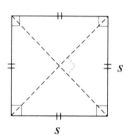

Area of square $= s \times s = s^2$

Perimeter $= s + s + s + s = 4s$

In a square:

- ✓ The diagonals bisect each other
- ✓ The diagonals are perpendicular to each other
- ✓ Each diagonal bisects the square into two isosceles right triangles

PARALLELOGRAM

A **parallelogram** is a quadrilateral in which the opposite sides run parallel to each other. Also, the opposite sides are of equal length, and the opposite angles are of equal measure (note the markings).

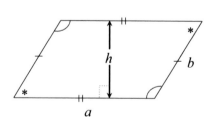

Area of parallelogram $= a \times h$

Area of parallelogram $\neq a \times b$

Perimeter $= 2a + 2b = 2(a + b)$

In a parallelogram:

- ✓ The diagonals (not shown) bisect each other
- ✓ Each diagonal bisects the parallelogram into two triangles

RHOMBUS

A **rhombus** is a quadrilateral with all four sides equal in length. The opposite angles are equal in measure.

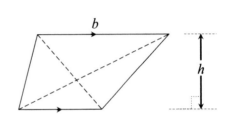

Area of rhombus $= \dfrac{(\text{diagonal } AC) \times (\text{diagonal } BD)}{2}$

Area of rhombus $= s \times h$

Area of rhombus $\neq s \times s$

Perimeter $= s + s + s + s = 4s$

In a rhombus:

- ✓ The diagonals bisect each other
- ✓ The diagonals are perpendicular to each other
- ✓ Each diagonal bisects the rhombus into two triangles

TRAPEZOID

A **trapezoid** is a quadrilateral in which two sides are parallel, and the other two sides are *not parallel*. The parallel sides are of unequal lengths. So are the unparallel sides.

Area of trapezoid $= \dfrac{(a + b)h}{2}$

Area of trapezoid $= \dfrac{(\text{sum of parallel sides}) \times \text{height}}{2}$

Perimeter $=$ sum of all four sides

In a trapezoid:

- ✓ The diagonals do not bisect each other
- ✓ The diagonals do not bisect the trapezoid into two triangles

ISOSCELES TRAPEZOID

This is a special type of trapezoid in which the non-parallel sides are of *equal length*. The formulas for area and perimeter are the same as those for a trapezoid (as above).

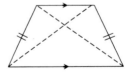

In an isosceles trapezoid:

- ✓ The diagonals are equal in length

Understanding what makes the above six quadrilaterals unique is important from mathematical reasoning point of view. Please review them carefully before you proceed further.

SOME FACTS ABOUT RECTANGLES

Infinitely many rectangles exist that have the same perimeter. Among them, the rectangle that is a square will have the *greatest* area. The figures below illustrate this point.

Rectangles with **perimeter = 100**

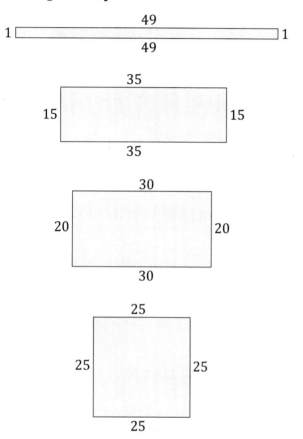

Infinitely many rectangles exist that have the same area. Among them, the rectangle that is a square will have the *least* perimeter. The figures below illustrate this point.

Rectangles with **area = 100**

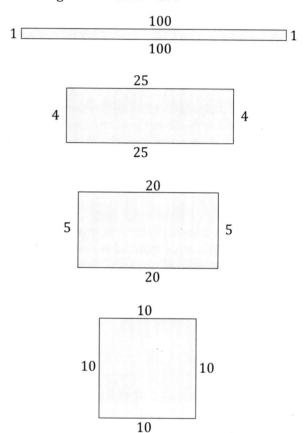

4.033 **How many sides does a polygon have if the average of its interior angles is 144°?**

The sum of the interior angles of a polygon with n sides is $180(n - 2)$.

The average of interior angles is $\dfrac{180(n-2)}{n}$. This is given as 144°. Therefore, equate the two.

$$144 = \frac{180(n - 2)}{n} \quad \Rightarrow \quad 144n = 180n - 360 \quad \Rightarrow \quad -36n = -360 \quad \Rightarrow \quad n = 10$$

4.034 **How many diagonals does a 12-sided polygon have?**

A 12-sided polygon has 12 vertices. A diagonal is formed by joining 2 non-adjacent vertices.

First, the # of unique line segments that can be formed by teaming up 2 vertices out of 12 is:

$$_{12}C_2 = \frac{12!}{2!(12-2)!} = \frac{12!}{2! \times 10!} = \frac{12 \times 11 \times 10!}{2 \times 10!} = 66$$

This number (66) also includes all the 12 sides. Therefore, subtract 12 from 66 to get only the number of diagonals, i.e. 66 – 12 = 54 diagonals.

4.035

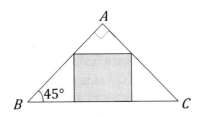

In the figure on the left, the shaded square inscribed within $\triangle ABC$ has an area of 64. What is the area of $\triangle ABC$?

$\angle C = 180 - (\angle A + \angle B) = 180 - (90 + 45) = 45°$

Therefore, $\triangle ABC$ is an isosceles right triangle.

Also, each of the three small triangles around the square is an isosceles right triangle.

Area of the square is 64, i.e. $\text{side}^2 = 64 \Rightarrow \text{side} = \sqrt{64} = 8$

This (8) is also the height of the square. We need to find the remaining height, i.e. the height of the small triangle on top of the square.

Figure on the right shows the top triangle drawn larger. Let's name it $\triangle ADE$. Drop an altitude AF onto side DE as shown. Then AF bisects $\triangle ADE$ into two isosceles right triangles. Also, DE = top side of square = 8

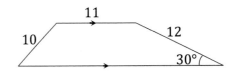

In $\triangle AFE$, $AF = FE$, where $FE = DF = \frac{1}{2}(DE) = \frac{1}{2}(8) = 4$

Therefore, $AF = 4$

Now, height of $\triangle ABC = AF +$ height of square $= 4 + 8 = 12$

Also, base of $\triangle ABC$ = base of square + bases of the two small isosceles triangles =

$8 + 8 + 8 = 24$

Area of $\triangle ABC = \dfrac{\text{base} \times \text{height}}{2} = \dfrac{24 \times 12}{2} = 144$

4.036 **Figure on the right is a trapezoid. Find its area. Find its perimeter.**

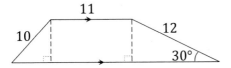

On the left is the figure redrawn with two altitudes forming two right triangles with a rectangle between them.

We need to find the height and the total length of base of the trapezoid.

The triangle on the right is a 30–60–90 triangle. Its height is its smallest side, which is half as long as the hypotenuse, i.e. ½ of 12, i.e. 6. This is also the height of the trapezoid and of the triangle on the left. Base of the 30–60–90 triangle can be found by the Pythagorean Theorem:

$\text{base}^2 + 6^2 = 12^2 \Rightarrow \text{base}^2 = 12^2 - 6^2 = 144 - 36 = 108 \Rightarrow \text{base} = \sqrt{108} = 6\sqrt{3}$

The triangle on the left has hypotenuse 10 and height 6. These numbers are the 5th and 3rd multiples of 2. Therefore, the base must be the 4th multiple of 2, i.e. 8. It is a 3-4-5 triangle.

Base of the trapezoid $= 8 + 11 + 6\sqrt{3} = 19 + 6\sqrt{3}$

Sum of the parallel sides of the trapezoid $= 11 + (19 + 6\sqrt{3}) = 30 + 6\sqrt{3}$

$$\text{Area of trapezoid} = \frac{\text{(sum of parallel sides)} \times \text{ht.}}{2} = \frac{(30 + 6\sqrt{3}) \times 6}{2} = 90 + 18\sqrt{3}$$

Perimeter = length of boundary = sum of all sides = $10 + 11 + 12 + (19 + 6\sqrt{3}) = 52 + 6\sqrt{3}$

4.037

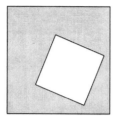

Figure on the left shows one square inside another. Area of the shaded region is 100 cm². If the side of the outer square is 5 cm longer than the side of the inner square, then what is the sum of the perimeters of the two squares?

In case you're clueless about where to start, please don't scratch your head too much, or your hair might fall off!

The tilted orientation of the inner square doesn't matter. The area of the shaded region is simply the area of the outer square, minus the area of the inner square.

Assume the inner square of side x cm. Then the area of the inner square would be x^2 cm².

The outer square has its side 5 cm longer than x, i.e. $(x + 5)$ cm. Area of the outer square would be: $(x + 5)^2 = (x + 5)(x + 5) = x^2 + 5x + 5x + 25 = x^2 + 10x + 25$

Area of the shaded region = (area of outer square) – (area of inner square) $\Rightarrow$

$100 = (x^2 + 10x + 25) - x^2$ $\Rightarrow$ $100 = 10x + 25$ $\Rightarrow$ $75 = 10x$ $\Rightarrow$ $x = 7.5$

Perimeter of inner square = $4x = 4(7.5) = 30$ cm

Perimeter of outer square = $4(x + 5) = 4(7.5 + 5) = 4(12.5) = 50$ cm

Sum of their perimeters = $30 + 50 = 80$ cm

4.038 Figure on the right shows the floor plan of an apartment I once lived in. It is drawn to scale. What fraction of the area of the apartment is taken up by the patio? Ignore wall thicknesses.

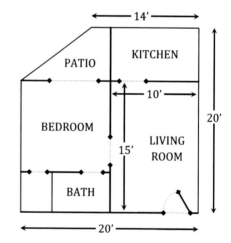

Simply subtracting one dimension from another would give us the unknown lateral or longitudinal dimensions.

The living room, bedroom and bath together form a 20' × 15' rectangle, the area of which is $20 \times 15 = \textbf{300 sq. feet.}$

The kitchen is a 10' × 5' rectangle, its area being $10 \times 5 = \textbf{50 sq. feet.}$

The patio is a trapezoid, with parallel sides being 4' and 10', and height being 5'.

$$\text{Area of patio} = \frac{\text{(sum of parallel sides)} \times \text{ht.}}{2} = \frac{(4 + 10) \times 5}{2} = \frac{14 \times 5}{2} = \textbf{35 sq. feet}$$

Total area of apartment = $300 + 50 + 35 = 385$ sq. feet

$$\text{Area of patio, as a fraction of the area of apartment} = \frac{\text{area of patio}}{\text{area of apartment}} = \frac{35}{385} = \frac{1}{11}$$

4.039

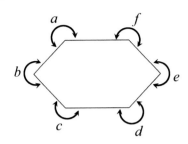

In the hexagon on the left, what is the sum of the marked angles *a* through *f*?

To better understand this example, let's consider only one angle at a time, say angle *a*. As shown below, it is the sum of the exterior angle in the counter-clockwise direction (say x_1), and 180°.

From the figure on the right, $a = 180 + x_1$

Similarly, let the exterior angles at *b*, *c*, *d*, *e* and *f* be x_2, x_3, x_4, x_5 and x_6, respectively.

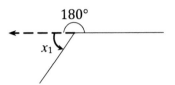

Then, the sum of the angles *a* through *f* would be:

$(180 + x_1) + (180 + x_2) + (180 + x_3) + (180 + x_4) + (180 + x_5) + (180 + x_6) =$

$1080 + (x_1 + x_2 + x_3 + x_4 + x_5 + x_6) = 1080 + 360 = \mathbf{1440°}$

In the above step, $(x_1 + x_2 + x_3 + x_4 + x_5 + x_6) = 360° =$ sum of exterior angles of a polygon.

4.040   Figure on the right shows parallelogram *PQRS*. Find (i) its area; (ii) the length of diagonal *PR*.

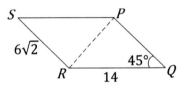

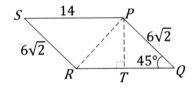

On the left is the figure redrawn with altitude *PT* perpendicular to side *RQ*. $\triangle PTQ$ is an isosceles right triangle, with $PT = TQ$, and $\angle TPQ = \angle Q = 45°$.

Using the Pythagorean Theorem in $\triangle PTQ$,

$(PT)^2 + (TQ)^2 = (PQ)^2 \quad \Rightarrow \quad (PT)^2 + (PT)^2 = (PQ)^2 \quad \Rightarrow \quad 2(PT)^2 = \left(6\sqrt{2}\right)^2 \quad \Rightarrow$

$2(PT)^2 = 72 \quad \Rightarrow \quad (PT)^2 = 36 \quad \Rightarrow \quad PT = \sqrt{36} = 6 = TQ$

(i)   Area of parallelogram *PQRS* = base *RQ* × height *PT* = 14 × 6 = 84

(ii) $RT = RQ - TQ = 14 - 6 = 8$ Notice that in $\triangle PTR$, $PT = 6$ and $RT = 8$. They are the 3rd and 4th multiples of 2. Therefore, hypotenuse $PR = $ (5th multiple of 2) = 10. $\triangle PTR$ is a 3–4–5 triangle. No need to use the Pythagorean Theorem. Save time!

4.041

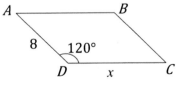

In the figure on the left, if the area of parallelogram *ABCD* is to be greater than $92\sqrt{3}$, then what is the minimum *prime integer value* of *x*?

On the right is the figure redrawn with altitude *ED* as shown. In the 30–60–90 $\triangle AED$, $AE = \frac{1}{2}(AD) = 4$

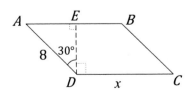

Using the Pythagorean Theorem,

$(AE)^2 + (ED)^2 = (AD)^2 \quad \Rightarrow \quad 4^2 + (ED)^2 = 8^2$

$\Rightarrow \quad (ED)^2 = 8^2 - 4^2 = 64 - 16 = 48$

$$\Rightarrow \quad ED = \sqrt{48} = \sqrt{16 \times 3} = 4\sqrt{3}$$

Area of parallelogram $ABCD$ = base DC × height ED $\Rightarrow$ $92\sqrt{3} = x \times 4\sqrt{3}$ $\Rightarrow$ $x = 23$

When $x = 23$, the area of parallelogram $ABCD$ will be $92\sqrt{3}$.

But the area is to be *greater than* $92\sqrt{3}$, which means x will have to be greater than 23, i.e. the very *next* prime number, i.e. $x = 29$.

4.042 **A rectangle has an area of 100. Which of the following could be possible?**

    (i)   **Its perimeter measures 35.**
    (ii) **Its diagonals measure 10 each.**
    (iii) **Its length is 13 times its width.**

  (i) Imagine the rectangle getting thinner and thinner. Then, to keep up with the area of 100, it will have to get longer and longer. Meaning, its perimeter will increase rapidly, the maximum perimeter being infinity (∞). The perimeter is at its minimum when the rectangle is a square. A square of area 100 will have its side equal to $\sqrt{100}$, or 10. Its perimeter will be $4 \times 10 = 40$, which is the minimum possible perimeter. Clearly, it cannot drop all the way to 35.

 (ii) When the rectangle of area 100 is a square, its sides measure 10 each. Each diagonal bisects the square into two isosceles right triangles. The length of each diagonal is $\sqrt{10^2 + 10^2} = \sqrt{100 + 100} = \sqrt{200} = \sqrt{100 \times 2} = 10\sqrt{2}$, which is greater than 10.

     When the rectangle is not a square, two equal sides will have their length greater than 10, whereas the other two will have their length less than 10. The diagonals, being hypotenuses of right triangles, will still be longer than 10. Therefore, whether the rectangle is a square or not, its diagonals will always measure more than 10.

(iii) If the rectangle gets thinner and thinner, then to maintain a constant area of 100, it will get longer and longer. Meaning, its length will become twice its width, then thrice, then 4 times, then 6½ times, then 8¾ times, then 11.473 times, **then 13 times**, then...... My point being, its length can be *any multiple* of its width, because there exist infinitely many configurations of a rectangle with a fixed area.

4.043 **In the figure on the right, *ABCD* is a rhombus with sides of length 13. If diagonal *BD* measures 10, then what is the length of diagonal *AC*?**

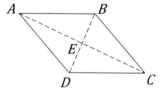

The diagonals of a rhombus are perpendicular to each other, and bisect each other.

In this example, diagonals BD and AC are perpendicular bisectors of each other at E.

Therefore, $BE = ED = \frac{1}{2}(BD) = \frac{1}{2}(10) = 5$

Using the Pythagorean Theorem in right $\triangle BEC$,

$(BE)^2 + (EC)^2 = (BC)^2$ $\Rightarrow$ $5^2 + (EC)^2 = 13^2$ $\Rightarrow$ $(EC)^2 = 13^2 - 5^2$ $\Rightarrow$

$(EC)^2 = 169 - 25 = 144$ $\Rightarrow$ $EC = \sqrt{144} = 12 = AE$

Length of diagonal $AC = AE + EC = 12 + 12 = 24$

PRACTICE EXAMPLES

4.044 If the sum of the interior angles of a polygon is 720°, then how many diagonals does the polygon have?

4.045 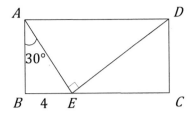 In the figure on the left, *ABCD* is a rectangle. Find the area of △*DEC*.

4.046 In the figure on the right, $\overleftrightarrow{l} \parallel \overleftrightarrow{m}$. If the area of the shaded region is 75, then how far are the lines from each other?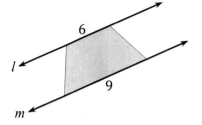

4.047 A rectangular swimming pool 50′ × 25′ is immediately surrounded by a marble pavement 4′ wide. What is the area of the marble pavement?

4.048 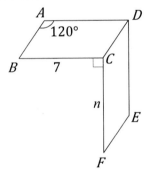 In the figure on the left, area of parallelogram *ABCD* is $35\sqrt{3}$. If area of parallelogram *CDEF* is greater than that, then what could be the least integer value of $\dfrac{n}{\sqrt{3}}$?

4.049 In the figure on the right, *PQRS* is a rectangle. $\overline{AB} \parallel \overline{PR} \parallel \overline{DC}$. What is the total area of the shaded regions?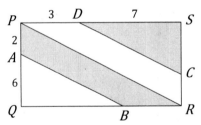

4.050 Three times the length of a rectangle is equal to five times its width. If the length and the width are both to be odd integers, then what is the least measure of its width?

4.051 A rectangle has a perimeter equal to 100. Which of the following could be true?

(i)   Its area equals 700.
(ii) Its length is 700 times its width.
(iii) The length of each of its diagonals is 25.

4.052 In the figure on the right, if all the angles measure 90°, then what is the perimeter of the figure?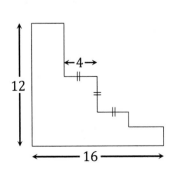

SPACE FOR CALCULATIONS

CIRCLES

Circles are the most commonly visible figures in our daily lives. Buttons, door knobs, dishes, wheels, cups and saucers, screws, lamps, pots etc. are all objects in which we see circles. Hence, circles have importance, and warrant academic exploration. Let's first study the terms associated with a circle.

A **circle** is the set of all coplanar points that are equidistant from a given point called **center**. The constant distance of each of those points from the center is called **radius** (plural: *radii*). A **chord** is a line segment joining any two points on a circle. The **diameter** is a chord passing through the center of the circle. It is also the longest possible chord, and is twice as long as the radius. A **tangent** is a line or line segment coplanar with the circle that intersects the circle at exactly one point. A **secant** is a line that intersects the circle at two points.

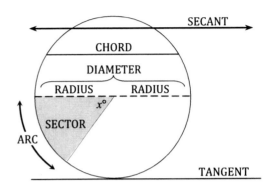

The **circumference** is the length of the boundary of the circle. An **arc** is a segment of the circumference of the circle. If the circumference is divided into two unequal parts, then the larger part is called **major arc**, and the smaller part is called **minor arc**. In the above figure, the arc shown is the minor arc, whereas the remaining part of the circumference would be the major arc. A **sector** is the region bounded by an arc and two radii. Just like major and minor arcs, there are **major and minor sectors**. In the figure, the shaded portion is the minor sector.

Let d = diameter r = radius C = circumference A = area of circle

Then, $d = 2r$ $C = 2\pi r$ $A = \pi r^2$ where $\pi \approx \dfrac{22}{7} \approx 3.14$

To find the length of an arc, use the following proportion:

$$\frac{\text{length of arc}}{\text{circumference}} = \frac{x}{360}$$ where x is the angle swept by the arc, as shown in the above diagram.

Clearly, the arc length is proportional to x; greater the degree measure of x, greater is the arc length.

To find the area of a sector, use the following proportion:

$$\frac{\text{area of sector}}{\text{area of circle}} = \frac{x}{360}$$

Just like arc length, area of sector also increases or decreases with x.

That's all there is to it when it comes to remembering formulas on circles for the GRE/GMAT. Now let's consider some important theorems and properties that will help us solve application type questions.

INSCRIBED ANGLE THEOREM

The angle made by two radii at the center of a circle is called a **central angle**. If two chords on a circle originate at the same point, then the angle they make at their common point of origin is called an **inscribed angle**.

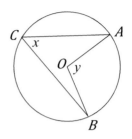

Select two random points A and B on the circle, and draw a central angle from them. From the same points A and B, draw an inscribed angle at a random point C on the circle.

Then, $\angle AOB = 2\angle ACB$ or $y = 2x$

THALES' THEOREM

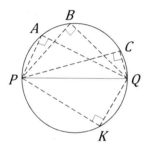

The diameter subtends a right angle at the circumference. *Subtends* means *faces*, or *is opposite to*.

As an example, pick any random point (A, or B, or C, or......) on the circumference of a circle as shown in the diagram on the left. Join the two ends of diameter PQ to that point. The angle formed at that point will always be a right angle (90°).

INTERSECTING CHORDS THEOREM

When two chords of a circle intersect each other, the product of the segments of one chord is equal to the product of the segments of the other chord.

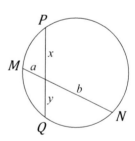

In the figure on the right, chord PQ intersects chord MN. a, b, x and y are the lengths of their segments.

Then, $ab = xy$

RADIUS PERPENDICULAR TO TANGENT

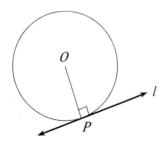

Consider $\overleftrightarrow{l}$ tangent to the circle at point P. If a radius PO is drawn from the point of tangency P, then that radius PO is perpendicular to the tangent $\overleftrightarrow{l}$ at point P.

i.e. $\overline{PO} \perp \overleftrightarrow{l}$.

CYCLIC AND BICENTRIC POLYGONS

A polygon is said to be **inscribed in a circle** if all of its vertices are points on the circumference of the circle. Such a polygon is referred to as a **cyclic polygon**. A circle is said to be **inscribed in a polygon** if every side of that polygon is a tangent to the circle. If a polygon can simultaneously inscribe one circle, and be inscribed in another circle, then it is called a **bicentric polygon**.

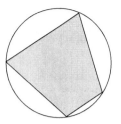

Quadrilateral
inscribed in a circle

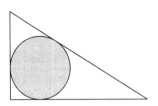

Circle inscribed
in a triangle

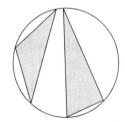

Two different triangles
inscribed in a circle

Every triangle, regardless of its configuration (acute, right, obtuse), can be inscribed in a circle. In other words, any three points anywhere in the universe will have exactly one circle passing through them.

Every regular polygon can be inscribed in a circle. Recall that a regular polygon is one that has all of its sides of equal length, and all of its interior angles of equal measure. Examples would be equilateral triangle, square, regular pentagon etc.

As a special case, **every rectangle** can be inscribed in a circle. The diagonals of the rectangle intersect at the center of the inscribing circle, and are the diameters of that circle.

If a polygon is irregular, then it may or may not be inscribed in a circle; that would vary individually case by case.

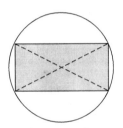

A rectangle
inscribed in a circle

4.053 **If the area of a circle is 81π, then what is its circumference?**

The radius is key to finding the circumference. Assume the radius as r.

Area of a circle $= \pi r^2 \quad \Rightarrow \quad 81\pi = \pi r^2 \quad \Rightarrow \quad 81 = r^2 \quad \Rightarrow \quad r = 9$

Circumference $= 2\pi r = 2\pi(9) = 18\pi$

4.054

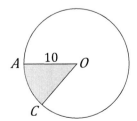

A circle has its center at O and radius 10. If the length of minor arc AC is 32, what is the area of the shaded sector?

$$\frac{\text{length of minor arc } AC}{\text{circumference}} = \frac{\angle AOC}{360°} = \frac{\text{area of shaded sector}}{\text{area of circle}} \quad \Rightarrow$$

$$\frac{32}{2\pi(10)} = \frac{\text{area of shaded sector}}{\pi(10)^2} \quad \Rightarrow \quad \text{area of shaded sector} = \frac{32 \times \pi(10)^2}{2\pi(10)} = 160$$

4.055 In the figure on the right, if the area of the shaded sector is 44, then what is the radius of the circle? Assume $\pi = \dfrac{22}{7}$.

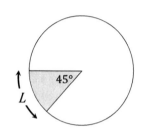

Assume the radius as r. The angle swept by the shaded sector is $360 - 220 = 140°$.

$$\frac{\text{area of shaded sector}}{\text{area of circle}} = \frac{\text{angle swept by shaded sector}}{360°}$$

$$\Rightarrow \quad \frac{44}{\pi r^2} = \frac{140}{360} \quad \Rightarrow \quad \frac{44}{\frac{22}{7}r^2} = \frac{7}{18} \quad \Rightarrow \quad \frac{44}{r^2} \times \frac{7}{22} = \frac{7}{18} \quad \Rightarrow \quad \frac{14}{r^2} = \frac{7}{18}$$

$$\Rightarrow \quad 7r^2 = 14 \times 18 \quad \Rightarrow \quad r^2 = \frac{14 \times 18}{7} = 2 \times 18 = 36 \quad \Rightarrow \quad r = \sqrt{36} = 6$$

4.056

In the figure on the left, if A is the area of the shaded sector, then what is L in terms of A?

Assume the radius as r.

$$\frac{\text{area of shaded sector}}{\text{area of circle}} = \frac{\text{angle swept by shaded sector}}{360°}$$

$$\Rightarrow \quad \frac{A}{\pi r^2} = \frac{45}{360} \quad \Rightarrow \quad \frac{A}{\pi r^2} = \frac{1}{8} \quad \Rightarrow \quad \pi r^2 = 8A \quad \Rightarrow \quad r^2 = \frac{8A}{\pi}$$

$$\frac{\text{length of an arc}}{\text{circumference}} = \frac{\text{angle swept by that arc}}{360°}$$

$$\Rightarrow \quad \frac{L}{2\pi r} = \frac{45}{360} \quad \Rightarrow \quad \frac{L}{2\pi r} = \frac{1}{8} \quad \Rightarrow \quad 2\pi r = 8L \quad \Rightarrow \quad r = \frac{4L}{\pi} \quad \Rightarrow \quad r^2 = \frac{16L^2}{\pi^2}$$

Equate these two values of r^2, and solve for L:

$$\frac{8A}{\pi} = \frac{16L^2}{\pi^2} \quad \Rightarrow \quad \pi \times 16L^2 = 8A \times \pi^2 \quad \Rightarrow \quad L^2 = \frac{8A\pi^2}{16\pi} = \frac{A\pi}{2} \quad \Rightarrow \quad L = \sqrt{\frac{A\pi}{2}}$$

4.057 The diagram on the right shows four shaded quarter circles drawn inside a square. What is the area of the region that is *not* shaded?

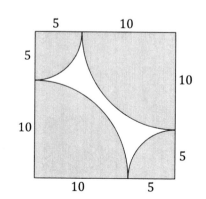

The area of the region that is not shaded is simply the area of the square, minus the areas of the four quarter circles. A quarter circle has ¼ the area of a full circle.

The two quarter circles with radius 5 will have a total

area of: $\dfrac{\pi \cdot 5^2}{4} + \dfrac{\pi \cdot 5^2}{4} = \dfrac{\pi \cdot 5^2}{2} = \dfrac{\pi \cdot 25}{2} = 12.5\pi$

The two quarter circles with radius 10 will have a total area of:

$$\frac{\pi \cdot 10^2}{4} + \frac{\pi \cdot 10^2}{4} = \frac{\pi \cdot 10^2}{2} = \frac{\pi \cdot 100}{2} = 50\pi$$

Area of the square with sides 15 is: $15 \times 15 = 225$

Therefore, area of the region that is not shaded = $225 - (12.5\pi + 50\pi) = 225 - 62.5\pi$

4.058

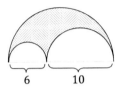

Two small semi-circles are inscribed within a larger semi-circle. What is the area of the shaded region?

Notice that the diameters of the three semi-circles are 6, 10 and 16, in increasing order. Therefore, their radii are 3, 5 and 8 in that order.

The area of the shaded region is the area of the largest semi-circle, minus the areas of the two smaller semi-circles.

Area of the shaded region $= \dfrac{\pi \cdot 8^2}{2} - \left(\dfrac{\pi \cdot 3^2}{2} + \dfrac{\pi \cdot 5^2}{2}\right) = \dfrac{64\pi}{2} - \left(\dfrac{9\pi}{2} + \dfrac{25\pi}{2}\right) = \dfrac{30\pi}{2} = 15\pi$

4.059 **What is the area of a circle that has a 32 × 24 rectangle inscribed in it?**

When a rectangle is inscribed in a circle, a diagonal of the rectangle (dotted) is the diameter of the circle.

In the figure, notice that in each of the right triangles, the hypotenuse is is the diagonal, which is the diameter. The triangles have their sides as the 3rd and 4th multiples of 8. Therefore, the hypotenuse must be the 5th multiple of 8, i.e. 40. The triangles you see are 3–4–5 right triangles.

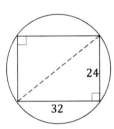

You would get the same result by using the Pythagorean Theorem, but it's better to avoid all those calculations, and save time on the GRE/GMAT.

Diameter = 40   Therefore, radius = 20

Area of the circle $= \pi \cdot 20^2 = 400\pi$

4.060

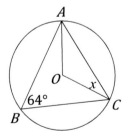

***O* is the center of the circle. What is the degree measure of *x*?**

Central $\angle AOC = 2(\text{inscribed } \angle ABC) = 2(64) = 128°$

$\triangle AOC$ is an isosceles triangle because $OA = OC =$ radius

Therefore, $\angle OAC = \angle OCA = x$

In $\triangle AOC$, $\angle A + \angle O + \angle C = 180°$ $\Rightarrow$ $x + 128 + x = 180$ $\Rightarrow$ $2x = 52$ $\Rightarrow$ $x = 26°$

4.061 A circular fountain 8 feet in diameter is to be surrounded by a mosaic pavement 1 foot wide. If laying the mosaic tiles costs \$35 per square foot, then how much will it cost to tile the pavement with mosaic tiles? Assume $\pi = \dfrac{22}{7}$.

Radius of the inner circle $= \dfrac{8}{2} = 4'$

Radius of the outer circle $= \dfrac{8+1+1}{2} = 5'$

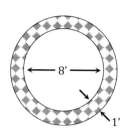

Area to be covered by mosaic tiles = area of outer circle − area of inner circle $= \pi \cdot 5^2 - \pi \cdot 4^2 = 25\pi - 16\pi = 9\pi$ ft^2

Cost $= \$35$ per ft$^2 \times 9\pi$ ft$^2 = 35 \times 9 \times \dfrac{22}{7} = 45 \times 22 = \$990$

4.062

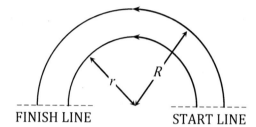

FINISH LINE   START LINE

Two athletes race along two semi-circular tracks. The inner and outer tracks have radii r and R. If the two athletes start and finish the race in the same amount of time, then what is the ratio of the speed of the runner in the outer track to that of the runner in the inner track?

Let's say the two athletes take time t to run the two semi-circular distances.

Distance run by the outer athlete $= \dfrac{\text{outer circumference}}{2} = \dfrac{2\pi R}{2} = \pi R$

Distance run by the inner athlete $= \dfrac{\text{inner circumference}}{2} = \dfrac{2\pi r}{2} = \pi r$

Speed of the outer athlete $= \dfrac{\text{distance run}}{\text{time}} = \dfrac{\pi R}{t}$

Speed of the inner athlete $= \dfrac{\text{distance run}}{\text{time}} = \dfrac{\pi r}{t}$

Ratio of their speeds $= \dfrac{\text{speed of outer athlete}}{\text{speed of inner athlete}} = \dfrac{\frac{\pi R}{t}}{\frac{\pi r}{t}} = \dfrac{\pi R}{t} \times \dfrac{t}{\pi r} = \dfrac{R}{r} = R : r$

4.063 A tire of diameter 2 feet starts rolling down a 30° incline. The vertical distance between the top and bottom of the incline is 44 feet. When the tire reaches the bottom of the incline, how many revolutions will it have made? Assume $\pi = \dfrac{22}{7}$.

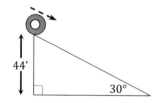

The configuration of the hill is in the form of a 30–60–90 triangle in which, the sloping side (hypotenuse) is twice as long as the vertical height (smallest side), i.e. $2 \times 44' = 88'$.

Radius of the tire is 1'. Assume the tire makes n revolutions while rolling down the hill.

Then, (circumference of tire) × (# of revolutions) = (distance rolled down)

$2\pi(1) \times n = 88 \quad \Rightarrow \quad n = \dfrac{88}{2\pi} = \dfrac{44}{\pi} = \dfrac{44}{\frac{22}{7}} = 44 \times \dfrac{7}{22} = 14$

4.064

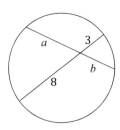

In the figure, if a and b are integers, then what's the maximum possible value of $(a + b)$?

Using the Intersecting Chords Theorem, we can say:

$$a \times b = 8 \times 3 \quad \Rightarrow \quad ab = 24$$

If a and b are to be integers, then we need to first find the different pairs of integers, the product of which is 24. They are:

$1 \times 24 = 24$ $2 \times 12 = 24$ $3 \times 8 = 24$ $4 \times 6 = 24$

The sum of which two numbers within the above pairs is maximum? Clearly, it's 1 and 24.

Therefore, a could be 1 and b could be 24, or a could be 24 and b could be 1.

Regardless of which one is which, the maximum value of $a + b = 25$.

Among the examples that we have solved up to this point, notice that more and more of them have involved concepts not just from geometry, but from algebra as well as basic arithmetic. Such is the nature of application oriented questions on the GRE/GMAT.

PRACTICE EXAMPLES

4.065   In the figure on the right, if the shaded sector covers 35% of the area of the circle, then what is the degree measure of x?

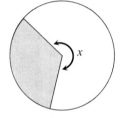

4.066

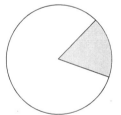

In the figure on the left, if the shaded sector covers 20% of the area of the circle, then what percent of the area of the non-shaded sector does the shaded sector cover?

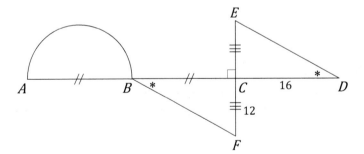

4.067   In the figure on the right, what is the ratio of length ED to the radius of the semi-circle?

4.068

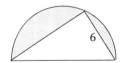

In the figure on the left, a triangle is inscribed in a semi-circle of radius 5. What is the area of the shaded region?

4.069   In the figure on the right, if the right triangle and the semi-circle have the same area, then what is d in terms of h?

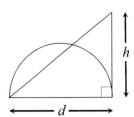

4.070

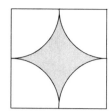

In the figure on the left, four quarter circles of radius 11 are drawn inside a square. What is the area of the shaded region?

4.071 The circumference of a circle is equal to the radius of another circle. What is the ratio of the area of the smaller circle to that of the larger circle?

4.072

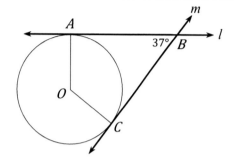

Figure on the left shows $\overrightarrow{l}$ and $\overleftrightarrow{m}$ tangential to the circle at A and C. O is the center of the circle. What is the degree measure of $\angle AOC$?

4.073 What is the area of a circle that is inscribed in a square that is inscribed in another circle of diameter 10?

4.074 In the figure on the right, O is the center of the circle. What is the radius of the circle?

4.075

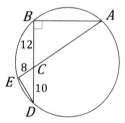

In the figure on the left, what is the area of $\triangle ABC$?

SOLIDS

Up to this point, we have only studied geometrical entities in one dimension (1-D) or two dimensions (2-D). Solids, which we shall now discuss, exist in three dimensions (3-D). A **solid** is an object that occupies space that is otherwise empty. A solid has definite volume and surface area. Let us consider the solids often tested on the GRE/GMAT. Get ready to stretch your imagination to its fullest!

RECTANGULAR SOLID

A **rectanglular solid** has all of its faces as rectangles.

Volume, $V = lwh$

Surface area, $A = 2lw + 2wh + 2hl = 2(lw + wh + hl)$

Diagonal, $d = \sqrt{l^2 + w^2 + h^2}$

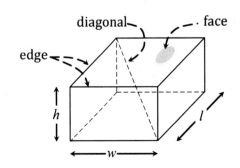

CUBE

A **cube** is a special type of rectangular solid in which, all the six faces are identical squares. All its edges are of equal length (e).

Volume, $V = e^3$

Surface area, $A = 6e^2$

Diagonal, $d = \sqrt{e^2 + e^2 + e^2} = \sqrt{3e^2} = e\sqrt{3}$

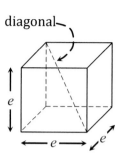

RIGHT CIRCULAR CYLINDER

A **right circular cylinder** is basically two circles (a bottom and a lid) joined by a curved surface perpendicular to the circles. If you take a pair of scissors and cut open the curved surface in the vertical direction, the curved surface will unwrap into a rectangle with height h and width equal to the circumference ($2\pi r$) of the lid (or of the bottom).

Volume, $V = \pi r^2 h$

Surface area, $A = \underbrace{2\pi r^2}_{\text{lid+bottom}} + \underbrace{2\pi rh}_{\text{curved surface}}$

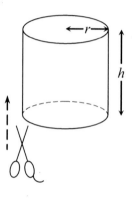

SPHERE

A **sphere** is a 3-D solid generated when a circle is revolved about its diameter.

Volume, $V = \dfrac{4}{3}\pi r^3$

Surface area, $A = 4\pi r^2$

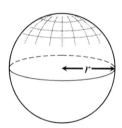

RIGHT CIRCULAR CONE

A **right circular cone** is a 3-D solid generated when a right triangle is revolved about one of its legs. The top most point at which the cone converges is called the **apex**. The circular bottom upon which the cone rests is called the **base**.

Volume, $V = \dfrac{1}{3}\pi r^2 h$

Surface area, $A = \underbrace{\pi r^2}_{\text{bottom}} + \underbrace{\pi rL}_{\text{curved surface}}$

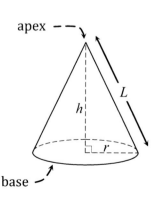

PRISM

A **prism** is a solid whose ends have the same shape and are parallel to each other. A rectangular solid is an example of a prism. So is a right circular cylinder. What characterizes a prism is that its ends can have *any* shape—polygonal, circular, elliptical, etc.—or no shape in particular. A prism can be imagined as a cut out section of an elongated solid with uniform cross-section.

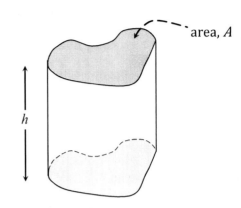

area, A

h

Volume, $V = A \cdot h$

PYRAMID

A **pyramid** is a solid with a polygonal base and triangular sides. The volume of a pyramid is $\frac{1}{3}$ the volume of a prism with the same polygonal base and the same vertical height.

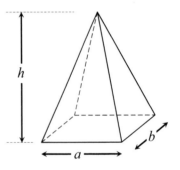

h

b

a

Volume, $V = \frac{1}{3}abh$

4.076 The edge of a cube is equal to the radius of a sphere. If the volume of the cube is 27, then what is the volume of the sphere?

Let e = edge of the cube, and r = radius of the sphere.

Volume of the cube $= 27 \quad \Rightarrow \quad e^3 = 27 \quad \Rightarrow \quad e = \sqrt[3]{27} = 3$

Given: $r = e = 3$

Volume of the sphere $= \frac{4}{3}\pi r^3 = \frac{4}{3}\pi(3)^3 = 4\pi(9) = 36\pi$

4.077 The diagonal of a cube is equal to the diameter of a sphere. What fraction of the volume of the sphere is the volume of the cube? Assume $\pi\sqrt{3} = 5.5$

Let e = edge of the cube, and d = diameter of the sphere

Diagonal of the cube $= \sqrt{e^2 + e^2 + e^2} = \sqrt{3e^2} = e\sqrt{3} = d$

Volume of the cube $= e^3$

Volume of the sphere $= \frac{4}{3}\pi\left(\frac{d}{2}\right)^3 = \frac{4}{3}\pi\left(\frac{d^3}{8}\right) = \frac{\pi d^3}{6} = \frac{\pi\left(e\sqrt{3}\right)^3}{6} = \frac{\pi e^3 \cdot 3\sqrt{3}}{6} = \frac{\pi e^3\sqrt{3}}{2}$

Volume of the cube, as a fraction of volume of the sphere $= \dfrac{\text{Volume of cube}}{\text{Volume of sphere}} =$

$$\frac{e^3}{\dfrac{\pi e^3 \sqrt{3}}{2}} = e^3 \times \frac{2}{\pi e^3 \sqrt{3}} = \frac{2}{\pi \sqrt{3}} = \frac{2}{5.5} = \frac{2}{5.5} \times \frac{10}{10} = \frac{20}{55} = \frac{4}{11}$$

Notice just how many different laws of fractions the above example tested.

4.078 A rectangular shoebox with dimensions 12" × 8" × 6" has several balls inside it. Each ball has a radius of 1". Exactly 4 balls are green. If one ball is picked at random from the box, then what is the *least* probability of it being green?

$$\text{P(ball picked = green)} = \frac{\text{\# of green balls}}{\text{total \# of balls}}$$

Given: # of green balls = 4

Clearly, the probability would be minimum when the denominator (total # of balls) is at its maximum. So we need to find the greatest number of balls the box can accommodate.

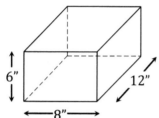

Radius of each ball = 1"   So, diameter = 2"

The box can hold: 6 balls along the 12" dimension
                                          4 balls along the 8" dimension
                                          3 balls along the 6" dimension

Maximum # of balls it can hold = 6 × 4 × 3 = 72

The least probability is: $\text{P(ball picked = green)} = \dfrac{\text{\# of green balls}}{\text{max. \# of balls}} = \dfrac{4}{72} = \dfrac{1}{18}$

4.079 Figure on the right shows a pot with hemispherical inner and outer surfaces. The annular surface is in the form of a circular ring. What is the total surface area of the pot if its inner and outer radii are 5" and 7"?

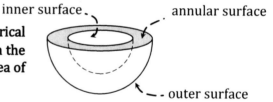

$$\text{Surface area of hemisphere} = \frac{\text{surface area of full sphere}}{2} = \frac{4\pi r^2}{2} = 2\pi r^2$$

Surface area of outer hemisphere = $2\pi(7)^2 = 2\pi(49) = 98\pi$

Surface area of inner hemisphere = $2\pi(5)^2 = 2\pi(25) = 50\pi$

Area of annular surface = area of larger circle – area of smaller circle =

$\pi(7)^2 - \pi(5)^2 = 49\pi - 25\pi = 24\pi$

Total surface area of the pot = area of (outer surface + inner surface + annular surface) =

$98\pi + 50\pi + 24\pi = \mathbf{172\pi}$

4.080

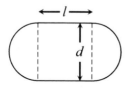

A cylindrical air tank has hemispherical ends. If the length of the cylindrical portion of the tank (*l*) is equal to the diameter of the hemispherical ends (*d*), then what is the total volume of the tank (*V*) in terms of *l* ?

It's given that the diameter d is equal to l.

The two hemispheres on the sides can be considered together as a full sphere. Its volume is:

$$\frac{4}{3}\pi(\text{radius})^3 = \frac{4}{3}\pi\left(\frac{d}{2}\right)^3 = \frac{4}{3}\pi\left(\frac{l}{2}\right)^3 = \frac{4}{3}\pi\left(\frac{l^3}{8}\right) = \frac{\pi l^3}{6}$$

The volume of the cylindrical portion between the two hemispheres is:

$$\pi\left(\frac{d}{2}\right)^2 l = \pi\left(\frac{l}{2}\right)^2 l = \pi\cdot\frac{l^2}{4}\cdot l = \frac{\pi l^3}{4}$$

Total volume of tank, V = volume of the two hemispheres + volume of cylindrical portion $\Rightarrow$

$$V = \frac{\pi l^3}{6} + \frac{\pi l^3}{4} = \pi l^3\left(\frac{1}{6}+\frac{1}{4}\right) = \pi l^3\left(\frac{4+6}{24}\right) = \frac{10\pi l^3}{24} = \frac{5\pi l^3}{12} \quad\Rightarrow\quad V = \frac{5\pi l^3}{12}$$

4.081 **If the angle at the apex of a right circular cone is 60°, then what is the ratio of the area of the base of the cone to the area of the curved surface of the cone?**

In the cone on the right, the vertical height (h), the radius (r) and the slant height (L) form a 30–60–90 right triangle in which, $L = 2r$.

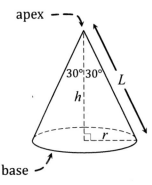

Area of the base $= \pi r^2$

Area of the curved surface $= \pi r L = \pi r(2r) = 2\pi r^2$

$$\frac{\text{Area of the base}}{\text{Area of curved surface}} = \frac{\pi r^2}{2\pi r^2} = \frac{1}{2} = 1:2$$

4.082 **The curved surface of a right circular cylinder has an area equal to that of the curved surface of a circular cone with height 12 and radius 9. If the height of the cylinder is 7½ times its radius, then what is its radius?**

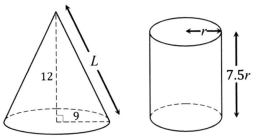

In the cone, 9, 12 and L form a right triangle in which, 9 and 12 are the 3rd and 4th multiples of 3. Therefore, L must be the 5th multiple of 3, i.e. 15.   The triangle is a 3–4–5 triangle.

Area of curved surface of the cone is:

$$\pi\cdot(\text{radius})\cdot(\text{slant height}) = \pi(9)(15) = 135\pi$$

Area of curved surface of the cylinder is:

$$(\text{circumference})\cdot(\text{height}) = (2\pi r)\cdot(7.5r) = 15\pi r^2$$

It's given that the two curved surfaces have equal areas:

$$15\pi r^2 = 135\pi \quad\Rightarrow\quad r^2 = \frac{135\pi}{15\pi} = 9 \quad\Rightarrow\quad r = \sqrt{9} = 3$$

4.083 **A pyramid has a rectangular base whose one side is twice the other. Its height is half the height of a right circular cone. If the cone and the pyramid have the same volume, then what is the ratio of the radius of the cone to the smaller side of the base of the pyramid?**

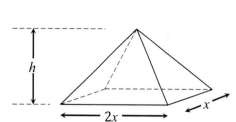

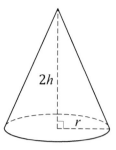

Assume the pyramid has a rectangular base of dimensions x and $2x$, and a vertical height h. Then, the vertical height of the cone would be $2h$. Assume the cone has a radius r.

$$\text{Volume of pyramid} = \frac{1}{3}(2x)(x)(h) = \frac{2x^2h}{3}$$

$$\text{Volume of cone} = \frac{1}{3}\pi r^2 h = \frac{\pi r^2 h}{3}$$

It's given that the two volumes are equal:

$$\frac{\pi r^2 h}{3} = \frac{2x^2h}{3} \quad \Rightarrow \quad \pi r^2 = 2x^2 \quad \Rightarrow \quad \frac{r^2}{x^2} = \frac{2}{\pi} \quad \Rightarrow \quad \frac{r}{x} = \sqrt{\frac{2}{\pi}} \quad \Rightarrow \quad r : x = \sqrt{2} : \sqrt{\pi}$$

PRACTICE EXAMPLES

4.084 Ice cream is scooped out of a cylindrical pot 6" in diameter and 6" tall, that is initially full. If the right circular cones in which it is filled (to the top) measure 2" in diameter and 4" in height, then *at least* how many cones would be required to use up all the ice cream from the pot?

4.085 The height of a right circular cone is equal to the diameter of a sphere. If the cone and the sphere have the same volume, then what is the ratio of the radius of the sphere to the radius of the cone?

4.086 A brick wall, 19 feet wide and 14 feet high, is to be constructed using bricks with the dimensions 12" × 6" × 4". What is the minimum number of bricks that would be required to construct the wall? (Note: It doesn't matter how the bricks are oriented. Also, ignore the volume taken up by the plaster.)

4.087 A cylindrical drum of height 3 feet has water in it, the water level being at 2'8". Four metal plates with dimensions 11" × 7" × 2" are gently released into the water. When they completely sink, the water level reaches exactly the top of the drum. What is the diameter of the drum? Assume $\pi = \frac{22}{7}$.

4.088 The edge of a cube is equal to the diameter of a sphere. What is the surface area (A) of the cube in terms of the surface area (a) of the sphere?

4.089 The rectangular floor of my dining room has dimensions 12' × 10'. It has 6" square tiles from end to end. The sides of the tiles are parallel to the walls of the room. Three tiles are defective. If I randomly step on a tile, then what's the probability that I will have stepped on a defective tile?

4.090 How many blocks are in the picture?

SPACE FOR CALCULATIONS

COORDINATE GEOMETRY

This section deals with plotting geometric entities on a 2-D plane. Each entity – a line, a triangle, a circle etc. – can be plotted on an **X-Y plane** (also known as the **coordinate plane**) using certain anchor points that lie on the entities. Each point has two coordinates that define its position on the X-Y plane: the **x-coordinate** (also known as abscissa) and the **y-coordinate** (also known as the ordinate). The x-coordinate is the horizontal distance from the vertical axis (the **y-axis**). The y-coordinate is the vertical distance from the horizontal axis (the **x-axis**).

Think of the x and y axes as two number lines – horizontal and vertical – that intersect at their 0. This point of intersection of the two axes is called the **origin**, whose coordinates are (0, 0). The two axes divide the coordinate plane into four **quadrants**: I, II, III and IV, numbered counter-clockwise. The x and y coordinates of a point can be positive or negative, depending upon which quadrant it is in. The following list shows the signs of the x and y coordinates in each quadrant.

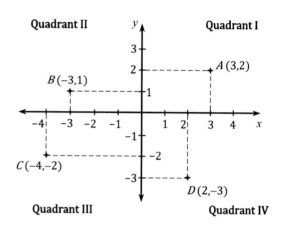

Quadrant	x-coordinate	y-coordinate
I	+	+
II	–	+
III	–	–
IV	+	–

DISTANCE FORMULA AND MIDPOINT FORMULA

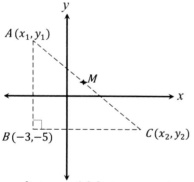

Consider three points A, B and C in a coordinate plane as shown on the left. The three line segments joining them form a right triangle whose base and height run parallel to the axes. Points A and B have the same horizontal distance from the y-axis. Therefore, their x coordinates are the same, i.e. $x_1 = -3$. y_1 could be anything. Let's assume that $y_1 = 8$.

Points B and C are at the same vertical distance from the x-axis. Therefore, their y coordinates are the same, i.e. $y_2 = -5$. x_2 could be anything. Let's assume that $x_2 = 6$.

Then, distance (d) between points $A(x_1, y_1)$ and $C(x_2, y_2)$ is given by the **distance formula**:

$$d = \sqrt{(y_2 - y_1)^2 + (x_2 - x_1)^2}$$

This is nothing but the Pythagorean Theorem interpreted in terms of coordinates: $|y_2 - y_1|$ being the length of the vertical leg AB, $|x_2 - x_1|$ being the length of the horizontal leg BC, and d being the length of the hypotenuse AC in right $\triangle ABC$. I've used absolute values because distances are always positive.

$$AC = \sqrt{(-5 - 8)^2 + \left(6 - (-3)\right)^2} = \sqrt{(-13)^2 + 9^2} = \sqrt{169 + 81} = \sqrt{250} = \sqrt{25 \times 10} = 5\sqrt{10}$$

Let M be the midpoint of $\overline{AC}$. Then coordinates of M are given by the **midpoint formula**:

$$M \equiv \left(\frac{x_1 + x_2}{2}, \frac{y_1 + y_2}{2}\right)$$

Notice that the coordinates of M are simply the arithmetic means of the respective coordinates of A and C.

$$M \equiv \left(\frac{-3+6}{2}, \frac{8+(-5)}{2}\right) \quad \Rightarrow \quad M \equiv \left(\frac{3}{2}, \frac{3}{2}\right) \quad \Rightarrow \quad M \equiv (1.5, 1.5)$$

SLOPE

Consider a horizontal line l_1 in the coordinate plane. $\overleftrightarrow{l_1}$ is not *sloping*. It is not *inclined*. In the language of geometry, slope is the measure of the *steepness* of an incline. $\overleftrightarrow{l_1}$ has no steepness. Therefore, slope of $\overleftrightarrow{l_1} = 0$. Let's designate the letter m to represent slope of a line in general. Meaning, let $m_1, m_2, m_3, \ldots\ldots$ be the slopes of $\overleftrightarrow{l_1}, \overleftrightarrow{l_2}, \overleftrightarrow{l_3}, \ldots\ldots$ As you can tell, the lines get steeper and steeper until $\overleftrightarrow{l_5}$, which is vertical. Its steepness is infinity (∞), i.e. $m_5 = \infty$.

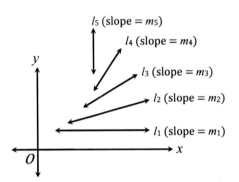

In the figure on the right: $m_1 < m_2 < m_3 < m_4 < m_5$

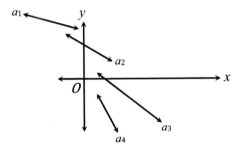

If a line or a line segment inclines upward from left to right, then it has a *positive slope*, as is the case with $\overleftrightarrow{l_2}$, $\overleftrightarrow{l_3}$ and $\overleftrightarrow{l_4}$ in the above diagram. If a line or a line segment inclines downward from left to right, then it has a *negative slope*, as is the case with $\overleftrightarrow{a_1}, \overleftrightarrow{a_2}, \overleftrightarrow{a_3}$ and $\overleftrightarrow{a_4}$ in the diagram on the left.

To calculate the exact value of slope, we would need to know the coordinates of two points on a line or line segment. Assume the points as $A(x_1, y_1)$ and $B(x_2, y_2)$. Then the slope (m) of $\overline{AB}$ would be:

$$m = \frac{y_2 - y_1}{x_2 - x_1}$$

SLOPES OF PERPENDICULAR LINES

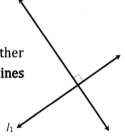

Consider $\overleftrightarrow{l_1} \perp \overleftrightarrow{l_2}$ as shown on the right. Let m_1 and m_2 be their slopes. If neither $\overleftrightarrow{l_1}$ nor $\overleftrightarrow{l_2}$ is vertical or horizontal, then the product of the slopes of such **lines perpendicular to each other** is (-1).

$$m_1 \cdot m_2 = -1 \quad \text{or} \quad m_1 = -\frac{1}{m_2} \quad \text{or} \quad m_2 = -\frac{1}{m_1}$$

EQUATION OF A LINE

Every line drawn on a 2-D coordinate system (*X-Y* plane) can be defined by a linear equation. We discussed linear equations in the chapter on algebra. Now let's try to reconcile them here.

A linear equation of the form $ax + by = c$, where *a*, *b* and *c* are real numbers, is said to be in its **standard form**.

A linear equation of the form $y = mx + b$, where *m* and *b* are real numbers, is said to be in its **slope-intercept form**. In the equation, *m* is the slope of the line, and *b* is the *y*-intercept (the *y*-coordinate of the point at which the line cuts the *y*-axis). Figure on the right makes this clear.

Notice that a linear equation can be converted from standard form to slope-intercept form by simply solving it for *y*:

$$y = -\frac{a}{b}x + \frac{c}{b}$$

The equation may look slightly different, but the format is essentially the same. The coefficient of the *x* term, i.e. $-\frac{a}{b}$ is nothing but the slope, and the constant term $\frac{c}{b}$ is the *y*-intercept.

A third form of a linear equation is the **two-point form**. This can be constructed when coordinates of *two points* on the line are known. Therefore, the name. Consider two points $A(x_1, y_1)$ and $B(x_2, y_2)$ that lie on $\overleftrightarrow{l}$. Then, equation of $\overleftrightarrow{l}$ is:

$$\frac{y - y_1}{y_1 - y_2} = \frac{x - x_1}{x_1 - x_2}$$

If you read the equation over and over, you will realize that the terms rhyme! So, is math now poetry?

Please note that only *x* and *y* are the variables in the two-point form. x_1, y_1, x_2 and y_2 are *constants*, i.e. they are the coordinates of points *A* and *B*. Please do not confuse them with *variables*.

A linear equation in the two-point form can be simplified and eventually expressed in either the standard form, or the slope-intercept form.

In algebra, when we were given two equations in *x* and *y*, and we solved them to find the values of *x* and *y*, what we really ended up doing was finding the *x* and *y* *coordinates* of the point of intersection of the two lines represented by the two equations. This is where algebra and geometry converge.

4.091 *ABCD* is a rhombus with sides *AD* and *BC* **parallel to the *x*-axis. Find the coordinates of the point of intersection of its diagonals.**

The diagonals of a rhombus bisect each other. To find their midpoint, we need to find the coordinates of either *C* or *D*.

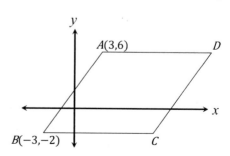

Then use the midpoint formula. Let's first work out the coordinates of C:

(y-coordinate of C) = (y-coordinate of B) = -2, because B and C are both at the same vertical distance from the x-axis.

(x-coordinate of C) = (x-coordinate of B) + (length BC).

In a rhombus, all four sides have equal lengths. Using the distance formula:

$$BC = BA = \sqrt{(y_2 - y_1)^2 + (x_2 - x_1)^2} = \sqrt{(-2 - 6)^2 + (-3 - 3)^2} = \sqrt{(-8)^2 + (-6)^2} =$$

$$\sqrt{64 + 36} = \sqrt{100} = 10$$

(x-coordinate of C) = $-3 + 10 = 7$

Using the midpoint formula, the midpoint of $A(3, 6)$ and $C(7, -2)$ has coordinates:

$$\left(\frac{x_1 + x_2}{2}, \frac{y_1 + y_2}{2}\right), \quad \text{i.e.} \quad \left(\frac{3 + 7}{2}, \frac{6 + (-2)}{2}\right), \quad \text{i.e.} \quad \left(\frac{10}{2}, \frac{4}{2}\right), \quad \text{i.e.} \quad (5, 2)$$

This is the point of intersection of the two diagonals.

4.092

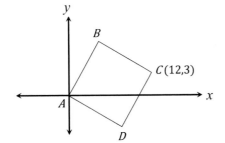

$ABCD$ is a square. What is the slope of diagonal BD?

The diagonals of a square are perpendicular to each other. Also, the product of slopes of two perpendicular lines (or line segments) is -1. To find the slope of BD, we need the coordinates of B and D, which we don't have. But we can find the slope of AC using the coordinates of $A(0, 0)$ and $C(12, 3)$.

$$\text{Slope of } AC = m_{AC} = \frac{y_2 - y_1}{x_2 - x_1} = \frac{3 - 0}{12 - 0} = \frac{1}{4}$$

$$m_{AC} \times m_{BD} = -1 \quad \Rightarrow \quad m_{BD} = \frac{-1}{m_{AC}} = \frac{-1}{\frac{1}{4}} = -1 \times \frac{4}{1} = -4$$

4.093 **If the point $(x, -3)$ lies on the line passing through $(-7, 2)$ and $(5, -6)$, then find x.**

It looks like the three points lie on the same line. Meaning, the slope calculated using the coordinates of any two points should be the same:

Slope using $(x, -3)$ and $(-7, 2)$ $\quad = \quad$ slope using $(-7, 2)$ and $(5, -6)$

$$\frac{y_2 - y_1}{x_2 - x_1} \quad = \quad \frac{2 - (-3)}{-7 - x} = \frac{-6 - 2}{5 - (-7)} \quad \Rightarrow \quad \frac{5}{-7 - x} = \frac{-8}{12} \quad \Rightarrow \quad -8(-7 - x) = 5 \times 12$$

$$\Rightarrow \quad 56 + 8x = 60 \quad \Rightarrow \quad 8x = 4 \quad \Rightarrow \quad x = \frac{1}{2}$$

4.094 **If $(-16, -5)$ is the center of a circle, and $(-4, 4)$ lies on the circle, then what is the circumference of the circle?**

A point lies *on the circle*, means that it lies on the circular boundary (see point P); *inside the circle* means within the boundary (point Q); and *outside the circle* means outside the boundary (point R).

Let Q actually be the center of the circle $(-16, -5)$. Then, using the distance formula:

Radius $PQ = \sqrt{\left(4 - (-5)\right)^2 + \left(-4 - (-16)\right)^2} = \sqrt{9^2 + 12^2} = \sqrt{81 + 144} = \sqrt{225} = 15$

Circumference $= 2\pi(\text{radius}) = 2\pi(15) = 30\pi$

4.095 **A circle has its center at (12, 5) and radius equal to 12.5. How many quadrants does the circle pass through?**

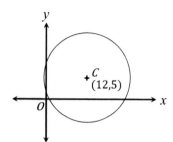

Draw the circle. You will see that it mostly exists in Quadrant I.

Its radius is 12.5, whereas its center is at a vertical distance of only 5 from the x-axis. Therefore, the circle extends below the x-axis into Quadrant IV.

Its center is also at a horizontal distance of 12 from the y-axis, when it's radius is actually 12.5. Therefore, the circle extends slightly to the left of the y-axis into Quadrant II as well.

To find out whether the circle extends into Quadrant III, let's find the distance OC. If the radius is greater than OC, then the circle extends beyond O into Quadrant III.

Using the distance formula, $OC = \sqrt{(y_2 - y_1)^2 + (x_2 - x_1)^2} = \sqrt{(12 - 0)^2 + (5 - 0)^2} =$

$\sqrt{12^2 + 5^2} = \sqrt{144 + 25} = \sqrt{169} = 13$

The radius (12.5) is not greater than OC (13). So the circle does not reach into Quadrant III.

The circle only passes through Quadrants I, II and IV, a total of 3 quadrants.

4.096 **Both circles in the figure are symmetric about the x-axis, and pass through the point A. The smaller circle passes through O as well. If $OB : OA = 2\frac{1}{2} : 4\frac{1}{2}$, then what is the area of the shaded region?**

From the fact that both circles are symmetric about the x-axis, OA and BA must be the diameters of the two circles. $OA = 18$

$BA = OB + OA \quad \Rightarrow \quad OB = BA - OA$

$\dfrac{OB}{OA} = \dfrac{2.5}{4.5} = \dfrac{25}{45} = \dfrac{5}{9} \quad \Rightarrow \quad 9(OB) = 5(OA) \quad \Rightarrow \quad 9(BA - OA) = 5(OA) \quad \Rightarrow$

$9(BA) - 9(OA) = 5(OA) \quad \Rightarrow \quad 9(BA) = 14(OA) \quad \Rightarrow \quad BA = \dfrac{14(OA)}{9} = \dfrac{14(18)}{9} = 28$

Area of the shaded region = area of larger circle − area of smaller circle

$= \pi\left(\dfrac{BA}{2}\right)^2 - \pi\left(\dfrac{OA}{2}\right)^2$

$= \pi\left(\dfrac{28}{2}\right)^2 - \pi\left(\dfrac{18}{2}\right)^2 = \pi(14)^2 - \pi(9)^2 = 196\pi - 81\pi = 115\pi$

4.097 **What are the coordinates of the point of intersection of the lines $y = 3x - 1$ and $y = -2x + 9$?**

Solve the equations simultaneously to find the x and y values, i.e. the x and y coordinates of their point of intersection.   First, equate the right hand sides of the two equations:

$3x - 1 = -2x + 9$ $\Rightarrow$ $5x = 10$ $\Rightarrow$ $x = 2$

Substitute this value of x in the first equation:

$y = 3(2) - 1 = 6 - 1 = 5$

The coordinates of the point of intersection are $(2, 5)$.

4.098 **Write the equation of the line passing through $(-10, -6)$ and $(4, 1)$ in the slope-intercept form.**

Simply use the two-point "poetry" form of a linear equation. Then solve it for y.

$$\frac{y - y_1}{y_1 - y_2} = \frac{x - x_1}{x_1 - x_2} \quad \Rightarrow \quad \frac{y - (-6)}{-6 - 1} = \frac{x - (-10)}{-10 - 4} \quad \Rightarrow \quad \frac{y + 6}{-7} = \frac{x + 10}{-14} \quad \Rightarrow$$

$$y + 6 = \frac{x + 10}{2} \quad \Rightarrow \quad y + 6 = \frac{1}{2}x + 5 \quad \Rightarrow \quad y = \frac{1}{2}x - 1$$

PRACTICE EXAMPLES

4.099 If $(-4, -1)$ is the center of the circle, and $(1, 2)$ is one endpoint of its diameter, then what is the other endpoint of the diameter?

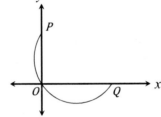

4.100 In the figure on the right, arc POQ is a part of a circle. If $OP = 9$ and $OQ = 14.6$, then what are the coordinates of the center of the circle?

4.101

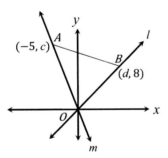

In the figure on the left:

Slope of $\overleftrightarrow{l} = \frac{4}{3}$

Slope of $\overleftrightarrow{m} = -2$

Write the slope-intercept equation of $\overline{AB}$.

4.102 Using the diagram on the right, write the slope-intercept equation of $\overleftrightarrow{l}$.

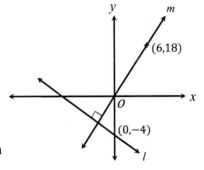

4.103 A circle has its center at $(-3, 7)$ and radius 5. Which of the following is (are) true?

(i) $(-6, 3)$ lies on the circle.

(ii) $(1, 6)$ lies outside the circle.

(iii) The circle passes through the fourth quadrant.

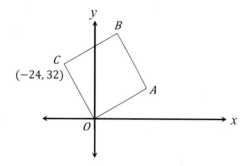

4.104 In the figure on the left, *OABC* is a square. What is the length of diagonal *AC*?

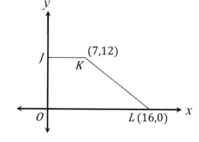

4.105 In the figure on the right, what is the perimeter of trapezoid *OJKL*?

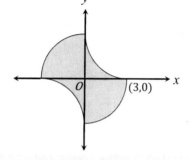

4.106 In the figure on the left, if each of the four arcs bounding the shaded region is a quarter circle, then what is the area of the shaded region?

CHAPTER 5

ODD TYPES OF QUESTIONS

On the GRE/GMAT, some questions don't quite seem to 'fit the mold;' they are atypical in nature designed to test your general understanding of mathematics as a subject. Although such questions may not be as frequent as the regular types we have studied thus far, it helps to know them well. Let's discuss each type, one by one.

MISSING DIGITS

Such questions involve a correctly worked out addition, subtraction, or multiplication, but a few digits may be missing. Their replacements will be letters. You will be expected to figure out the values of those letters. You can always guess and come up with the right answers, but it might take you forever to make guess after guess. The questions are set up to test your knowledge of some basic rules of arithmetic. Let's understand how it all works with some examples.

5.001 **Find *A*, *K* and *C* in the addition shown below.**

$$\begin{array}{ccccc} & 3 & K & A & 8 \\ + & A & C & C & 1 \\ \hline 1 & C & K & A & K \\ \hline \end{array}$$

First, find out what you can by simply performing the basic operation (addition). For example, in the bottom row, the rightmost K is $8 + 1 = \mathbf{9}$. Fill in 9 for all the *K*s, and rewrite.

$$\begin{array}{ccccc} & 3 & 9 & A & 8 \\ + & A & C & C & 1 \\ \hline 1 & C & 9 & A & 9 \\ \hline \end{array}$$

Once again, perform the addition. The highlighted column clearly shows that C must be **0**, i.e. $9 + 0 = 9$. Please don't think that C could also be 10, 20, 30,...... which would still give 9 for that digit. No! That's because C is a *single digit*. So it can't be 10 or 20 or 30...... Replace all *C*s with 0, and rewrite.

$$\begin{array}{ccccc} & 3 & 9 & A & 8 \\ + & A & 0 & 0 & 1 \\ \hline 1 & 0 & 9 & A & 9 \\ \hline \end{array}$$

The highlighted cells tell us that $3 + A = 10$, i.e. $A = \mathbf{7}$.

5.002 In the correctly worked out addition below, what could be the minimum value of $P(\neq Q)$?

$$
\begin{array}{r r}
P & 3 \\
Q & 5 \\
P & 1 \\
+ \quad Q & 8 \\
\hline
2 \quad 5 & 7
\end{array}
$$

Notice that the arrangement of P and Q doesn't matter. Both Ps could be above and both Qs could be below, or it could be the opposite, or they could be dispersed in any way. Regardless, neither P nor Q can be zero, because the number in each row is a *2-digit number*.

Now, look for any carry-overs that may exist. For example, you can readily add up all the digits on the right: $3 + 5 + 1 + 8 = 17$.   7 is written in the final row, which means that 1 is the carried-over digit above P.

Therefore, $P + Q + P + Q + 1 = 25$ $\Rightarrow$ $2(P + Q) = 24$ $\Rightarrow$ $P + Q = 12$

To find the minimum possible value of P, we would need to consider the maximum possible value of Q. Since Q is a *single digit*, its value could be 9 at the most.

Therefore, $P + 9 = 12$ $\Rightarrow$ $P = 3$

5.003 Find A, B and C in the multiplication shown below.

$$
\begin{array}{r r r r r}
 & 3 & 6 & A & \\
\times & & & B & C \\
\hline
 & B & 3 & 0 & \\
C & A & A & A & + \\
\hline
C & 6 & C & 8 & 0
\end{array}
$$

Perform basic addition. Watch out for carry-overs. Look at the 8 in the bottom row:

$3 + A = 8$ $\Rightarrow$ $A = 5$ We can be sure of that because there's no carry-over digit above 3. Nor can $3 + A$ equal 18, 28, 38...... because A is a *single digit*, it can't be 15, 25, 35...... Replace all As with 5, and rewrite.

$$
\begin{array}{r r r r r}
 & 3 & \boxed{6} & 5 & \\
\times & & & B & C \\
\hline
 & B & \boxed{3} & \boxed{0} & \\
C & 5 & 5 & 5 & + \\
\hline
C & 6 & C & 8 & 0
\end{array}
$$

Now look at the highlighted digits. C multiplied by 5 should give 0, 10, 20, 30...... Meaning, C could be 0, 2, 4, 6, or 8.

Next, look at the boxed digits.   If $C = 0$, then $0 \times 6 = 0$.   If $C = 2$, then $2 \times 6 = 12$, plus the carried-over 1 from 2×5, giving $12 + 1 = 13$. That brings down the $\boxed{3}$. Try $C = 4, 6$, or 8; the numbers won't match up. Hence, $C = 2$. Replace all the Cs with 2, and rewrite.

$$
\begin{array}{r r r r r}
 & 3 & 6 & 5 & \\
\times & & & B & 2 \\
\hline
 & B & 3 & 0 & \\
2 & 5 & 5 & 5 & + \\
\hline
2 & 6 & 2 & 8 & 0
\end{array}
$$

Notice the carried-over 1 above the highlighted digits. That's because only $1 + 5 = 6$.

The very next column to the *right* of the highlighted digits means: $B + 5 = 12$, i.e. $B = 7$.

PRACTICE EXAMPLES

5.004 Find A, B, and C in the addition scheme below.

$$
\begin{array}{ccccc}
 & 2 & B & 4 & A \\
 & A & 9 & B & C \\
+ & 5 & 3 & 6 & 8 \\
\hline
A & C & C & B & 9 \\
\end{array}
$$

5.005 In the correctly worked out addition below, find the maximum possible value of $Z (\neq X \neq Y)$.

$$
\begin{array}{cccc}
 & X & 2 & 4 \\
 & Y & 9 & 7 \\
+ & Z & 5 & 5 \\
\hline
1 & 0 & 7 & 6 \\
\end{array}
$$

5.006 In the multiplication below, find A, B and C.

$$
\begin{array}{cccccc}
 & & 1 & A & 3 & \\
 & \times & & 9 & A & \\
\hline
 & & 5 & B & C & \\
 & 1 & C & 8 & B & + \\
\hline
 & 1 & 3 & A & A & 2 \\
\end{array}
$$

SPECIAL OPERATORS

We are familiar with the regular arithmetic operators like $+$, $-$, $\times$ and $\div$. But sometimes a question may involve strange looking symbols unfamiliar in the mathematical context. The operation pertaining to such symbols will be defined in terms of the operators we know. Let's see how to handle such questions using the following example.

5.007 $a \odot b = 3a + 2b$ and $a \otimes b = a - 2b$.   Find $(2 \odot 3) \otimes 4$.

The symbols $\odot$ and $\otimes$ are defined in terms of the arithmetic operations we know. When two numbers are $\odot$ed, it means the result is $3 \times$ (1st number) $+ 2 \times$ (2nd number). Likewise, the definition for $\otimes$ is: (1st number) $- 2 \times$ (2nd number). So let's perform $2 \odot 3$ first, because it is in parentheses, and then $\otimes$ that result with 4:

$2 ☺ 3 = 3(2) + 2(3) = 6 + 6 = 12$

$12 ☯ 4 = 12 - 2(4) = 12 - 8 = 4$

Therefore, $(2 ☺ 3) ☯ 4 = 4$

PRACTICE EXAMPLE

5.008 $x▼y = x^2y - 1$ and $x▲y = 2x - 3y$.   Find $\dfrac{5▼1}{10▲5}$.

SPACE FOR CALCULATIONS

SPECIAL DEFINITIONS

A non-mathematical definition can be *made up* using some mathematical operators or properties. The definition does not have to be true in real life. It's a fun way to test your understanding of the math concepts. Here's an example:

5.009   An item has a "fair price" if the price is a prime number which, when divided by 3, leaves a remainder of 2. What is the arithmetic mean of all the "fair prices" from $10 to $20?

What's really being asked is: which prime numbers from 10 to 20 will leave a remainder of 2 when divided by 3? Well, what are the prime numbers between 10 and 20? Just 11, 13, 17 and 19. 11 and 17 are the only ones that leave a remainder of 2 when divided by 3. Their arithmetic mean is: $\dfrac{11+17}{2} = \dfrac{28}{2} = \$14$

PRACTICE EXAMPLE

5.010   A speed is said to be "preferred speed" if it is a multiple of 2, 3 and 4. What is the average speed of all the "preferred speeds" between 0 mph and 50 mph, inclusive?

SPACE FOR CALCULATIONS

SEQUENCES

A sequence is an ordered list of numbers. Each number within a sequence follows a certain set pattern. Examples on sequences can be solved simply by interpreting words. No formulas are needed.

5.011   In a sequence, each term is obtained by multiplying the previous term by 2, and then subtracting 2 from the result. If the 1st term is 8, then what is the units digit of the 43rd term?

From the given formation:

1st term:	8	=	8
2nd term:	2(8) – 2 = 16 – 2	=	14
3rd term:	2(14) – 2 = 28 – 2	=	26
4th term:	2(26) – 2 = 52 – 2	=	50
5th term:	2(50) – 2 = 100 – 2	=	98
6th term:	2(98) – 2 = 196 – 2	=	194

Notice the pattern in the units digit: 8, 4, 6, 0, ...

Every 4th term in the sequence has 0 in its units place. The 40th term will have it too. The 43rd term will have 6 in its units place.

5.012 A sequence starts with the number 13. After that, each odd numbered term is obtained by adding 4 to the previous term, and each even numbered term is obtained by tripling the previous term. What is the sum of the units digits of the first 20 terms?

1st term:	13	**Odd numbered term**
2nd term:	3(13) = 39	**Even numbered term**
3rd term:	39 + 4 = 43	**Odd numbered term**
4th term:	3(43) = 129	**Even numbered term**

Notice the pattern in the units digit: 3, 9, 3, 9, 3, 9, ...

The sum of the units digits of the first 20 terms = $(3 + 9) \times 10 = 12 \times 10 = 120$

5.013 $a, 5, 7, b, c, 35, 67, 131, ...$

In the above sequence, each term is 3 less than twice the previous term. What is the average of a, b and c?

Consider the first two terms:	$a, 5$
Then, 5 is 3 less than twice a:	$5 = 2a - 3$ $\Rightarrow$ $8 = 2a$ $\Rightarrow$ $a = 4$
Consider the next two terms:	$7, b$
Then, b is 3 less than twice 7:	$b = 2(7) - 3 = 14 - 3 = 11$
Consider the next two terms:	$c, 35$
Then, 35 is 3 less than twice c:	$35 = 2c - 3$ $\Rightarrow$ $38 = 2c$ $\Rightarrow$ $c = 19$

Average of a, b and $c = \dfrac{a + b + c}{3} = \dfrac{4 + 11 + 19}{3} = \dfrac{34}{3} = 11.\overline{3}$

PRACTICE EXAMPLES

5.014 In a sequence, each term is obtained by multiplying the previous term by 3, and then adding 1 to the result. If the 1st term is 1, then what is the arithmetic mean of the units digits of the first 100 terms?

5.015 A sequence starts with the number 3. After that, each odd numbered term is obtained by taking the absolute value of the previous term, and each even numbered term is obtained by subtracting 7 from the previous term. What is the 30th term of the sequence?

5.016 $2, 0, -8, -40, -168, ...$

In the above sequence, each term is q more than p times the previous term. What is $2p + q$?

FUNCTIONS

A function shows the relationship between two variables, one variable being dependent, whereas the other being independent. Let's say I go out to buy a few loaves of bread. The amount of money (M) I spend will depend upon the number of loaves (L) I purchase. Greater the number of loaves, more will be the money I spend.

Mathematically, $M = f(L)$, pronounced "M equals f of L," meaning **M is a function of L.** The letter f stands for function; f is not multiplied by L. Greater the value of L, greater will M turn out to be. M depends on L, i.e. M is the dependent variable, and L is the independent variable. Similarly, any other letter (g, h, j etc.) can also be used to denote a function.

Let's say a loaf of bread costs $2. Then, $M = 2 \times L$. Now, the function $f(L)$ is defined as $2L$.

$M = f(L) = 2L$

So, if you're asked to evaluate $f(7)$, substitute 7 for L in $f(L)$:

$M = f(7) = 2 \times 7 = 14$

5.017 $f(x) = 2x + 7$ and $g(x) = x - 5$. Find $\dfrac{f(g(2))}{g(f(2))}$.

When evaluating a function of a function, e.g. $f(g(x))$, *first evaluate the innermost function. Then move outwards.* Using this method, let's find $f(g(2))$ and $g(f(2))$ separately, and then divide the first one by the second.

$g(2) = 2 - 5 = -3$ $\qquad f(g(2)) = f(-3) = 2(-3) + 7 = -6 + 7 = 1$

$f(2) = 2(2) + 7 = 4 + 7 = 11$ $\qquad g(f(2)) = g(11) = 11 - 5 = 6$

Therefore, $\dfrac{f(g(2))}{g(f(2))} = \dfrac{1}{6}$

PRACTICE EXAMPLE

5.018 $f(x) = x^2 + 1$, $g(x) = x^2 - 6$, and $h(x) = 2x - 1$. Find $f(g(h(-1))) \cdot |g(h(0))|$

GRAPHED FUNCTIONS

This is the graphical version of the functions we just studied in the last section. You will be given an X-Y plot showing one or more functions. You will be expected to pick the value of the function at a certain value of x. That value of the function is simply the y-value on the curve (or line) corresponding to the x-value. Once again, if there are several functions within one another, always start with the innermost function and proceed outward.

5.019 **Using the graph on the right, find the value of $f(g(2))$.**

Let's first evaluate $g(2)$, meaning for the value of $x = 2$, find $g(x)$. From the $g(x)$ curve, at $x = 2$, $\ \ y = g(2) = 3$.

Next, $f(g(2)) = f(3)$. From the slanted line, at $x = 3$, $\ \ y = f(3) = 2$

Therefore, $f(g(2)) = 2$

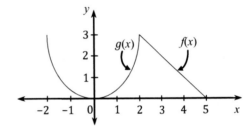

Note: Figure drawn to scale

PRACTICE EXAMPLE

5.020

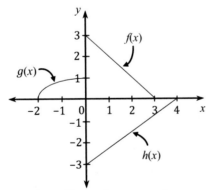

Note: Figure drawn to scale

Using the plot on the left, find the value of $f(g(h(4)))$.

CHAPTER 6

VISUAL DATA ANALYSIS

Visual data analysis, also known as data interpretation, involves analyzing visual information in the form of tables, charts, or graphs, and answering questions on it. Typically, on the GRE/GMAT, there may be up to three questions asked on the same set of data. Questions on data interpretation also test your knowledge of basic arithmetic: fractions, decimals, percentages, ratios and proportions etc.

There are several commonly used forms of visually representing information:

Line graphs involve one or more lines plotted on the same plane. Accordingly, the same vertical scale may be used for all lines, or two different vertical scales may be used on either side. If two scales are used, then it will clearly be mentioned which scale applies to which line.

Bar charts involve several vertical (or horizontal) bars placed parallel to one another along either the vertical axis or the horizontal axis. Bar charts are often used to compare the growth or decline of one or more quantities simultaneously. Bar charts can also be stacked type in which, two or more bars are placed on top of one another.

Pie charts (also known as **circle graphs**) are circles best used when expressing percentage break-up of a quantity. A pie chart is a circle divided into sectors, each sector proportional to the percentage of the quantity it represents.

Finally, **tables** may be used either individually or several at a time. Tables are used when actual numbers are to be presented for calculations.

The above four forms of data are frequently used in conjunction with one another on the GRE/GMAT.

While answering questions on visual data analysis, there are some things you should remember:

> ➤ Do not assume that a figure is drawn to scale unless it is specifically mentioned as such.

> ➤ Approximate numbers whenever you can by using simple rounding off techniques.

> ➤ When figures are *drawn to scale*, look for visual patterns. If a portion of a pie looks bigger, then it is bigger. If a bar looks taller, then it does represent a greater quantity. In most cases, you will not have to perform tedious calculations.

These points can best be understood with examples starting on the next page.

Questions 6.001 – 6.005 are based on the following tables:

SCHEDULE OF FIXED HEALTH INSURANCE PREMIUMS:
WYSIWYG INSURANCE COMPANY

	Ages	Monthly ($)	Annual ($)
Individual	18 – 50	80	960
	51 – 64	120	1,440
Single parent	18 – 50	100	1,200
	51 – 64	140	1,680
Two parents and children, if any	18 – 50	120	1,440
	51 – 64	150	1,800

Additionally available riders* (ages 18-64)

Type of Rider	Benefit	Monthly Premium ($)
A	$1,000 upon first time hospitalization	5
B	$2,000 upon first time hospitalization	10
C	$1,000 per subsequent hospitalization	15

* A *rider* is a feature added to an insurance policy at an additional cost.

6.001 For a 40 year old individual, the annual premium for a health insurance policy with riders A and C is

(A) $960
(B) $980
(C) $1,200
(D) $1,220
(E) $1,440

Look up the 18 – 50 age bracket for an *individual* from the table on the left. Be careful to select not the monthly premium, but the *annual* premium. That's $960. From the table on the right, look up the total premium for riders A and C. That's $5 + $15 = $20 *per month*, i.e. its annual equivalent would be $20 × 12 = $240. Finally, the total annual premium for the policy and the riders A and C would be $960 + $240 = $1,200. The answer is (C).

6.002 Nikolai bought health insurance on his 41st birthday. He did not purchase any riders with his policy. He never got married, and had no children. If he kept his policy unchanged until it discontinued on his 65th birthday, then how much did he pay for his health insurance over all those years?

(A) $23,040
(B) $29,760
(C) $30,240
(D) $35,520
(E) $36,000

Nikolai's annual premium changed from $960 (18 – 50 age bracket) to $1,440 (51 – 64 age bracket) on his 51ˢᵗ birthday. From his 41ˢᵗ birthday to his 51ˢᵗ birthday, i.e. for 10 years, he paid $960 × 10 = $9,600. From his 51ˢᵗ birthday to his 65ᵗʰ birthday, i.e. for 14 years, he paid $1,440 × 14 = $20,160. Thus, he paid a total of $9,600 + $20,160 = $29,760 in insurance premiums over all those years. The answer is (B).

6.003 At 27, Martha decides to marry Stewart. They both have health insurance policies. How much will they save annually if they get a joint insurance plan?

(A) $0
(B) $480
(C) $960
(D) $1,240
(E) It cannot be determined from the information given

Stewart's age is not known. Nor is it known whether they both have any riders with their policies. It is also not known whether either of them has any children, so we don't know whether to place them initially in the individual or single parent category. Therefore, this question cannot be answered with the available information. The answer is (E).

PRACTICE QUESTIONS

6.004 If Judy was hospitalized three times in her life, then how much more money could she have received in benefits had she bought riders B and C along with her insurance policy?

(A) $2,000
(B) $3,000
(C) $4,000
(D) $5,000
(E) $6,000

6.005 At 28, Stacy was unmarried, and had no children. She had health insurance with no riders. At 30, she married Donald, 31. If they added riders A and C to their newly joint insurance policy, and equally split the premium among themselves, then Stacy's new monthly premium

(A) Decreased by $30
(B) Remained unchanged
(C) Increased by $60
(D) Decreased by $10
(E) Increased by $20

Questions 6.006 – 6.010 are based on the following pie charts:

2009 SALES REVENUES:
ORION MOTOR COMPANY (OMC)

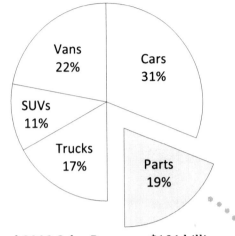

Total 2009 Sales Revenue: $101 billion

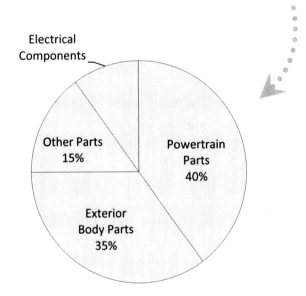

6.006 In 2009, the ratio of OMC's revenue from the sales of electrical components to that of its exterior body parts was

(A) 1 : 5

(B) 2 : 7

(C) 1 : 2

(D) 3 : 7

(E) None of the above

The second pie chart shows the percentage break-up of the revenue from parts sales. Directly use percentages; do not waste your time calculating the actual revenues. In percentage terms, revenue from the sales of electrical components would be $100\% - (40\% + 35\% + 15\%) = 10\%$. For exterior body parts, it is 35%.

$$\text{Electrical} : \text{Exterior} = \frac{\text{Electrical}}{\text{Exterior}} = \frac{10}{35} = \frac{2}{7} = 2 : 7$$

The answer is (B).

6.007 In 2009, OMC's revenue from car sales exceeded its revenue from SUV sales by

(A) $20.0 billion
(B) $20.1 billion
(C) $20.2 billion
(D) $20.3 billion
(E) $20.4 billion

Look up the first pie chart. Car sales: 31%, SUV sales: 11%. There's no need to calculate each revenue out of $101 billion, and then subtract one from the other. Instead, directly take the difference of the two percentages: $31\% - 11\% = 20\%$.

$$20\% \text{ of } \$101 \text{ billion} = \frac{20}{100} \times 101 = \frac{2}{10} \times 101 = \frac{202}{10} = 20.2 = \$20.2 \text{ billion}$$

To save time, I have conveniently avoided writing *billion* too many times, but I hope you get the point: Simply find 20% of 101. Those many billions. The answer is (C).

6.008 If cars built by OMC were priced from $10,000 to $30,000 a piece, then which of the following could <u>not</u> be the number of cars sold by OMC in 2009?

(A) 1 million
(B) 1½ million
(C) 2 million
(D) 2½ million
(E) 3 million

Looking at the answer choices hints that we are looking at a range of values, i.e. the maximum and minimum number of cars sold.

The revenue from car sales was 31% of $101 billion. If you want, you can calculate its actual value at $31.31 billion. But it's not necessary; just recognize that it was more than $30 billion.

If all the cars sold were the cheapest ones at $10,000 each, then the greatest number of cars sold would be

$$\frac{\text{revenue from car sales}}{\text{price of each car}} = \frac{(> \$30,000,000,000)}{\$10,000} = (> 3 \text{ million}) \text{ cars}$$

If all the cars sold were the most expensive ones at $30,000 each, then the least number of cars sold would be

$$\frac{\text{revenue from car sales}}{\text{price of each car}} = \frac{(> \$30,000,000,000)}{\$30,000} = (> 1 \text{ million}) \text{ cars}$$

In either case, *more than* 1 million cars were sold. Answer choice (A) is the only number of cars that could not have been sold. The answer is (A).

PRACTICE QUESTIONS

6.009 In 2009, if 25% of OMC's revenue from the sales of powertrain parts came from the sales of engines, then what was its approximate revenue from engine sales?

    (A) $2 billion
    (B) $5 billion
    (C) $8 billion
    (D) $25 billion
    (E) $40 billion

6.010 Which of the following can be inferred from the pie charts?

    (i)   In 2009, OMC sold about half as many trucks as it sold vans and SUVs combined.
    (ii) In 2009, at the most 15% of OMC's revenue from the sales of parts came from the sales of interior body parts.
    (iii) In 2009, OMC's combined revenue from the sales of powertrain parts and vans exceeded its revenue from the sales of trucks and SUVs.

    (A) (i) only
    (B) (ii) only
    (C) (iii) only
    (D) (i) and (ii) only
    (E) (ii) and (iii) only

SPACE FOR CALCULATIONS

Questions 6.011 – 6.015 are based on the graph from the next page:

6.011 In which year was the GDP equally split between the manufacturing and service sectors?

    (A) 1985
    (B) 1990
    (C) 1995
    (D) 2000
    (E) None of the above

Be very alert in answering this question. While considering the answer choices one by one, be sure to notice the two different vertical scales on the graph. For the year 1985, the GDP from manufacturing sector (dotted line) was valued at $30 billion (on the left hand scale), while that from the service sector (solid line) also was valued at $30 billion (on the right hand scale). The answer is (A). Similarly, look up the other answer choices; none of them will have the same value for both the sectors.

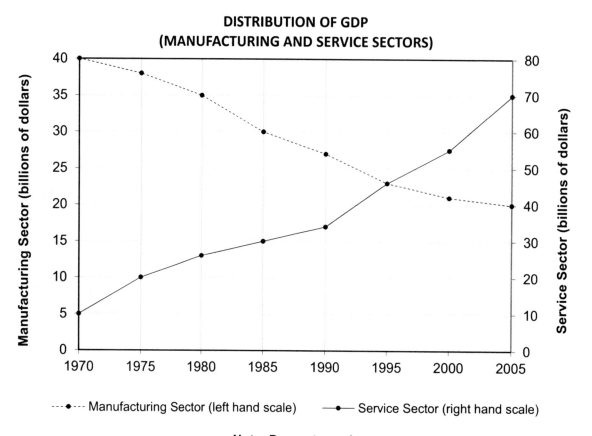

DISTRIBUTION OF GDP
(MANUFACTURING AND SERVICE SECTORS)

--•-- Manufacturing Sector (left hand scale)   —•— Service Sector (right hand scale)

Note: Drawn to scale.

6.012 In the year 1985, 25% of the GDP came from automobile manufacturing. What percent of the manufacturing sector's contribution to GDP came from automobile manufacturing, in 1985?

(A) 20%

(B) 25%

(C) $33\frac{1}{3}\%$

(D) 50%

(E) $66\frac{2}{3}\%$

In 1985, the total GDP was 30 + 30 = $60 billion. 25% of that was from automobile manufacturing, i.e. 25% of 60 = ¼ of 60 = $15 billion.

$$\frac{\text{GDP from automobile mfg.}}{\text{GDP from mfg.}} \times 100\% = \frac{\$15 \text{ billion}}{\$30 \text{ billion}} \times 100\% = \frac{1}{2} \times 100\% = 50\%$$

The answer is (D).

6.013 Which of the following can be inferred from the graph?
- (i)   In 1980, the contribution by manufacturing sector to GDP was greater than the contribution by service sector to GDP by over $10 billion.
- (ii) In 1975, the GDP was more than $60 million.
- (iii) In the 1990s, in terms of GDP, the service sector overtook the manufacturing sector.

- (A) (i) only
- (B) (ii) only
- (C) (iii) only
- (D) (ii) and (iii) only
- (E) (i), (ii) and (iii)

- (i)   In 1980, manufacturing sector contributed $35 billion (left hand scale) to the GDP. Service sector contributed slightly more than $25 billion (right hand scale). The graph is drawn to scale, so you should believe it the way it looks for service sector. The difference between the two contributions is not more than, but *less than* $10 billion. This statement is therefore not true.
- (ii) In 1975, manufacturing sector contributed a bit less than $40 billion to the GDP, whereas service sector contributed $20 billion to the GDP. The result is a bit less than $60 billion. But it is still greater than $60 *million* ! This statement is about million, not billion. It is therefore true.
- (iii) In 1985 itself, the manufacturing and service sectors were each at $30 billion. After that, the manufacturing sector continued to decline, whereas the service sector continued to grow. Meaning, the service sector overtook the manufacturing sector sometime before 1990, not *in the 1990s*. This statement is therefore not true.

The answer is (B).

PRACTICE QUESTIONS

6.014 Which of the following 5-year periods saw the greatest percentage increase in the contribution of the service sector to the GDP?
- (A) 1970 – 1975
- (B) 1975 – 1980
- (C) 1990 – 1995
- (D) 1995 – 2000
- (E) 2000 – 2005

6.015 For the 5-year period from 1975 to 1980, what was the approximate ratio of the growth (in dollars) of the service sector to the decline (in dollars) of the manufacturing sector?

(A) 1 : 1
(B) 5 : 8
(C) 1 : 2
(D) 7 : 15
(E) 2 : 1

SPACE FOR
CALCULATIONS

Questions 6.016 – 6.020 are based on the following graph:

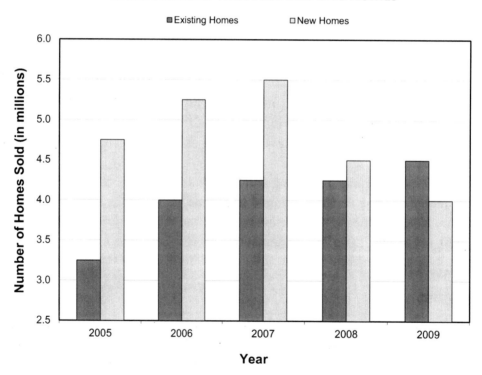

SALES FIGURES: NEW AND EXISTING HOMES

Note: Drawn to scale.

6.016 Which of the following years had the least total number of homes sold?

(A) 2005
(B) 2006
(C) 2007
(D) 2008
(E) 2009

Imagine the excess portion of each taller column cut in half and placed on top of the corresponding shorter column. The two columns would then even out in each case. Those are the pairs we would add up and compare. Or we could simply compare their leveled out levels. Which one do you think would be the shortest?

Think logically: The levels for 2006 and 2007 would be considerably above the 4.0 line. The levels for 2008 and 2009 would be at least half way between the 4.0 and 4.5 lines. Only the level for 2005 will be very close to the 4.0 line, and will therefore be the least. This is the beauty of graphs drawn to scale! The answer is (A). You can always do the actual additions and compare the results, but it might take you much longer.

6.017 Which of the following years had the highest ratio of the number of new homes sold to the number of existing homes sold?

(A) 2005
(B) 2006
(C) 2007
(D) 2008
(E) 2009

2008 has almost the same new and existing home sales, and is therefore out of the question. 2009 had greater existing homes sold than new homes sold. Between 2005, 2006 and 2007 it's clear that 2005 would have the greatest numerator *as compared with the denominator*, and has therefore the highest ratio. The answer is (A).

6.018 If 50% of all the new homes sold in 2007 were brick houses, then approximately what was the difference between the number of existing homes sold and the number of new brick houses sold, in 2007?

(A) 150,000
(B) 250,000
(C) 1,000,000
(D) 1,250,000
(E) 1,500,000

The number of existing homes sold in 2007 was approximately 4.25 million. (It's almost half-way between the 4.0 and 4.5 lines.) The number of new homes sold was 5.5 million, 50% of which would be $\frac{1}{2} \times 5.5$ million = 2.75 million. That was the number of brick houses sold in 2007. The difference between the two numbers would be approximately 4.25 – 2.75 = 1.5 million. The answer is (E).

PRACTICE QUESTIONS

6.019 If there has always been a 5% sales tax on every new home sold, and a 2% sales tax on every existing home sold, then approximately what was the total amount of taxes paid on the sales of new and existing homes in 2009?

(A) $290,000

(B) $305,000

(C) $2,900,000

(D) $3,050,000

(E) It cannot be determined from the information given.

6.020 In the graph, if the percent change in the number of existing homes sold from year to year is equal to the percent change in the value of each existing home from year to year, then what would be the value of my home in 2009 if I bought it as an existing home for $100,0000 back in 2006?

(A) $104,500

(B) $112,000

(C) $145,000

(D) $112,500

(E) It cannot be determined from the information given.

SPACE FOR CALCULATIONS

Questions 6.021 – 6.025 are based on the graphs from the next page:

6.021 In which of the following years were the oil exports of Zigzagistan the highest?

(A) 2003

(B) 2004

(C) 2005

(D) 2006

(E) 2007

From the bar graph, the choice is between 2004 and 2006. Actual calculations are not necessary. The light gray bar (exports) spans exactly 40 million barrels for 2006. For 2004, it spans just a little more than 40 million barrels. The answer is (B).

6.022 In 2006, what was the approximate oil export to the U.S., as a percentage of the exports to India and China combined?

(A) 14 million barrels

(B) 75%

(C) 40%

(D) 83%

(E) 35%

OIL PRODUCTION, REPUBLIC OF ZIGZAGISTAN
(millions of barrels)

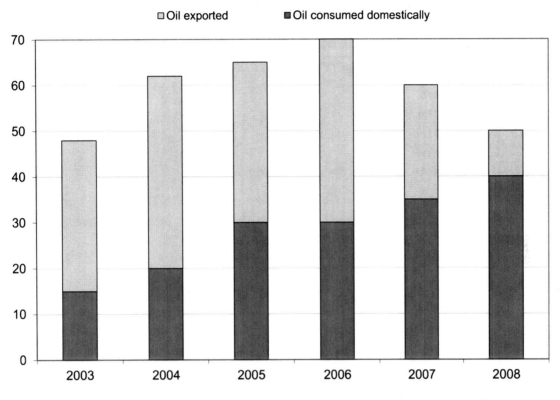

Note: Drawn to scale.

OIL EXPORTED, 2006

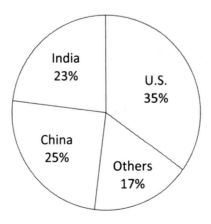

There's no need to actually calculate the oil exported to the U.S., India and China. From the pie chart, directly take the percentage values.

$$\frac{\text{Oil exported to the U.S.}}{\text{Oil exported to India and China}} \times 100\% = \frac{35\%}{23\% + 25\%} \times 100 = \frac{35}{48} \times 100 \approx \frac{3}{4} \times 100 \approx 75\%$$

I reduced 35 and 48 by 12, i.e. 3 times (approximately) and 4 times. The answer is (B).

6.023 In 2005, the GDP of Zigzagistan was $6 billion, 20% of which came from its oil exports. What was the approximate average price of exported oil, in 2005?

(A) $25 per barrel
(B) $35 per barrel
(C) $40 per barrel
(D) $50 per barrel
(E) It cannot be determined from the information given.

20% of $6 billion is $1.2 billion, i.e. revenue from oil exports. From the bar graph, the amount of oil exported in 2005 was 35 million barrels. The average price of exported oil would be:

$$\frac{\text{Revenue from oil exports}}{\text{Amount of oil exported}} = \frac{\$1,200,000,000}{35,000,000 \text{ barrels}} = \frac{12 \times 100}{35} \approx \frac{100}{(< 3)} \approx (> \$33.\overline{3}/\text{barrel})$$

In the above steps, 12 reduces 35 slightly less than 3 times. Therefore, the denominator below 100 would be slightly less than 3. Therefore, the value of the fraction would be slightly greater than $33.\overline{3}$. The closest answer choice is $35/barrel. The answer is (B).

PRACTICE QUESTIONS

6.024 In which of the following years did Zigzagistan experience the greatest percentage increase in its domestic oil consumption over the previous year?

(A) 2004
(B) 2005
(C) 2006
(D) 2007
(E) 2008

6.025 In 2008 and 2009, Zigzagistan experienced the same percentage drop in its oil production over the previous year. Also, in 2008 and 2009, its domestic oil consumption increased by the same amount over the previous year. Which of the following must be true?

(i) In 2009, Zigzagistan did not produce enough oil to meet its domestic demand.
(ii) In 2009, the domestic oil demand in Zigzagistan exceeded 45 million barrels.
(iii) In 2009, Zigzagistan experienced a 60% drop in its oil exports.

(A) (i) only
(B) (ii) only
(C) (iii) only
(D) (i) and (iii) only
(E) (i), (ii) and (iii)

Questions 6.026 – 6.030 are based on the graph and the table from the next page:

6.026 Which of the following highways cost the most to construct?

(A) H-1
(B) H-4
(C) H-5
(D) H-6
(E) H-7

For each highway, multiply the total cost per mile (from the table) with the approximate number of miles (from the graph).

H-1 cost a bit more than $3 \times 400 = \$1200$ million

H-4 cost about $5 \times 250 = \$1250$ million

H-5 cost $3 \times 400 = \$1200$ million

H-6 cost $8 \times 200 = \$1600$ million

H-7 cost a bit less than $3 \times 350 = \$1050$ million

The answer is (D). Don't waste time converting millions to billions. Just make comparisons.

6.027 What is the approximate ratio of the total cost of construction of highway H-2 to that of highway H-8?

(A) $2:1$
(B) $5:6$
(C) $5:3$
(D) $2:5$
(E) $3:2$

Cost of construction of highway H-2 $= 2.5 \times 200 = \$500$ million

Cost of construction of highway H-8 $\approx 3 \times 100 \approx \$300$ million

Ratio of cost of construction of H-2 to that of H-8 is $500:300$, or simply $5:3$.

The answer is (C).

LENGTHS OF HIGHWAYS (MILES)

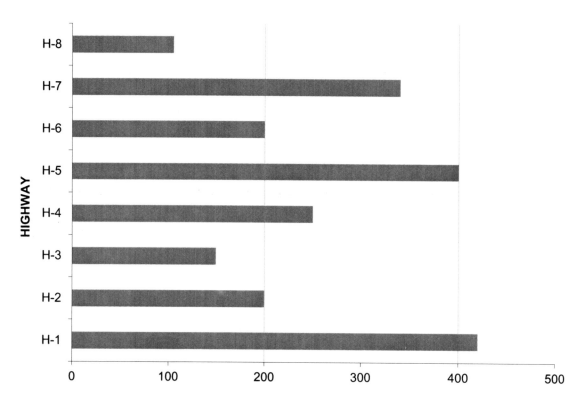

Note: Drawn to scale.

	COST OF CONSTRUCTION PER MILE ($, MILLIONS)			
Highway	Materials	Labor	Overheads	Total
H-1	1.6	1.0	0.4	3.0
H-2	1.4	0.7	0.4	2.5
H-3	0.8	0.5	0.2	1.5
H-4	2.3	1.7	1.0	5.0
H-5	2.0	0.5	0.5	3.0
H-6	3.8	2.0	2.2	8.0
H-7	2.1	0.5	0.4	3.0
H-8	2.1	0.5	0.4	3.0

6.028 Which of the following can be inferred from the given information?

(i) The total cost of materials for the construction of highway H-2 was a lot more than $28 million.

(ii) Just under 13% of the total cost of construction of highway H-3 was attributed to the overheads.

(iii) The total costs of construction of highways H-4 and H-5 were approximately the same.

(A) (i) only
(B) (ii) only
(C) (iii) only
(D) (i) and (iii) only
(E) (i), (ii) and (iii)

(i) For highway H-2, look for the cost of *materials* per mile. Do not accidentally consider the *total cost* per mile. Total cost of materials for the construction of H-2 $= 1.4 \times 200 = \$280$ million. Therefore, this statement is true.

(ii) If overheads comprise a certain percent of the total cost *per mile*, then that percentage will remain the same regardless of the number of miles constructed. So, consider only the per mile figures for highway H-3:

$$\frac{\text{Overheads per mile}}{\text{Total cost per mile}} \times 100 = \frac{0.2}{1.5} \times 100 = \frac{20}{1.5} = \frac{200}{15} = \frac{40}{3} = 13.\overline{3}\% = \text{just } over \text{ } 13\%$$

Therefore, this statement is not true.

(iii) Cost of construction of highway H-4 $\approx 5 \times 250 = \$1250$ million
Cost of construction of highway H-5 $= 3 \times 400 = \$1200$ million
They are approximately the same. This statement is therefore true.

The answer is (D).

PRACTICE QUESTIONS

6.029 Which of the following highways had the cheapest labor per mile of construction, as a fraction of the total cost per mile of construction?

(A) H-1
(B) H-2
(C) H-4
(D) H-6
(E) H-8

6.030 Half the length of highway H-5 is due for repairs. If the cost of materials for repairs is 50% less than the cost of materials for construction, then what would be the total cost to repair the due portion of highway H-5?

(A) $400 million
(B) $600 million
(C) $800 million
(D) $1.0 billion
(E) $1.2 billion

CHAPTER 7

QUANTITATIVE COMPARISON

Quantitative Comparison is specific to the GRE, not to the GMAT. However, I recommend studying this chapter even if you're taking the GMAT because these questions involve mathematical reasoning, not just numerical computations. If you're taking the GRE, then I advise studying the chapter on Data Sufficiency as well, even though that chapter is specific to the GMAT. Data Sufficiency also requires mathematical reasoning.

Questions on Quantitative Comparison involve two columns: Column A and Column B. Using your quantitative reasoning skills, you are to decide which quantity among the two columns is greater. Your answer would be:

(A) if the quantity in Column A is greater

(B) if the quantity in Column B is greater

(C) if the two quantities are equal

(D) if the relationship cannot be determined from the information given

In order to correctly analyze such questions, it is important to remember certain key points:

➤ Do not assume that numbers are always *integers*. 'Number' is a generic term; it could mean an integer or a fraction (or a decimal, which is essentially a fraction).

➤ Always consider positive, negative, zero and one. For example, $x^2 = 25$ means $x = \pm 5$, not just +5. Similarly, be alert about whether a number could be zero as well, or one as well. The properties of zero and one are quite frequently tested on this section.

➤ Pay particular attention to the > versus the ≥ sign, and the < versus the ≤ sign.

➤ Recognize the difference between parenthesis (brackets) and modulus signs (absolute value signs).

➤ Recognize the difference between positive and non-negative numbers. Recall that *positive numbers* are all numbers greater than zero. Zero itself is not included. On the other hand, *non-negative numbers* include all positive numbers and the number zero as well.

Let's analyze some **commonly tested number properties** before we move on to solving examples:

Statement: $\frac{a}{2}$ is an integer.

What it means: If $\frac{a}{2}$ is an integer, then a must be an even integer. Here's how:

$\frac{a}{2} = $ integer $\Rightarrow a = 2 \times$ integer $=$ even integer, because 2 multiplied by any integer $=$ even integer

Statement: $\frac{a}{2}$ is *not* an integer.

What it means: If $\frac{a}{2}$ is not an integer, then a could be an integer or a fraction. Here's how:

If $a = 1$, then $\frac{a}{2} = \frac{1}{2} \neq$ an integer.   But if $a = \frac{1}{3}$, then too $\frac{a}{2} = \frac{\frac{1}{3}}{2} = \frac{1}{6} \neq$ an integer.

Statement: $\frac{a}{2}$ is an odd integer.

What it means: If $\frac{a}{2}$ is an odd integer, then a must be an even integer. Here's why:

Let $\frac{a}{2} = 3$.   Then, $a = 2 \times 3 = 6$.   Let $\frac{a}{2} = 11$.   Then, $a = 2 \times 11 = 22$.

In general, let $\frac{a}{2} = k$.   Then, $a = 2 \times k$.   2 multiplied by *any integer* will result in an even integer.

Statement: $\frac{a}{2}$ is an even integer.

What it means: If $\frac{a}{2}$ is an even integer, then a must be an even integer. Here's why:

Let $\frac{a}{2} = 4$.   Then, $a = 2 \times 4 = 8$.   Let $\frac{a}{2} = 10$.   Then, $a = 2 \times 10 = 20$.

In general, let $\frac{a}{2} = k$.   Then, $a = 2 \times k$.   2 multiplied by *any integer* will result in an even integer.

Statement: $2a$ is an integer.

What it means: If $2a$ is an integer, then a could be an integer of a fraction. Here's how:

$2a = $ integer $\Rightarrow a = \dfrac{\text{integer}}{2} =$ integer or fraction, depending upon even or odd numerator.

Statement: $2a$ is *not* an integer.

What it means: 2 multiplied by any integer a (odd or even) will always result in an integer. If $2a$ is not an integer, then a must be a fraction.

Statement: $2a$ is an odd integer.

What it means: If $2a$ is an odd integer, then a can only be a fraction. Here's how:

$2a = $ odd integer $\Rightarrow a = \dfrac{\text{odd integer}}{2} =$ fraction   (because odd integers can't be divided by 2)

Statement: $2a$ is an even integer.

What it means: If $2a$ is an even integer, then a has to be an integer (odd or even).

Statement: $\frac{a}{b}$ is an integer.

What it means: If $\frac{a}{b}$ is an integer, then a is a multiple of b. a and b could both be integers, or they could both be fractions, or a could be an integer whereas b could be a fraction. But if a is a fraction, and b is an integer, then the result will not be an integer. Here's how:

Both are integers: $a = 6$, $\quad b = 3$. Then, $\dfrac{a}{b} = \dfrac{6}{3} = 2 =$ an integer.

Both are fractions: $a = \dfrac{1}{2}$, $\quad b = \dfrac{1}{4}$. Then, $\dfrac{a}{b} = \dfrac{\frac{1}{2}}{\frac{1}{4}} = \dfrac{1}{2} \times \dfrac{4}{1} = 2 =$ an integer.

Only a is an integer: $a = 2$, $\quad b = \dfrac{1}{7}$. Then, $\dfrac{a}{b} = \dfrac{2}{\frac{1}{7}} = 2 \times \dfrac{7}{1} = 14 =$ an integer.

Only b is an integer: $a = \dfrac{1}{3}$, $\quad b = 5$. Then, $\dfrac{a}{b} = \dfrac{\frac{1}{3}}{5} = \dfrac{1}{15} \neq$ an integer.

Statement: $x^2 > 0$ (This has been the most easily forgotten property in my teaching experience).

What it means: x^2 is positive. But x could be either positive or negative. Here's how:

Let $x^2 = 36$. Then $x = \sqrt{36} = \pm 6$ because $6 \times 6 = 36$ and $(-6) \times (-6) = 36$ as well.

Statement: $a^b = 1$

What it means: It could mean three cases, considered individually. Here's each of them:

Case 1: We know that $x^0 = 1$, where $x \neq 0$. Meaning, if $a^b = 1$, then a could be any real number other than zero, and b could be zero. Please recall that *real numbers* are all numbers (integers, fractions, decimals, positive and negative).

Case 2: We also know that $1^n = 1$, where n could be any integer. Meaning, if $a^b = 1$, then a could be 1, and b could be any integer.

Case 3: Finally, $(-1)^n = 1$, where n could be any *even integer*. Meaning, if $a^b = 1$, then a could be (-1), and b could be any even integer.

You will be able to answer most Quantitative Comparison questions without actually solving the problem.

7.001	Column A	Column B
	The smallest prime factor of 876	The smallest prime factor of 875

There's no need to prime factorize 875 or 876 all the way. The smallest prime factor of any even number is 2. The smallest prime factor of any odd number ($\neq 1$) is at least 3, because odd numbers are not divisible by 2 at all. Therefore, Column B is greater. The answer is (B).

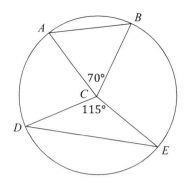

7.002 *C* is the center of the circle.

Column A

Area of Δ*ABC*

Column B

Area of Δ*CDE*

The two triangles are isosceles because two of their sides have the same length (radius). Imagine the two triangles, each resting on one of its radii.

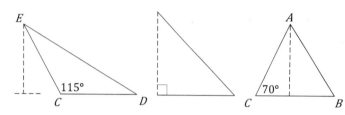

As ∠*C* increases from 70° to 90°, the triangle will get taller with each degree increase. When ∠*C* exceeds 90°, the triangle will get shorter with each degree increase. The triangle will have its greatest height (dotted) when ∠*C* = 90°.

Notice that each of these triangles has the same base (= radius).

$$\text{Area of a triangle} = \frac{\text{base} \times \text{height}}{2}$$

The base is constant; therefore, greater the height, greater will be the area. The closer ∠*C* is to 90°, the taller will the triangle be. 70° is closer to 90° than is 115°. Therefore, Δ*ABC* has a greater area. The answer is (A).

7.003

Column A

The standard deviation of:

161, 163, 164, 165, 167

Column B

The standard deviation of:

54, 58, 64, 69, 74

Standard deviation is the measure of the dispersion of the numbers, given by:

$$\sigma = \sqrt{\frac{\Sigma(\mu - x)^2}{n}}$$

x is each of the numbers deviating from the mean *μ*. Notice that the more spread out the numbers are, the greater will the numerator be, and therefore the greater the standard deviation *σ*. Column B has numbers with a greater spread. The answer is (B).

7.004 In the figure on the right, *b* > *d*.

Column A
a

Column B
c

$a + b = 180°$

$d + c = 180°$

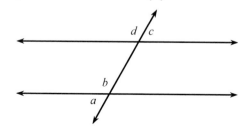

Therefore, $a + b = d + c$

If the two horizontal lines were parallel, then $b = d$ because b and d would be the corresponding angles of two parallel lines cut by a transversal.

But $b > d$. Therefore, to keep the sum ($a + b = d + c = 180°$) constant at $180°$, a would have to be less than c. The answer is (B).

7.005

Column A	Column B
61% of 23	23% of 61

$$61\% \text{ of } 23 = \frac{61}{100} \times 23 = \frac{61 \times 23}{100}$$

$$23\% \text{ of } 61 = \frac{23}{100} \times 61 = \frac{23 \times 61}{100}$$

There's no difference between the two. The answer is (C).

7.006

Column A	Column B
$\dfrac{2^8}{(-2)^8}$	$\dfrac{2^{-8}}{-2^{-8}}$

Please do not perform any calculations. Instead, notice that $(-2)^8$ will be a positive number because of an even exponent for a negative number. Also, -2^{-8} will be a negative number because the negative sign in front of 2 is not within parenthesis. So it is the equivalent of *negative of* 2^{-8}. Meaning, the negative sign will stay; the end result will be negative. **Any negative number is always smaller than any positive number.** The answer is (A).

7.007

Column A	Column B
The product of the slopes of the diagonals of a rhombus.	1

The diagonals of a rhombus are perpendicular to each other. Meaning, the product of their slopes is -1. However, it is not stated whether either of the diagonals is vertical or horizontal. If either of them is horizontal, then its slope is 0, whereas the slope of the other diagonal (vertical) would be infinity (∞). Zero multiplied by infinity is undefined. Therefore, the value of Column A is uncertain. The answer is (D).

7.008   **Ruhi is a high school student. Most students in her class do not like algebra.**

Column A	Column B
The probability that Ruhi likes Algebra.	50%

Most students in her class do not like algebra. It means that more than 50% of students in her class do not like algebra, i.e. less than 50% of the students in her class *like* algebra. If you randomly pick one student from her class (Ruhi, or anyone else), then the probability that that student likes algebra would be less than 50%. The answer is (B).

7.009 $0 < x < 1$

Column A	Column B
$x \cdot \sqrt[3]{x}$	$\dfrac{1}{x}$

If $x < 1$, then $\sqrt{x}, \sqrt[3]{x}, \sqrt[4]{x}, \sqrt[5]{x}$ etc. will each be less than 1. Meaning, x and $\sqrt[3]{x}$ are both less than 1. Their product will certainly be less than 1, i.e. $x \cdot \sqrt[3]{x} < 1$.

On the other hand, $\dfrac{1}{x}$ will be greater than 1, i.e. $\dfrac{1}{x} > 1$. The answer is (B).

7.010   A right circular cone has the same height as that of a pyramid with square base.

<u>Column A</u>   <u>Column B</u>

The volume of the cone   The volume of the pyramid

Volume of a cone $= \dfrac{\pi r^2 h}{3}$,   and   Volume of a pyramid $= \dfrac{abh}{3}$

It's only mentioned that the two heights (h) are equal. There is no information about the radius (r) of the cone, or the side ($a = b$) of the square base of the pyramid. Therefore, the relationship cannot be determined with the information given. The answer is (D).

7.011   <u>Column A</u>   <u>Column B</u>
$$\frac{2}{3} \times \frac{4}{5} \times \frac{6}{7}$$ $$\frac{4}{5} \times \frac{6}{7} \times \frac{8}{9}$$

Notice that $\dfrac{4}{5} \times \dfrac{6}{7}$ appears in both columns. Let's call it x (>0). Now, the comparison is between $\dfrac{2}{3}x$ and $\dfrac{8}{9}x$, or simply $\dfrac{2}{3}$ and $\dfrac{8}{9}$.

Figure on the right shows the technique on comparing fractions from the chapter on arithmetic (page 21). Thus, $\dfrac{8}{9} > \dfrac{2}{3}$. The answer is (B).

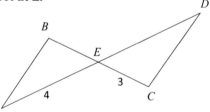

7.012   In the figure on the right, $AB \parallel CD$, and BC and AD intersect at E.

<u>Column A</u>   <u>Column B</u>
Length of side BE Length of side ED

$\angle B = \angle C$ because they are alternate interior angles of parallel line segments cut by a transversal. For the same reason, $\angle A = \angle D$.

$\angle BEA = \angle DEC$ because they are vertically opposite to each other.

Therefore, $\triangle ABE \sim \triangle DCE$. Their sides are in proportion:

$BE : EA = CE : ED$ $\Rightarrow$ $\dfrac{BE}{EA} = \dfrac{CE}{ED}$ $\Rightarrow$ $\dfrac{BE}{4} = \dfrac{3}{ED}$ $\Rightarrow$ $(BE) \cdot (ED) = 12$

The product of lengths of BE and ED is constant at 12. Meaning, the individual lengths could be in any combination. It's not possible to tell which one would be greater. The answer is (D).

7.013 m and n are non-negative numbers, and $(m - n)^2 = m^2 - n^2$.
<u>Column A</u>   <u>Column B</u>

m n

Expand the left side of the given equation and solve:

$$m^2 - 2mn + n^2 = m^2 - n^2 \quad \Rightarrow \quad -2mn + n^2 = -n^2 \quad \Rightarrow \quad 2n^2 - 2mn = 0 \quad \Rightarrow$$

$$2(n^2 - mn) = 0 \quad \Rightarrow \quad n^2 - mn = 0 \quad \Rightarrow \quad n(n - m) = 0 \quad \Rightarrow \quad n = 0 \ \text{ or } \ n = m$$

Meaning, n can either be 0 or m. (Keep in mind that non-negative numbers include 0 as well.) It is not possible to establish which one is greater. Therefore, the answer is (D).

7.014 The population of a country increases by 7% every decade.

Column A

The time it will take for the population of the country to increase from 20 million to 30 million.

Column B

The time it will take for the population of the country to increase from 30 million to 40 million.

This is a case of exponential growth; the growth being slow initially, and speeding up later. It will therefore take more time for the population to grow from 20 to 30 million than it will from 30 to 40 million. The answer is (A).

7.015 $a \, ፠ \, b = \dfrac{1}{a} + \dfrac{1}{b}$ and $a \, ⇸ \, b = \dfrac{1}{a} \times \dfrac{1}{b}$

Column A

$$\dfrac{1}{19} \, ፠ \, \dfrac{1}{14}$$

Column B

$$\dfrac{1}{19} \, ⇸ \, \dfrac{1}{14}$$

$$\dfrac{1}{19} \, ፠ \, \dfrac{1}{14} = \dfrac{1}{\frac{1}{19}} + \dfrac{1}{\frac{1}{14}} = 19 + 14$$

$$\dfrac{1}{19} \, ⇸ \, \dfrac{1}{14} = \dfrac{1}{\frac{1}{19}} \times \dfrac{1}{\frac{1}{14}} = 19 \times 14$$

Clearly, 19×14 is greater than $19 + 14$. The actual calculations are not necessary. Therefore, the answer is (B).

7.016 In the figure on the right, $a = c$.

Column A	Column B
b	d

The four lines enclose a quadrilateral. The sum of the interior angles of a quadrilateral is 360°.

$$a + b + c + d = 360 \quad \Rightarrow \quad a + b + a + d = 360 \quad \Rightarrow \quad b + d = 360 - 2a$$

Even if we knew the actual value of a, we could still not determine whether b or d is greater. b and d could be in any combination, as long as they add up to $360 - 2a$. The answer is (D).

7.017

Column A

The remainder when a prime number greater than 2 is divided by 2

Column B

The remainder when a prime number greater than 3 is divided by 3

Every prime number greater than 2 is an odd number. It will always leave a remainder of 1 when divided by 2.

A prime number greater than 3 when divided by 3 will leave a remainder of 1 or 2. For example, 5 divided by 3 leaves a remainder of 2, whereas 7 divided by 3 leaves a remainder of 1. Therefore, it's not possible to generally say which column is greater. The answer is (D).

7.018

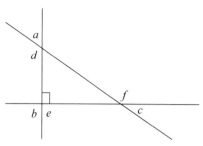

Column A	Column B
$a + b + c$	The arithmetic mean of a, b, c, d, e and f

a, b and c are the vertically opposite angles to the three interior angles of the triangle. Their sum must be the same as the sum of the interior angles: $a + b + c = 180°$

Observe that: $a + d = 180°$ $b + e = 180°$ $f + c = 180°$

Their arithmetic mean $= \dfrac{a + b + c + d + e + f}{6} = \dfrac{180 + 180 + 180}{6} = 90°$

The answer is (A).

7.019   In the figure on the right, $x \leq y$.

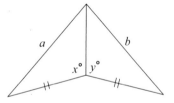

Column A	Column B
$(a + b)(a - b)$	0

$x \leq y$ means $x = y$ or $x < y$.

If $x = y$, then $a = b$. Therefore, $(a + b)(a - b) = (b + b)(b - b) = (2b)(0) = 0$

If $x < y$, then $a < b$. Therefore, $(a + b)(a - b) = $ (a positive number)(a negative number) $= $ (a negative number), i.e. $(a + b)(a - b) < 0$.

Meaning, $(a + b)(a - b) \leq 0$.   It is therefore not possible to generalize which column is greater. The answer is (D).

7.020 $f(x) = x^2$ and $g(x) = \sqrt[3]{x}$

Column A	Column B
$f(g(8))$	$g(f(8))$

$g(8) = \sqrt[3]{8} = 2$ $\Rightarrow$ $f(g(8)) = f(2) = 2^2 = 4$

$f(8) = 8^2 = 64$ $\Rightarrow$ $g(f(8)) = \sqrt[3]{64} = 4$

The two columns are equal. The answer is (C).

PRACTICE EXAMPLES (SET 1)

7.021

Column A	Column B
The minor angle between the hour and minute hands of a clock at 3:30	The minor angle between the hour and minute hands of a clock at 5:45

7.022

Column A	Column B
27% of 3800	38×27

7.023 $0 \leq x \leq 1$

Column A	Column B
x^2	$\dfrac{1}{x^2}$

7.024   In the figure on the right, $AB \parallel CD$ and $x > y$.

Column A	Column B
Length of diagonal AC	Length of diagonal BD

7.025 $abc = 0$ and $\dfrac{a}{c} = \dfrac{-1}{\pi}$

Column A	Column B
ac	b

7.026   A rectangle is inscribed in a circle.

Column A	Column B
Three times the length of a diagonal of the rectangle	The circumference of the circle

7.027

Column A	Column B
$\dfrac{1}{3} + \dfrac{1}{5} + \dfrac{1}{7}$	$\dfrac{1}{4} + \dfrac{1}{6} + \dfrac{1}{8}$

7.028   The two rectangles have identical dimensions.

Column A	Column B
The degree measure of x	The degree measure of y

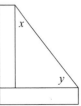

7.029 At an amusement park, the number of men, women and children present are in the ratio 3 : 7 : 16.

Column A	Column B

For every man, the number of women present at the park

For every woman, the number of children present at the park

7.030

Column A

$$\frac{27 \times 26!}{25!}$$

Column B

$$\frac{27!}{25 \times 24!}$$

7.031 A cylindrical can has a hemispherical cap. The height of the can, the outer diameter of the can, and the outer diameter of the cap are all equal in measure.

Column A

The outer (curved) surface area of the cap

Column B

The outer (curved) surface area of the can

7.032 k and m are positive integers.

Column A

The minimum value of $|3k - 7|$

Column B

The minimum value of $|5m - 9|$

7.033 $0.5 \leq x \leq 0.6$ and $0.8 \leq y \leq 0.9$

Column A

The value of x when rounded off to the nearest units place

Column B

The value of y when rounded off to the nearest tens place

7.034

Column A

$p + s$

Column B

$q + r$

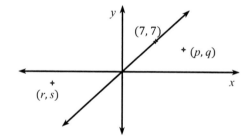

7.035 A car travels m miles in h hours, followed by n miles in t hours, where $2m > 3n$, and $2t > 3h$.

Column A

The average speed of the car while traveling the first m miles

Column B

The average speed of the car while traveling the next n miles

7.036

Column A	Column B
The sum of the unique prime factors of 35	The sum of the unique prime factors of 36

7.037 The circle is tangential to the coordinate axes, and has a rectangle inscribed in it.

Column A	Column B
Area of the rectangle	40

7.038 Sarah and Polly are both teenagers, and Sarah is 2 years older than Polly.

Column A	Column B
Polly's age, as a fraction of Sarah's age today	Polly's age 2 years ago, as a fraction of Sarah's age back then

7.039

Column A	Column B
$(0.638)^2$	$\dfrac{1}{(0.638)^2}$

7.040

Column A	Column B
The number of 4-digit integers that can be formed by using only odd digits	The number of 4-digit integers that can be formed by using only even digits

SPACE FOR CALCULATIONS

PRACTICE EXAMPLES (SET 2)

7.041 $ab = -24$, where a and b are integers.

<u>Column A</u>	<u>Column B</u>
The greatest value of $|a - b|$	The greatest value of $|a| - |b|$

7.042 AB is the diameter of the circle.

<u>Column A</u>	<u>Column B</u>
Area of $\triangle ADB$	Area of $\triangle ACB$

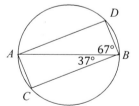

7.043   I was born in 1975.

<u>Column A</u>	<u>Column B</u>
The probability that I was born in January 1975	The probability that I was born in February 1975

7.044

Column A	Column B
25 × 52	26 × 51

7.045 $\left(\dfrac{u+v}{-2}\right)^2 < \left(\dfrac{u-v}{-2}\right)^2$

Column A	Column B
uv	0

7.046 At Company X, at least 25% of the employees are male. Of the female employees, at the most two-thirds are married.

Column A	Column B
The number of male employees at Company X	The number of female employees that are married, at Company X

7.047 $p^2 = -64q$ and $r^3 = 64s$, where p, q, r and s are non-zero numbers.

Column A	Column B
q	rs

7.048 $\dfrac{a}{b} = \dfrac{c}{d}$ where a, b, c and d are non-zero integers.

Column A	Column B
$\dfrac{2a - 2b}{3a + 3b}$	$\dfrac{4c - 4d}{5c + 5d}$

7.049 k and c are non-zero numbers.

Column A	Column B
$\dfrac{k}{c}$	$\dfrac{k + 1.22}{c + 1.22}$

7.050 Rectangle X has an area of 36. Rectangle Y has a perimeter of 36.

Column A	Column B
The minimum perimeter of Rectangle X	The maximum area of Rectangle Y

7.051 3 men can build a fence in 8 days, working at a uniform rate.

Column A	Column B
The time it will take 4 men to build 2 such fences, working at the same rate as above	The time it will take 14 men to build 7 such fences, working at the same rate as above

7.052 $0 < x < 1$

Column A	Column B
$\sqrt[3]{\sqrt[7]{x}}$	$\sqrt[5]{\sqrt[9]{x}}$

7.053 O is the center of the circle. C bisects minor arc AB, and D bisects major arc AB.

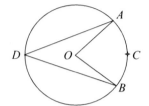

Column A	Column B
Area of the loop $AOBCA$	Area of the loop $AOBDA$

7.054

Column A	Column B
$\dfrac{3}{7} + \dfrac{4}{9} + \dfrac{5}{11} + \dfrac{6}{13}$	$\dfrac{3}{7} \times \dfrac{4}{9} \times \dfrac{5}{11} \times \dfrac{6}{13}$

7.055

Column A	Column B
The diagonal of a rectangular box with edges 3, 4 and 5	The diagonal of a cube with edges 4

7.056   A sandwich costs $3, and a burger costs $5. You are to buy at least one of each in the money you're given. You are to spend exactly all the money.

Column A	Column B
Given $20, the maximum number of items (sandwiches + burgers) you could buy	Given $21, the maximum number of items (sandwiches + burgers) you could buy

---

7.057

Column A	Column B
100×10^{-13}	$0.1^2 \times 0.1^5 \times 0.1^3$

7.058   The equation of $\overleftrightarrow{l}$ is $y = px + q$. Also, $|a|^b = 1$, and $b = 3a$.

Column A	Column B
p	q

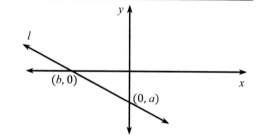

7.059   In a class of 50 students, there are students majoring in English, Chemistry, both, and neither. 30 students are majoring in English, and 27 in Chemistry.

Column A	Column B
The number of students majoring in Chemistry, but not English	The number of students majoring in English, but not Chemistry

7.060 x is a prime number less than 100.

Column A	Column B

The number of distinct digits possible in the units place of $2x$

The number of distinct digits possible in the units place of x^2

CHAPTER 8

DATA SUFFICIENCY

Data Sufficiency is specific to the GMAT, not to the GRE. However, I recommend studying this chapter even if you're taking the GRE because these questions involve mathematical reasoning, not just numerical computations. If you're taking the GMAT, then I advise studying the chapter on Quantitative Comparison as well, even though that chapter is specific to the GRE. Quantitative Comparison also requires mathematical reasoning.

Questions on Data Sufficiency involve some initial information, accompanied by two statements numbered (1) and (2). You are to decide whether the two statements provide enough information for you to answer the question. Your answer would be:

(A) if statement (1) ALONE is sufficient, but statement (2) is not sufficient

(B) if statement (2) ALONE is sufficient, but statement (1) is not sufficient

(C) if BOTH statements TOGETHER are sufficient, but NEITHER statement ALONE is sufficient

(D) if EACH statement ALONE is sufficient

(E) if statements (1) and (2) TOGETHER are NOT sufficient

In order to correctly analyze such questions, it is important to remember certain key points:

➤ Do not assume that numbers are always *integers*. 'Number' is a generic term; it could mean an integer or a fraction (or a decimal, which is essentially a fraction).

➤ Always consider positive, negative, zero and one. For example, $x^2 = 25$ means $x = \pm 5$, not just +5. Similarly, be alert about whether a number could be zero as well, or one as well. The properties of zero and one are quite frequently tested on this section.

➤ Pay particular attention to the > versus the ≥ sign, and the < versus the ≤ sign.

➤ Recognize the difference between parenthesis (brackets) and modulus signs (absolute value signs).

➤ Recognize the difference between positive and non-negative numbers. Recall that *positive numbers* are all numbers greater than zero. Zero itself is not included. On the other hand, *non-negative numbers* include all positive numbers and the number zero as well.

Let's analyze some **commonly tested number properties** before we move on to solving examples:

Statement: $\frac{a}{2}$ is an integer.

What it means: If $\frac{a}{2}$ is an integer, then a must be an even integer. Here's how:

$\frac{a}{2}$ = integer $\Rightarrow$ $a = 2 \times$ integer = even integer, because 2 multiplied by any integer = even integer

Statement: $\frac{a}{2}$ is *not* an integer.

What it means: If $\frac{a}{2}$ is not an integer, then a could be an integer or a fraction. Here's how:

If $a = 1$, then $\frac{a}{2} = \frac{1}{2} \neq$ an integer. But if $a = \frac{1}{3}$, then too $\frac{a}{2} = \frac{\frac{1}{3}}{2} = \frac{1}{6} \neq$ an integer.

Statement: $\frac{a}{2}$ is an odd integer.

What it means: If $\frac{a}{2}$ is an odd integer, then a must be an even integer. Here's why:

Let $\frac{a}{2} = 3$. Then, $a = 2 \times 3 = 6$. Let $\frac{a}{2} = 11$. Then, $a = 2 \times 11 = 22$.

In general, let $\frac{a}{2} = k$. Then, $a = 2 \times k$. 2 multiplied by *any integer* will result in an even integer.

Statement: $\frac{a}{2}$ is an even integer.

What it means: If $\frac{a}{2}$ is an even integer, then a must be an even integer. Here's why:

Let $\frac{a}{2} = 4$. Then, $a = 2 \times 4 = 8$. Let $\frac{a}{2} = 10$. Then, $a = 2 \times 10 = 20$.

In general, let $\frac{a}{2} = k$. Then, $a = 2 \times k$. 2 multiplied by *any integer* will result in an even integer.

Statement: $2a$ is an integer.

What it means: If $2a$ is an integer, then a could be an integer of a fraction. Here's how:

$2a$ = integer $\Rightarrow$ $a = \frac{\text{integer}}{2}$ = integer or fraction, depending upon even or odd numerator.

Statement: $2a$ is *not* an integer.

What it means: 2 multiplied by any integer a (odd or even) will always result in an integer. If $2a$ is not an integer, then a must be a fraction.

Statement: $2a$ is an odd integer.

What it means: If $2a$ is an odd integer, then a can only be a fraction. Here's how:

$2a$ = odd integer $\Rightarrow$ $a = \frac{\text{odd integer}}{2}$ = fraction   (because odd integers can't be divided by 2)

Statement: $2a$ is an even integer.

What it means: If $2a$ is an even integer, then a has to be an integer (odd or even).

Statement: $\frac{a}{b}$ is an integer.

What it means: If $\frac{a}{b}$ is an integer, then a is a multiple of b. a and b could both be integers, or they could both be fractions, or a could be an integer whereas b could be a fraction. But if a is a fraction, and b is an integer, then the result will not be an integer. Here's how:

Both are integers: $a = 6$, $\quad b = 3$. Then, $\dfrac{a}{b} = \dfrac{6}{3} = 2 =$ an integer.

Both are fractions: $a = \dfrac{1}{2}$, $b = \dfrac{1}{4}$. Then, $\dfrac{a}{b} = \dfrac{\frac{1}{2}}{\frac{1}{4}} = \dfrac{1}{2} \times \dfrac{4}{1} = 2 =$ an integer.

Only a is an integer: $a = 2$, $b = \dfrac{1}{7}$. Then, $\dfrac{a}{b} = \dfrac{2}{\frac{1}{7}} = 2 \times \dfrac{7}{1} = 14 =$ an integer.

Only b is an integer: $a = \dfrac{1}{3}$, $\quad b = 5$. Then, $\dfrac{a}{b} = \dfrac{\frac{1}{3}}{5} = \dfrac{1}{15} \neq$ an integer.

Statement: $x^2 > 0$ (This has been the most easily forgotten property in my teaching experience).

What it means: x^2 is positive. But x could be either positive or negative. Here's how:

Let $x^2 = 36$. Then $x = \sqrt{36} = \pm 6$ because $6 \times 6 = 36$ and $(-6) \times (-6) = 36$ as well.

Statement: $a^b = 1$

What it means: It could mean three cases, considered individually. Here's each of them:

Case 1: We know that $x^0 = 1$, where $x \neq 0$. Meaning, if $a^b = 1$, then a could be any real number other than zero, and b could be zero. Please recall that *real numbers* are all numbers (integers, fractions, decimals, positive and negative).

Case 2: We also know that $1^n = 1$, where n could be any integer. Meaning, if $a^b = 1$, then a could be 1, and b could be any integer.

Case 3: Finally, $(-1)^n = 1$, where n could be any *even integer*. Meaning, if $a^b = 1$, then a could be (-1), and b could be any even integer.

You will be able to answer most Data Sufficiency questions without actually solving the problem.

8.001 **Is c an integer?**

(1) $\dfrac{c}{12}$ is an integer.

(2) $\dfrac{c}{1/12}$ is an integer.

Statement (1): If $\dfrac{c}{12}$ is an integer, then clearly c is divisible by 12, i.e. c must be an integer.

This statement is sufficient to answer the question.

Statement (2): $\dfrac{c}{1/12} = c \times \dfrac{12}{1} = 12c$ If $12c$ is an integer, then c could be an integer

or a fraction. For example, c could be ¼, in which case $12 \times ¼ = 3 =$ integer. Or c could be any integer, in which case, $12c$ would also be an integer. This statement is not sufficient to answer the question.

If one of the statements is sufficient by itself, while the other is not, then do not waste your time considering them together. Meaning, your answer cannot be (C).

Here, the answer is (A).

8.002 **35% of the students in a school are male. How many students are female?**

(1) The school has 500 students.
(2) The school has 49 male students.

Statement (1): 65% of the students must be female. 65% of 500 would give us the number of female students. This statement is sufficient to answer the question.

Statement (2): Assume $x =$ the total number of students. Then, 35% of $x = 49$. From this, x can be found out. Then, 65% of $x =$ number of female students. This statement is sufficient to answer the question.

Each statement is independently sufficient to answer the question. The answer is (D).

8.003 **Is $y < 0$?**

(1) $x = y^2$
(2) $x = y^3$

Neither statement gives any clue about the value of y. So let's try to solve for y by using both the statements together. Equate the two values of x:

$y^2 = y^3 \quad \Rightarrow \quad y^2 - y^3 = 0 \quad \Rightarrow \quad y^2(1 - y) = 0$ Here, we have a zero product.

If $y^2 = 0$, then $y = 0$.

If $1 - y = 0$, then $y = 1$.

Meaning, y could be 0 or 1, but never negative, i.e. $y \not< 0$.

Statements (1) and (2) are together sufficient to answer the question, although neither of them is independently sufficient. The answer is (C).

8.004 **In a class of 80 students, there are students majoring in English, Chemistry, both, and neither. How many students are majoring in English?**

(1) 24 students are majoring in Chemistry but not English.
(2) 11 students are majoring in both English and Chemistry.

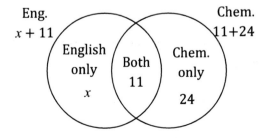

The above figure on the right is a Venn diagram drawn using the information given. In order to find how many students are majoring in English $(x + 11)$, we would need to know x.

Alternatively, we would need to know how many students (say y) are majoring in neither English nor Chemistry, so that $11 + x + 24 + y = 80$. Knowing y in the equation, we could find x, and therefore $x + 11$. But neither statement provides that information.

Statements (1) and (2) are together not sufficient to answer the question. The answer is (E).

8.005 **The circle on the right has its center at _O._ What is the value of x?**

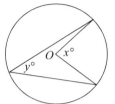

(1) $y = 41$
(2) $x + y = 108$

x is the degree measure of a central angle. y is the degree measure of an inscribed angle. In the configuration shown, central angle is twice as large as inscribed angle. Meaning, $x = 2y$.

Statement (1): y is given. x can be found out using $x = 2y$. This statement is sufficient to answer the question.

Statement (2): $x = 2y$ also means $y = \frac{1}{2}x$. Substituting $y = \frac{1}{2}x$ in this statement would give $x + \frac{1}{2}x = 108$. Solving for x gives its value. This statement is sufficient to answer the question.

Each statement is independently sufficient to answer the question. The answer is (D).

8.006 **Is x an odd integer?**

(1) x, when divided by 5, leaves 2 as a remainder.
(2) x, when divided by 4, leaves 3 as a remainder.

Statement (1): This means $x =$ (a multiple of 5) $+ 2$. A multiple of 5 could be odd or even, i.e. 5, 10, 75, 400 etc. Adding 2 would accordingly result in an odd or even number. This statement is not sufficient to answer the question.

Statement (2): This means $x =$ (a multiple of 4) $+ 3$. A multiple of 4 is always even. Adding 3 would always make it odd. This statement is sufficient to answer the question.

The answer is (B).

8.007 **Is Karen the only niece?**

(1) Karen is the only child.
(2) Karen is the only grandchild.

Statement (1): Karen may be the only child. But it's possible Karen has an uncle who has one or more daughters; or maybe an aunt with one or more daughters. In that case, Karen is not the only niece. Or maybe Karen's uncles/aunts have no daughters, in which case, she is the only niece. This statement is not sufficient to answer the question.

Statement (2): This statement means that even if Karen has uncles/aunts, none of them has any children. But Karen could have a sister, in which case, she's not the only niece. This statement does not say whether she has any sisters. This statement is not sufficient to answer the question.

The statements together mean that Karen has no sisters, and no cousins. But we don't know if she has any uncles/aunts. So, maybe she is nobody's niece. Statements (1) and (2) are together not sufficient to answer the question. The answer is (E).

8.008 If $a^b = c$, what is the value of b?

(1) $c = 1$
(2) $a = 1$

Statement (1): If $c = 1$, then either a = any non-zero real number and $b = 0$, or $a = 1$ and b = any integer, or $a = -1$ and b = any even integer. Meaning, b could be any integer (that includes zero). This statement is not sufficient to answer the question.

Statement (2): If $a = 1$, then we cannot find the value of b without knowing the value of c. Remember, if $a = 1$, then c could be ± 1. As an example, $1^{\frac{1}{2}} = \sqrt{1} = \pm 1$. This statement is not sufficient to answer the question.

The two statements together mean that $1^b = 1$. But it also means that b could be any integer. Statements (1) and (2) are together not sufficient to answer the question. The answer is (E).

8.009 A dealership has c cars, t trucks and v vans, and no other vehicles. If $c : t : v = 5 : 4 : x$, what is the minimum number of vans the dealership could have?

(1) If a vehicle is randomly selected, there's a 20% chance that it will be a van.
(2) $t : c = (c + t) : (c + t + v)$

Statement (1): The probability of randomly selecting a van is equal to the proportion in which vans are present in the group of vehicles. So, 20% of the vehicles are vans. This is the value of x, and can be converted into a fraction, and accommodated into the ratio $5 : 4 : x$. When the ratio is reduced to its lowest terms, the value of x would be the minimum number of vans. This statement is sufficient to answer the question.

Statement (2):

$$t : c = (c + t) : (c + t + v) \quad \Rightarrow \quad \frac{t}{c} = \frac{c + t}{c + t + v} \quad \Rightarrow \quad \frac{4}{5} = \frac{5 + 4}{5 + 4 + x}$$

The value of x can be worked out from the above equation. When the ratio $5 : 4 : x$ is reduced to its lowest terms, the value of x would be the minimum number of vans. This statement is sufficient to answer the question.

Each statement is independently sufficient to answer the question. The answer is (D).

8.010 In the figure on the right, $ABCD$ is a rectangle of area 400. What is the value of y?

(1) $x + y = 50$
(2) $x - y = 30$

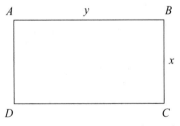

Please do not be tempted to think that simply solving the two equations simultaneously will lead to the value of y, and that the answer must be (C).

The area is given as 400. Therefore, $xy = 400$.

Statement (1): $x + y = 50 \quad \Rightarrow \quad x = 50 - y \qquad$ Substitute in $xy = 400$ to get:

$(50 - y)y = 400 \quad \Rightarrow \quad 50y - y^2 - 400 = 0 \quad \Rightarrow \quad y^2 - 50y + 400 = 0 \quad \Rightarrow$

$y^2 - 40y - 10y + 400 = 0 \quad \Rightarrow \quad y(y - 40) - 10(y - 40) = 0 \quad \Rightarrow$

$(y - 40)(y - 10) = 0 \quad \Rightarrow \quad y = 40 \quad \text{or} \quad y = 10$

That gives two different values of y. This statement is not sufficient to answer the question.

Statement (2): $x - y = 30$ $\Rightarrow$ $x = 30 + y$ Substitute in $xy = 400$ to get:

$(30 + y)y = 400$ $\Rightarrow$ $30y + y^2 = 400$ $\Rightarrow$ $y^2 + 30y - 400 = 0$ $\Rightarrow$

$y^2 + 40y - 10y - 400 = 0$ $\Rightarrow$ $y(y + 40) - 10(y + 40) = 0$ $\Rightarrow$

$(y + 40)(y - 10) = 0$ $\Rightarrow$ $y = -40$ or $y = 10$

The length y can never be negative (-40). Therefore, y must be 10. This statement is sufficient to answer the question.

The answer is (B).

8.011 What is the standard deviation of $-7, 4, 0, 1, -15, x, 0$ and 8?

(1) The mean of the numbers is -1.
(2) The median of the numbers is 0.5.

Statement (1): Add all the eight numbers, and divide the result by 8. Equate that to the mean (-1). Solve for x. Once x is known, the standard deviation of all the eight numbers can be found out by using the formula for it. This statement is sufficient to answer the question.

Statement (2): Arrange the numbers in increasing order: $-15, -7, 0, 0, 1, 4, 8, x$ Where would x be in the list? The list has an even number of elements; the median would be the average of the middle two numbers. But how can we be sure that x is at the far end of the list? What if $x = 1$? Then it would be one of the middle two numbers, the median being 0.5. The value of x would need to be known to be able to determine the standard deviation. This statement is not sufficient to answer the question.

The answer is (A).

8.012 What is the price of a gallon of milk?

(1) A gallon of milk and 2 loaves of bread cost \$7.
(2) 2 gallons of milk and 3 loaves of bread cost \$12.

Assume G as the price of a gallon of milk, and L as the price of a loaf of bread.

Statement (1): $G + 2L = 7$ There are two unknowns (G and L) in the equation. It's not possible to find G without knowing L. This statement is not sufficient to answer the question.

Statement (2): $2G + 3L = 12$ Once again, two unknowns. Can't find G without knowing L. This statement is not sufficient to answer the question.

Statements (1) and (2) together form a pair of linear equations. They can be solved simultaneously to find the values of G and L. Statements (1) and (2) are together sufficient to answer the question, although neither of them is independently sufficient. The answer is (C).

8.013 Is $\sqrt{a + b} = \sqrt{a} + \sqrt{b}$?

(1) $a = 2$
(2) $a - b = 2$

Neither statement, if considered alone, leads us anywhere. So, consider them together. Substitute $a = 2$ from statement (1) into statement (2):

$$a - b = 2 \quad \Rightarrow \quad 2 - b = 2 \quad \Rightarrow \quad -b = 0 \quad \Rightarrow \quad b = 0$$

In that case, $\sqrt{a + b} = \sqrt{2 + 0} = \sqrt{2}$ and $\sqrt{a} + \sqrt{b} = \sqrt{2} + \sqrt{0} = \sqrt{2}$

They are both equal. Thus, statements (1) and (2) are together sufficient to answer the question, although neither of them is independently sufficient. The answer is (C).

8.014 **In the diagram on the right, is $AC = BD$?**

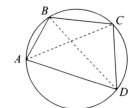

(1) $BC = AD$
(2) $AB = BC$

Statement (1): If $BC = AD$, then the quadrilateral is either a rectangle, or a square, or an isosceles trapezoid. In each case, diagonal $AC =$ diagonal BD. This statement is sufficient to answer the question.

Statement (2): If $AB = BC$, then the position of point D will determine what type of a quadrilateral $ABCD$ will be. Accordingly, the diagonals AC and BD may or may not be of the same length. This statement is not sufficient to answer the question.

The answer is (A).

8.015 **Is m a negative integer?**

(1) $m = 2n^2 + 3$
(2) $|m|$ is a prime number.

Statement (1): Whether the value of n is positive or negative, n^2 will always be positive. So, $2n^2 + 3$ will always be positive, i.e. m is positive. m will turn out to be an integer only if n is an integer. This statement doesn't say whether n is an integer. This statement is not sufficient to answer the question.

Statement (2): A prime number is always a positive integer. $|m|$ being a prime number only means that m is either a positive or a negative integer. But this statement does not say whether m is positive or negative. This statement is not sufficient to answer the question.

The two statements together mean that m is positive, and that m is an integer. Thus, statements (1) and (2) are together sufficient to answer the question, although neither of them is independently sufficient. The answer is (C).

8.016 **In the figure on the right, is $x > y$?**

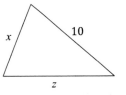 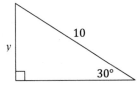

(1) $x = 6$
(2) $z = 8$

The triangle on the right is a 30–60–90 right triangle. In such triangles, the side facing the 30° angle is half as long as the hypotenuse. Therefore, $y = \frac{1}{2}(10) = 5$.

Statement (1): Directly compare x with y. This statement is sufficient to answer the question.

Statement (2): Knowing the value of z will not help us find x because it is not given whether the triangle on the left is a right triangle. Therefore, we don't have a value of x to compare with y. This statement is not sufficient to answer the question.

The answer is (A).

8.017 If a, b and c are integers, and $a^b = b^c$, what is the value of b?

(1) $a = c$
(2) $c = 0$

Statement (1): If $a = c$, then $a^b = b^a$. This equality means that $a = b =$ any non-zero number. This statement is not sufficient to answer the question.

Statement (2): If $c = 0$, then $b \neq 0$ because 0^0 is undefined. Therefore, $a^b = b^0 = 1$. Also, b would have to be an integer, for $a^b = 1$. Meaning, a would have to be 1 so that $1^{b \neq 0} = 1$. That is not enough to tell the value of b. This statement is not sufficient to answer the question.

Statements (1) and (2) together mean that $0^b = b^0$. If $b \neq 0$, then $b^0 = 1$. It means $0^b = 1$. But zero raised to *nothing* gives 1. So the value of b would be undefined. The answer is (E).

8.018 If $\dfrac{p}{q} + 17 = 29$, what is the value of $0.2p - 5.45$?

(1) $q = 16.4$
(2) $q = 3p - 1.08$

Statement (1): To tell the value of $0.2p - 5.45$, the only thing we need to know is the value of p. This statement gives us the value of q, which can be substituted into the given equation to find the value of p. This statement is sufficient to answer the question.

Statement (2): This statement is a linear equation in p and q. The given equation is also a linear equation in p and q. The two equations can be solved simultaneously to find the value of p and q. Then the value of p can be used to find the value of $0.2p - 5.45$. This statement is sufficient to answer the question.

Each statement is independently sufficient to answer the question. The answer is (D).

8.019 Derrick can wash his car in 20 minutes. If Derrick and Sean work together, how long will it take them to wash the car?

(1) Sean can wash the car in 45 minutes.
(2) Sean can wash the car and a truck in 1½ hours.

Statement (1): This statement gives us Sean's rate of washing a car. Knowing the two individual rates of work, we can find their group rate of work, and therefore, the time it will take them to wash the car together. This statement is sufficient to answer the question.

Statement (2): This statement cannot be used to determine Sean's individual rate of work when it comes to washing a car only. This statement is not sufficient to answer the question.

The answer is (A).

8.020 Will $(a \pm b)$ when divided by 4 result in an even number?

(1) a is a multiple of 12.
(2) b is a multiple of 28.

An easy way to answer this question is to not delve too deep into number properties, and not overanalyze the problem. The question is about $a \pm b$. Meaning, to answer the question, we will need some information about a, and some about b. Clearly, we will need both the statements. What we don't know is whether the two statements together will be sufficient to answer the question. Our answer choice would be either (C) or (E).

Consider just the 1st and 2nd multiples of 12, and from each, subtract 28.

If $a = 12$, and $b = 28$, then $a - b = 12 - 28 = -16$. $-16 \div 4 = -4 =$ an even number.

If $a = 24$, and $b = 28$, then $a - b = 24 - 28 = -4$. $-4 \div 4 = -1 =$ an odd number.

That's enough to say that statements (1) and (2) are together not sufficient to answer the question. The answer is (E).

PRACTICE EXAMPLES (SET 1)

8.021 Is $\dfrac{p}{qr}$ an integer?

(1) p is an odd multiple of r, and $r \neq 0$
(2) $\dfrac{p}{r}$ is an even multiple of q, and $q \neq 0$

8.022 What was the price of Company X's stock at the end of January 2007?

(1) Company X's stock price increased by $16 during January 2007.
(2) Company X's stock price increased by 12.78% during January 2007.

8.023 If x, y, a and b are positive integers, is $\dfrac{x+a}{y+b} > \dfrac{x}{y}$?

(1) $x < y$ and $a = b$
(2) $x < y$ and $a > b$

8.024 The average of 4 positive integers is 45½. What is the largest among them?

(1) The average of the 3 smaller integers is 42.
(2) The 4 integers are consecutive integers.

8.025 What is the radius of a semi-circle?

(1) The perimeter of the semi-circle is 32.
(2) The length of the semi-circular arc is 20.

8.026 If $x \, \mathcal{H} \, y = xy + 1$, and $x \, \boxed{\circ} \, y = 2 - 3x$, what is the value of $(A \, \mathcal{H} \, B) \boxed{\circ} \, C$?

(1) $A = 12 = 2B$
(2) $C = -B$

8.027 I was born in 1976. What's the probability that I'm 34 years old?

(1) Today is March 20, 2010.
(2) June 15 is my birthday.

8.028 In the diagram on the right, what is the length of segment MN?

(1) $d = 4$
(2) $b = \sqrt[3]{c + 4}$

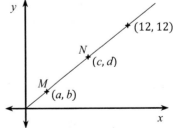

8.029 A fuel mixture contains 70% alcohol and 30% gasoline by volume. After some gasoline is added to it, the new mixture contains alcohol and gasoline in the ratio 2 : 3 by volume. How much gasoline is added?

(1) The original mixture had 15 gallons of fuel.
(2) The final mixture had 25 gallons of fuel.

8.030 What is the minimum value of $|pq - r|$?

(1) $q = \dfrac{r}{p}$

(2) $p \cdot |q| = r$

8.031 x is an integer. Is $x \geq 14$?

(1) x is not a prime number.
(2) $12 < x < 18$

8.032 Lisa is taller than Amy, who is shorter than Ken, who is 5'6" tall. Is Dixon taller than Lisa?

(1) Dixon is shorter than Mike, who is taller than Ken.
(2) Lisa is ½" taller than Mike.

8.033 In $\triangle PQR$ on the right, is $\angle R = 106°$?

(1) The product of slopes of sides PQ and QR is -1.
(2) PQR is an isosceles triangle, with $\angle Q = 37°$.

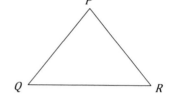

8.034 $3^{ax-b} = \dfrac{1}{3^c}$ and $x \neq 0$. What is the value of a?

(1) $b = c$
(2) $b^c = 1$

8.035 After driving for 4 hours, Steve has completed $\dfrac{2}{3}$ the trip. How many miles long is the total trip?

(1) Steve's average speed was 55 mph during the 4 hours he drove.
(2) If Steve maintains an average speed of 60 mph during the remainder of the trip, he can finish the trip in a total of 7 hours.

8.036 If the two sides forming the 90° angle in a right triangle are of integer lengths a and b, is the area of the triangle an even integer?

(1) a and b are consecutive integers.
(2) $a + b > 5$

8.037 Is k an odd number?

(1) $2k$ is an even number.
(2) $\dfrac{k}{2}$ is NOT a whole number.

8.038 Mr. Appleworm eats 8 apples on Wednesdays. Is it possible he eats 5 apples on Mondays?

(1) He eats at least one apple a day, and a total of 17 apples a week.
(2) He never eats more than 8 apples a day.

8.039 A corporation's revenues for the years 1997, 1998, 1999 and 2000 were $27M, $19M, $24M and $30M, respectively. Did its revenues exceed $27M in Year X?

(1) The revenue for Year X was between the revenues for 1997 and 2000, inclusive.
(2) The average revenue for 1997, 2000 and Year X was $27M.

8.040 Does the point (p, q) lie vertically below $\overleftrightarrow{l}$?

(1) $p = 5$
(2) $q < p$

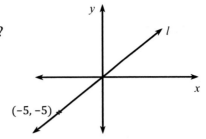

PRACTICE EXAMPLES (SET 2)

8.041 A box has red, yellow and blue cubical blocks in it, with each block having a different letter written boldly on it. How many blocks are in the box?

(1) The lettered blocks can be arranged in a straight line in 20! unique sequences.
(2) If a block is randomly selected from the box, there's a 45% probability that it is either red or blue.

8.042 In the figure on the right, is $x = 90°$?

(1) $AC = BD$
(2) $BC = AD$

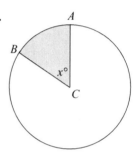

8.043 How old is Sue?

(1) 6 years from now, Sue's age will be 2½ times her son's age then.
(2) 4 years ago, Sue's age was 10 times her son's age back then.

8.044 C is the center of the circle on the right. What is the length of minor arc AB?

(1) Area of the shaded sector is 496.
(2) $x = 51$

8.045 Is mn an integer?

(1) n is a non-zero integer.
(2) $\dfrac{m}{n}$ is an integer.

8.046 Bank X pays me 3% simple annual interest on my investment of $x. Bank Y pays me 4% simple annual interest on my investment of $y. What percent of my total annual interest comes from Bank X?

(1) $x + y = 2000$
(2) $2x : 3y = 7 : 4$

8.047 In the figure on the right, ABC is a right triangle. What is the area of the shaded triangle?

(1) $x = 10$
(2) $y = 5\sqrt{3}$

8.048 If $ab^3 < cb$, is $b > 1$?

(1) $ab^2 < c$
(2) $b^2 > b$

8.049 What was the population of Country C in 1970?

(1) From 1900 to 2000, the population of Country C increased by 20% every 25 years.
(2) The population of Country C was 27 million in 1983.

8.050 Is x an even integer?

(1) $\dfrac{x}{100} - 0.02 < 5$

(2) $\dfrac{x}{100} - 0.002 \geq 5$

8.051 In the adjacent figure, is *ABC* an isosceles triangle?

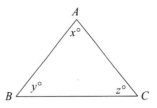

(1) $\dfrac{1}{x} = \dfrac{1}{z}$

(2) $\dfrac{2}{x} = \dfrac{3}{y} = \dfrac{2}{z}$

8.052 When the digits of a 2-digit number are reversed, the resulting number is greater than the original number. What is the original number?

(1) One of the digits is 4 times the other.
(2) The resulting number is greater than the original number by 54.

8.053 If *a* and *b* are whole numbers, what is the value of $a^2 - b^2$?

(1) *a* is three more than *b*.
(2) $|a + b| = 7$

8.054 Is $x^2 + x > \sqrt[3]{x} + \dfrac{1}{x}$?

(1) *x* is a non-zero integer.
(2) $0 < x < 1$

8.055 An airplane has 40 business class seats. How many economy class seats does it have?

(1) The percentage of business class seats occupied by men is equal to the percentage of economy class seats in the plane.
(2) There are 16 women traveling in business class.

8.056 How many sides does a polygon have?

(1) The sum of its interior angles is 360°.
(2) The sum of its exterior angles is 360°.

8.057 If $x(2x + 7)(x - 3) = 0$, what is the value of *x*?

(1) *x* is a non-negative number.
(2) *x* is a positive number.

8.058 In the figure on the right, *ABCD* is a square. Is the area of $\triangle ADE$ greater than the area of $\triangle CDE$?

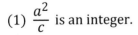

(1) $x = 115$
(2) $y = 155$

8.059 Is $\dfrac{ab}{c}$ an integer?

(1) $\dfrac{a^2}{c}$ is an integer.

(2) b^2 is an integer.

8.060 What is the standard deviation of the four numbers *a*, *b*, *c* and *d* if their average is 11?

(1) Their median is 8.
(2) The sum of their squares is 476.

SPACE FOR CALCULATIONS

SOLUTIONS TO PRACTICE EXAMPLES

CHAPTER 1: ARITHMETIC

1.005 $5 - 3 - 6 \Rightarrow -4$

1.006 $-2 - 11 - 1 \Rightarrow -14$

1.007 $-2 + 3 + 4 - 5 \Rightarrow 0$

1.008 $13 + 6 - 3 - 7 \Rightarrow 9$

1.013 $2 \times 4 \times 3 \times 3 \Rightarrow 72$

1.014 $-\dfrac{24}{24} \Rightarrow -1$

1.015 $-24 - 3 \Rightarrow -27$

1.016 $-2 + 2 \Rightarrow 0$

1.022 $2 \times 3 \times 3 \times 5$

1.023 $7 \times 3 \times 2$

1.024 $3 \times 2 \times 2 \times 2$

1.025 $7 \times 7 \times 3$. Smallest prime factor is 3.

1.026 $17 \times 3 \times 5 \times 2$. Greatest prime is 17.

1.027 Factors of 125: 1, 5, 25, 125. Factors of 50: 1, 2, 5, 10, 25, 50. HCF is 25. LCM $= \dfrac{125 \times 50}{25} = 250$.

1.028 Factors of 36: 1, 2, 3, 4, 6, 9, 12, 18, 36. Factors of 24: 1, 2, 3, 4, 6, 8, 12, 24. GCD is 12. LCM $= \dfrac{36 \times 24}{12} = 72$.

1.029 Factors of 81: 1, 3, 9, 27, 81. Factors of 54: 1, 2, 3, 6, 9, 18, 27, 54. GCF is 27. LCM $= \dfrac{81 \times 54}{27} = 162$.

1.030 Factors of 49: 1, 7, 49. Factors of 98: 1, 2, 7, 14, 49, 98. HCF is 49. LCM $= \dfrac{49 \times 98}{49} = 98$.

1.036 $\dfrac{12 \times 2 + 11}{12} \Rightarrow \dfrac{35}{12}$

1.037 $\dfrac{7 \times 5 + 4}{7} \Rightarrow \dfrac{39}{7}$

1.038 $\dfrac{7 \times 7 + 3}{7} \Rightarrow \dfrac{52}{7}$

1.039 $\dfrac{5 \times 1 + 4}{5} \Rightarrow \dfrac{9}{5}$

1.040 $\dfrac{3 \times 13 + 2}{3} \Rightarrow \dfrac{41}{3}$

1.046 $51 \div 50$ is 1. Remainder 1. Fractional part $\dfrac{1}{50}$. Result: $1\dfrac{1}{50}$.

1.047 $7 \div 5$ is 1. Remainder 2. Fractional part $\dfrac{2}{5}$. Result: $1\dfrac{2}{5}$.

1.048 $36 \div 11$ is 3. Remainder 3. Fractional part $\dfrac{3}{11}$. Result: $3\dfrac{3}{11}$.

1.049 $23 \div 7$ is 3. Remainder 2. Fractional part $\dfrac{2}{7}$. Result: $3\dfrac{2}{7}$.

1.050 $32 \div 3$ is 10. Remainder 2. Fractional part $\dfrac{2}{3}$. Result: $10\dfrac{2}{3}$.

1.056 $1 + 7 + \left(\dfrac{4}{5} + \dfrac{2}{5}\right) \Rightarrow 8 + \dfrac{6}{5} \Rightarrow 8 + 1 + \dfrac{1}{5} \Rightarrow 9\dfrac{1}{5}$

1.057 $6 - 4 + \left(\dfrac{5}{6} - \dfrac{1}{6}\right) \Rightarrow 2 + \dfrac{4}{6} \Rightarrow 2\dfrac{4}{6} \Rightarrow 2\dfrac{2}{3}$

1.058 $7\dfrac{12}{12} - 2\dfrac{11}{12} \Rightarrow 7 - 2 + \left(\dfrac{12}{12} - \dfrac{11}{12}\right) \Rightarrow 5\dfrac{1}{12}$

1.059 $9 - 6 + \left(\dfrac{3}{8} - \dfrac{5}{7}\right) \Rightarrow 3 + \left(\dfrac{21-40}{56}\right) \Rightarrow 3 - \dfrac{19}{56} \Rightarrow 2 + \dfrac{56}{56} - \dfrac{19}{56} \Rightarrow 2 + \dfrac{37}{56} \Rightarrow 2\dfrac{37}{56}$

1.060 $10 - 13 + \left(\dfrac{4}{5} - \dfrac{7}{9}\right) \Rightarrow -3 + \left(\dfrac{36-35}{45}\right) \Rightarrow -3 + \dfrac{1}{45} \Rightarrow -\left(\dfrac{3}{1} - \dfrac{1}{45}\right) \Rightarrow -\dfrac{135-1}{45} \Rightarrow -\dfrac{134}{45} \Rightarrow -2\dfrac{44}{45}$

1.066 $\dfrac{\overset{2}{\cancel{16}}\,^{1}}{\cancel{15}\,_{1}} \times \dfrac{\cancel{3}^{1}}{\cancel{8}^{1}} \times \dfrac{\cancel{5}^{1}}{\cancel{4}^{2}} \Rightarrow \dfrac{1}{2}$ The 3 and 5 *together* reduce the 15.

1.067 $\dfrac{\cancel{22}^{11}}{3} \times \dfrac{5}{\cancel{2}^{1}} \Rightarrow \dfrac{55}{3}$

1.068 $\dfrac{\cancel{8}^{4}}{\cancel{5}^{1}} \times \dfrac{\cancel{8}^{1}}{3} \times \dfrac{\cancel{5}^{1}}{\cancel{16}^{2}\,_{1}} \Rightarrow \dfrac{4}{3}$

1.069 $\dfrac{3}{\cancel{14}^{2}} \times \dfrac{\cancel{2}}{5} \times \dfrac{\cancel{21}^{3}}{1} \Rightarrow \dfrac{9}{5}$

1.070 $\dfrac{3}{2} \times \dfrac{2}{3} \times \dfrac{5}{4} \times \dfrac{4}{5} \times \ldots \ldots \dfrac{99}{98} \times \dfrac{98}{99}$ Cancel the 2s, 3s, 4s, 5s, … … 99s. Answer: 1.

1.075 $\dfrac{11}{\cancel{13}^{\,1}} \times \dfrac{\cancel{\cancel{39}}^{\,1}}{\cancel{12}^{\,4}} \Rightarrow \dfrac{11}{4}$

1.076 $\dfrac{\cancel{29}^{\,1}}{\cancel{5}^{\,1}} \times \dfrac{\cancel{10}^{\,2}}{\cancel{58}^{\,2}} \Rightarrow \dfrac{2}{2} \Rightarrow 1$

1.077 $\dfrac{\cancel{14}^{\,2}}{1} \times \dfrac{8}{\cancel{7}^{\,1}} \Rightarrow 16$

1.078 $\dfrac{11}{2} \times \dfrac{5}{34} \Rightarrow \dfrac{55}{68}$

1.085 $\dfrac{\cancel{3}^{\,1}}{8} \times \dfrac{7}{\cancel{6}^{\,2}} \Rightarrow \dfrac{7}{16}$

1.086 $\dfrac{\frac{20}{1} - \frac{4}{5}}{12} \Rightarrow \dfrac{\frac{100-4}{5}}{12} \Rightarrow \dfrac{\cancel{96}^{\,8}}{5} \times \dfrac{1}{\cancel{12}^{\,1}} \Rightarrow \dfrac{8}{5}$

1.087 $\dfrac{\frac{8+3}{12}}{\frac{11}{6}} \Rightarrow \dfrac{\cancel{11}^{\,1}}{\cancel{12}^{\,2}} \times \dfrac{\cancel{6}^{\,1}}{\cancel{11}^{\,1}} \Rightarrow \dfrac{1}{2}$

1.088 $\dfrac{\frac{7+18}{63}}{\frac{5}{1}} \Rightarrow \dfrac{\cancel{25}^{\,5}}{63} \times \dfrac{1}{\cancel{5}^{\,1}} \Rightarrow \dfrac{5}{63}$

1.089 $\dfrac{\frac{4}{1} + \frac{5}{7}}{\frac{11}{14}} \Rightarrow \dfrac{\frac{28+5}{7}}{\frac{11}{14}} \Rightarrow \dfrac{\cancel{33}^{\,3}}{\cancel{7}^{\,1}} \times \dfrac{\cancel{14}^{\,2}}{\cancel{11}^{\,1}} \Rightarrow 6$

1.090 $\dfrac{\cancel{3}^{\,1} \times \frac{1}{\cancel{6}^{\,2}} \times \cancel{\frac{1}{6}}}{4 \times \cancel{\frac{1}{6}}} \Rightarrow \dfrac{\frac{1}{2}}{4} \Rightarrow \dfrac{1}{2} \times \dfrac{1}{4} \Rightarrow \dfrac{1}{8}$

1.097 $\dfrac{23}{30}$ is greater.

1.098 Yes.

1.099 $\dfrac{21}{16}$ is smaller.

1.100 $\dfrac{5}{4}$ is the smallest.

1.101 $\dfrac{1}{4}, \dfrac{4}{17}, \dfrac{3}{13}$

1.102 $\dfrac{12}{7}, \dfrac{9}{4}, \dfrac{5}{2}$

1.105

(i) $\dfrac{x}{y}$ is a *proper* fraction. The same positive number is added to both x and y. Therefore, the resulting fraction is greater.

(ii)   Adding 3.1 to the numerator and to the denominator will increase the value of the fraction. Instead, adding 3.2 to the numerator will further increase the value of the fraction.

(iii)   Adding $\dfrac{1}{6}$ to the numerator and to the denominator will increase the value of the fraction. Instead, adding $\dfrac{1}{5} \left(> \dfrac{1}{6} \right)$ to the numerator will further increase the value of the fraction.

(iv)   Original fraction is *proper*. Its value is less than 1. The flipped fraction is *improper*. Its value is greater than 1.

(v)   Regardless of the individual values of x and y, the value of $x - y$ will always be negative, and that of $x + y$ will always be positive. Therefore, $\dfrac{x-y}{x+y}$ is negative. *Any negative number is always less than any positive number.*

1.106

(i) The *improper* fraction is now a *proper* fraction. Its value has dropped below 1 from its original value above 1.

(ii) $p-(-q)$ is the same as $p + q$. Both $p + q$ and $p - q$ are positive. The numerator has decreased, whereas the denominator has increased. The result is smaller than $\dfrac{p}{q}$.

(iii) $\dfrac{p}{q}$ is an *improper* fraction. The same positive number is added to both p and q. Therefore, the resulting fraction is smaller.

(iv) Adding 4.8 to the numerator and to the denominator will reduce the value of the fraction. Instead, adding 5.3 (> 4.8) to the denominator will further reduce the value of the fraction.

(v) Adding $\dfrac{11}{12}$ to the numerator and to the denominator will reduce the value of the fraction. Instead, adding $\dfrac{12}{13} \left(> \dfrac{11}{12} \right)$ to the denominator will further reduce the value of the fraction.

1.112 $6 \times 14 = 7 \times b \Rightarrow \dfrac{6 \times 14}{7} = \dfrac{7 \times b}{7} \Rightarrow b = 12$

1.113 $4 \times n = 5 \times 8 \Rightarrow \dfrac{4 \times n}{4} = \dfrac{5 \times 8}{4} \Rightarrow n = 10$

1.114 $16 \times 5 = y \times 8 \Rightarrow \dfrac{16 \times 5}{8} = \dfrac{y \times 8}{8} \Rightarrow 10 = y$

1.115 $e \times 51 = 17 \times 5 \Rightarrow \dfrac{e \times 51}{51} = \dfrac{17 \times 5}{51} \Rightarrow e = \dfrac{5}{3}$

1.116 $39 \times d = 12 \times 13 \Rightarrow \dfrac{39 \times d}{39} = \dfrac{12 \times 13}{39} \Rightarrow d = 4$

1.121

$$
\begin{array}{ccccc}
 & & \overset{1}{ } & \overset{1}{ } & & \\
 & 1. & 7 & 9 & 1 \\
 1 & 3. & 0 & 0 & 0 \\
+ & 0. & 8 & 4 & 3 \\
\hline
 1 & 5. & 6 & 3 & 4 \\
\end{array}
$$

1.122

$$
\begin{array}{cccccc}
 & & & \overset{5}{\cancel{6}} & \overset{14}{\cancel{4}} & \\
 & 8. & 7 & 6 & 4 & 8 & 2 \\
- & 8. & 1 & 2 & 9 & 0 & 0 \\
\hline
 & 0. & 6 & 3 & 5 & 8 & 2 \\
\end{array}
$$

1.123

$$
\begin{array}{cccc}
 & & \overset{4}{\cancel{5}} & \overset{9}{\cancel{0}} & \overset{10}{\cancel{0}} \\
 & 7. & 5 & 0 & 0 \\
- & 3. & 2 & 8 & 3 \\
\hline
 & 4. & 2 & 1 & 7 \\
\end{array}
$$

1.124

$$
\begin{array}{cccc}
 & \overset{0}{ } & \overset{9}{\cancel{0}} & \overset{10}{\cancel{0}} \\
 3 & 1. & 0 & 0 \\
- & & 0. & 5 & 4 \\
\hline
 3 & 0. & 4 & 6 \\
\end{array}
$$

1.129

$$
\begin{array}{cccc}
 & & \overset{4}{ } & \overset{1}{ } & \\
 & & 4. & 9 & 3 \\
 & \times & 1 & 0. & 5 \\
\hline
 & & 2 & 4 & 6 & 5 \\
\overset{1}{ } & & 0 & 0 & 0 & + \\
 & 4 & 9 & 3 & + & + \\
\hline
 & 5 & 1. & 7 & 6 & 5 \\
\end{array}
$$

1.130

$$
\begin{array}{ccccc}
 & & 0. & 0 & 7 & 1 \\
 & \times & & 4. & 7 & 5 \\
\hline
 & & & & \overset{1}{ } \\
 & & \overset{1}{ } & 3 & 5 & 5 \\
 & \overset{1}{ } & 4 & 9 & 7 & + \\
 & 2 & 8 & 4 & + & + \\
\hline
 & 0. & 3 & 3 & 7 & 2 & 5 \\
\end{array}
$$

1.131 $4.1400000 \times 10{,}000, \Rightarrow 41{,}400$

1.132 $4.980000 \times 100, \Rightarrow 498$

1.139 $\frac{0.688}{1.2} \times \frac{1000}{1000} \Rightarrow \frac{688}{12} \times \frac{1}{100} \Rightarrow 57.\overline{3} \times \frac{1}{100} \Rightarrow 0.57\overline{3}$

1.140 $\frac{3.1}{7} \times \frac{10}{10} \Rightarrow \frac{31}{7} \times \frac{1}{10} \Rightarrow 4.42 \times \frac{1}{10} \Rightarrow 0.442$

1.141 $\frac{4}{25} \times \frac{10}{10} \Rightarrow \frac{40}{25} \times \frac{1}{10} \Rightarrow 1.6 \times \frac{1}{10} \Rightarrow 0.16$

1.142 $0008{,}76 \div 100 \Rightarrow 0.0876$

1.143 $\frac{5.491}{0.13} \times \frac{1000}{1000} \Rightarrow \frac{5491}{13} \times \frac{1}{10} \Rightarrow 422.3 \times \frac{1}{10} \Rightarrow 42.23$

1.144 $\frac{77}{0.15} \times \frac{100}{100} \Rightarrow \frac{7700}{15} \Rightarrow 513.\overline{3}$

1.149 $0.475 \times \frac{1000}{1000} \Rightarrow \frac{475}{1000} \Rightarrow \frac{19}{40}$

1.150 $21 + \left(0.25 \times \frac{100}{100}\right) \Rightarrow 21 + \frac{25}{100} \Rightarrow 21\frac{1}{4}$

1.151 $0.1\frac{1}{2} \Rightarrow 0.15 \Rightarrow 0.15 \times \frac{100}{100} \Rightarrow \frac{15}{100} \Rightarrow \frac{3}{20}$

1.152 $26.025 \Rightarrow 26 + \left(0.025 \times \frac{1000}{1000}\right) \Rightarrow 26 + \frac{25}{1000} \Rightarrow 26\frac{1}{40}$

1.158 $\frac{10.71}{100} \Rightarrow 0.1071$

1.159 $\frac{0.897}{100} \Rightarrow 0.00897$

1.160 $\frac{97\frac{1}{3}}{100} \Rightarrow \frac{97.\overline{3}}{100} \Rightarrow 0.97\overline{3}$

1.161 $\frac{11.4\frac{5}{6}}{100} \Rightarrow \frac{11.48\overline{3}}{100} \Rightarrow 0.11483\overline{3}$

1.162 $\frac{0.84\frac{3}{4}}{100} \Rightarrow \frac{0.8475}{100} \Rightarrow 0.008475$

1.168 $0.706 \times 100\% \Rightarrow 70.6\%$

1.169 $0.085 \times 100\% \Rightarrow 8.5\%$

1.170 $4.11 \times 100\% \Rightarrow 411\%$

1.171 $9 \times 100\% \Rightarrow 900\%$

1.172 $5.45 \times 100\% \Rightarrow 545\%$

1.178 $75 \times \frac{1}{100} \Rightarrow \frac{75}{100} \Rightarrow \frac{3}{4}$

1.179 $250 \times \frac{1}{100} \Rightarrow \frac{250}{100} \Rightarrow \frac{5}{2}$

1.180 $5\frac{4}{5}\% \Rightarrow \frac{5\times5+4}{5}\% \Rightarrow \frac{29}{5} \times \frac{1}{100} \Rightarrow \frac{29}{500}$

1.181 $8.7 \times \frac{1}{100} \Rightarrow \frac{8.7}{100} \Rightarrow \frac{8.7}{100} \times \frac{10}{10} \Rightarrow \frac{87}{1000}$

1.182 $0.61 \times \frac{1}{100} \Rightarrow \frac{0.61}{100} \Rightarrow \frac{0.61}{100} \times \frac{100}{100} \Rightarrow \frac{61}{10{,}000}$

1.188 $\frac{7}{8} \times 100\% \Rightarrow \frac{700}{8}\% \Rightarrow 87.5\%$

1.189 $\frac{19}{\overset{1}{25}} \times \overset{4}{\cancel{100}}\% \Rightarrow 76\%$

1.190 $\frac{\overset{5}{\cancel{15}}}{\overset{6}{\cancel{18}}} \times 100\% \Rightarrow \frac{500}{6}\% \Rightarrow 83.\overline{3}\%$

1.191 $3\frac{1}{2} \times 100\% \Rightarrow 3.5 \times 100\% \Rightarrow 350\%$

1.192 $\frac{\overset{3}{\cancel{39}}}{\overset{5}{\cancel{65}}} \times \overset{20}{\cancel{100}}\% \Rightarrow 60\%$

1.199 $\dfrac{4\frac{2}{3}}{21} \Rightarrow \dfrac{\frac{3\times4+2}{3}}{21} \Rightarrow \frac{\overset{2}{\cancel{14}}}{3} \times \frac{1}{\underset{3}{\cancel{21}}} \Rightarrow \frac{2}{9}$

1.200 $1\frac{2}{3} \Rightarrow \frac{\frac{3\times1+2}{3}}{\frac{6\times4+5}{6}} \Rightarrow \frac{5}{3^1} \times \frac{6^2}{29} \Rightarrow \frac{10}{29}$

1.201 $\frac{\frac{7}{6}}{14} \Rightarrow \frac{7^1}{6} \times \frac{1}{14^2} \Rightarrow \frac{1}{12}$

1.202 x represents 3, and y represents 4. $2x$ would represent 6, and $3y$ would represent 12. Therefore, $2x : 3y = 6 : 12$. It can be further reduced by 2 to $1 : 2$.

1.203 dime : quarter = 10 ¢ : 25¢, or 10 : 25. It can be further reduced by 5 to 2 : 5.

1.204 $\frac{4}{7} = \frac{c}{28} \Rightarrow 7 \times c = 4 \times 28 \Rightarrow c = \frac{4\times28}{7} \Rightarrow c = 16$

1.208 b represents 2 in the first proportion, and 3 in the second. The LCM of 2 and 3 would be the common number for b. Raise the first proportion by 3, and the second by 2. You should get $a : b = 21 : 6$, and $c : b = 10 : 6$. Combine them to get $a : b : c = 21 : 6 : 10$.

1.209 x represents 3 in the first proportion, and 4 in the second. Their LCM is 12. Raise the first proportion by 4, and the second by 3. You should get $w : x = 8 : 12$, and $x : y = 12 : 27$. Combine them to get $w : x : y = 8 : 12 : 27$.

1.214 The numbers are not uniformly spaced out. So, do the actual math. $\frac{12+0+3+10+5}{5} \Rightarrow 6$

1.215 $\frac{3+y+19+1}{4} = 5 \Rightarrow \frac{y+23}{4} = 5 \Rightarrow y + 23 = 20 \Rightarrow y = -3$

1.216 1 to 15 are 15 integers, equally spaced out. Their midpoint is 8, which is their mean.

1.217 Average is $\frac{m+n}{2} \Rightarrow \frac{8}{2} \Rightarrow 4$

1.226 Note: -5^2 means $-(5^2)$ or $(-1) \times 5^2$, not $(-5)^2$, because the negative sign is outside the brackets. $\frac{(-1)\times5^2\times5^{-4}}{5^7} \Rightarrow (-1)(5^{2-4-7}) \Rightarrow (-1)(5^{-9}) \Rightarrow -\frac{1}{5^9}$

1.227 $2^{3\times4\times5} \Rightarrow 2^{60}$

1.228 $\frac{64^{13}}{64^{12}} \Rightarrow 64^{13-12} \Rightarrow 64^1 \Rightarrow 64$ It was convenient to convert 4^3 to 64, rather than the other way round. So I did it that way. But if you want, you can convert 64 to 4^3. You should get the same answer.

1.229 $\frac{(2^3)^{-12}\times2^{40}}{((2^2)^2)^2} \Rightarrow \frac{2^{3\times(-12)+40}}{2^{2\times2\times2}} \Rightarrow \frac{2^4}{2^8} \Rightarrow 2^{4-8} \Rightarrow 2^{-4} \Rightarrow \frac{1}{2^4} \Rightarrow \frac{1}{16}$

1.230 $x^{15} = (x^5)^3 = 3^3 = 27$

1.231 $(5^2)^m = 5^3 \Rightarrow 5^{2m} = 5^3$ The bases are equal. The exponents must be equal too. $2m = 3 \Rightarrow m = \frac{3}{2}$

1.232 $3^2 \times 3^{11} \times 3^3 = 3^b \Rightarrow 3^{2+11+3} = 3^b \Rightarrow 3^{16} = 3^b \Rightarrow 16 = b$

1.233 Prime factorization of 6: 2×3; 7 is already prime; prime factn. of 8: $2 \times 2 \times 2$; prime factn. of 18: $2 \times 3 \times 3$. Unique primes are 2, 3, 7. Highest power of 2 is 2^3, of 3 is 3^2, of 7 is 7. LCM $= 2^3 \times 3^2 \times 7 = 8 \times 9 \times 7 = 504$

1.234 Prime factorization of 12: $2 \times 2 \times 3$; of 18: $2 \times 3 \times 3$; of 24: $2 \times 2 \times 2 \times 3$; of 30: $2 \times 3 \times 5$; of 36: $2 \times 2 \times 3 \times 3$. Unique primes: 2, 3, 5. Highest power of 2 is 2^3, of 3 is 3^2, of 5 is 5. LCM $= 2^3 \times 3^2 \times 5 = 8 \times 9 \times 5 = 360$

1.239 $\sqrt{5} - 4\sqrt{5\times(3\times3)} + 3\sqrt{5\times(2\times2)} \Rightarrow \sqrt{5} - 3\times4\sqrt{5} + 2\times3\sqrt{5} \Rightarrow \sqrt{5} - 12\sqrt{5} + 6\sqrt{5} \Rightarrow -5\sqrt{5}$

1.240 $\frac{3\sqrt{5\times2}}{\sqrt{7\times(2\times2)}} \times \frac{\sqrt{2}}{\sqrt{5\times3}} \Rightarrow \frac{3\sqrt{5}\sqrt{2}\sqrt{2}}{2\sqrt{7}\sqrt{5}\sqrt{3}} \Rightarrow \frac{3}{\sqrt{7}\sqrt{3}} \Rightarrow \frac{\sqrt{3}\sqrt{3}}{\sqrt{7}\sqrt{3}} \Rightarrow \sqrt{\frac{3}{7}}$

1.245 $8 - \frac{7^2}{7} \Rightarrow 8 - 7 \Rightarrow 1$

1.246 4

1.247 $\frac{8-3\sqrt{100-96}}{4} \Rightarrow \frac{8-3\times2}{4} \Rightarrow \frac{2}{4} \Rightarrow \frac{1}{2}$

1.248 12

1.254 3.78

1.255 $14 - 23 \Rightarrow -9$

1.256 $11 + 1 \Rightarrow 12$

1.257 $\frac{-37+20}{12+22} \Rightarrow \frac{-17}{34} \Rightarrow -\frac{1}{2}$

1.258 $\frac{8\times3}{(-6)\times(-2)} \Rightarrow \frac{24}{12} \Rightarrow 2$

1.261 $-3w$, being negative, is less than w. $+\sqrt{w}$ is greater than w, because w is a proper fraction, i.e. its value is less than one. w^3 is less than w, whereas $2w$ is greater than w.

1.262 z^4 is the smallest, because a proper fraction times itself times itself times itself is going to be a much smaller fraction. $\frac{1}{z}$ is now an improper fraction, greater than one. Therefore, the ascending order is: $z^4, z, \frac{1}{z}$.

1.267 8.82×10^{-1}

1.268 9.469×10^9

1.269 $-4 \times 200 \times 200 \Rightarrow -4 \times 2 \times 100 \times 2 \times 100 \Rightarrow -16 \times 100 \times 100 \Rightarrow -1.6 \times 10^5$

1.270 $\frac{60,000}{300} \Rightarrow \frac{600}{3} \Rightarrow 200 \Rightarrow 2.0 \times 10^2$

CHAPTER 2: ALGEBRA

2.005 $\frac{20+5-(-4)}{3^2-(-4)(5)} \Rightarrow \frac{29}{29} \Rightarrow 1$

2.006 $\frac{3-4}{3+4-5} \Rightarrow -\frac{1}{2}$

2.007 $\frac{a+t+y}{3} \Rightarrow \frac{3-4+5}{3} \Rightarrow \frac{4}{3}$

2.008 $\frac{1}{\sqrt{60+64}} \Rightarrow \frac{1}{\sqrt{124}} \Rightarrow \frac{1}{2\sqrt{31}}$

2.014 $25ma - mp$

2.015 $\frac{3caz}{21} + \frac{cza}{21} - 11bdy \Rightarrow \frac{4caz}{21} - 11bdy$

2.016 $\frac{9xp^2}{n^3} + \frac{4ac}{b}$

2.017 ax

2.018 Adding the three polynomials, we get only b. This b is 8, given. $b^2 + 3 = 8^2 + 3 = 67$

2.024 $-3lm^4n^4$

2.025 $-6a^3p^3q^3$

2.026 $4uic^2 - 4u^2c^2$

2.027 $8f - afy + 8yi - aiy^2$

2.028 Rewrite in terms of the shorter term first. It's easier that way. $a^3 + a^2b - a^2x + ax + bx - x^2$

2.034 $\frac{-7q}{p}$

2.035 $-\frac{6w}{29x}$

2.036 $\frac{by^7}{3x^2}$

2.037 $\frac{k^3b^4}{-k^2b^3} - \frac{c^3k^2b^3}{-k^2b^3} \Rightarrow c^3 - kb$

2.038 $\frac{mx^2n}{-2m^2n^2x^2} + \frac{3xn^2m}{-2m^2n^2x^2} - \frac{8m^2xn}{-2m^2n^2x^2} \Rightarrow -\frac{1}{2mn} - \frac{3}{2xm} + \frac{4}{xn}$

2.042 $\frac{1}{(4x^{-2}y^7e^5)^3} \Rightarrow \frac{1}{64x^{-6}y^{21}e^{15}} \Rightarrow \frac{x^6}{64y^{21}e^{15}}$

2.043 $\frac{-p^6q^4r^{24}}{-p^{12}q^3r^{24}} \Rightarrow \frac{q}{p^6}$

2.044 $\left(\frac{a^3b^2}{d^4e^4f^3}\right)^2 \times \left(\frac{e^3f^2}{a^2bc^2}\right)^3 \Rightarrow \frac{a^6b^4e^9f^6}{a^6b^3c^6d^8e^8f^6} \Rightarrow \frac{be}{c^6d^8}$

2.053 $(6x)^2 - y^2 \Rightarrow (6x + y)(6x - y)$

2.054 $c^2 - (4d)^2 \Rightarrow (c + 4d)(c - 4d)$

2.055 $1 - (9z)^2 \Rightarrow (1 + 9z)(1 - 9z)$

2.056 $p^2 - \left(\frac{q}{rs}\right)^2 \Rightarrow \left(p + \frac{q}{rs}\right)\left(p - \frac{q}{rs}\right)$

2.057 $\left\{\left(\frac{1}{32} + 8\right) + \left(\frac{1}{32} - 8\right)\right\} \cdot \left\{\left(\frac{1}{32} + 8\right) - \left(\frac{1}{32} - 8\right)\right\} \Rightarrow \left\{\frac{1}{32} + 8 + \frac{1}{32} - 8\right\}\left\{\frac{1}{32} + 8 - \frac{1}{32} + 8\right\} \Rightarrow \frac{2}{32} \cdot 16 \Rightarrow 1$

2.058 $\left\{\left(\frac{1}{4} + \frac{1}{x}\right) + \left(\frac{1}{4} - \frac{1}{x}\right)\right\} \cdot \left\{\left(\frac{1}{4} + \frac{1}{x}\right) - \left(\frac{1}{4} - \frac{1}{x}\right)\right\} \Rightarrow \left\{\frac{1}{4} + \frac{1}{x} + \frac{1}{4} - \frac{1}{x}\right\} \cdot \left\{\frac{1}{4} + \frac{1}{x} - \frac{1}{4} + \frac{1}{x}\right\} \Rightarrow \frac{2}{4} \cdot \frac{2}{x} \Rightarrow \frac{1}{x}$

2.059 $(t^4)^2 - (w^4)^2$
$(t^4 + w^4)(t^4 - w^4)$
$(t^4 + w^4)((t^2)^2 - (w^2)^2)$
$(t^4 + w^4)(t^2 + w^2)(t^2 - w^2)$
$(t^4 + w^4)(t^2 + w^2)(t + w)(t - w)$

2.060 $d^2 - c^2 = 11$
$(d + c)(d - c) = 11$
$(d + c)(4) = 11$
$d + c = \frac{11}{4}$
$\frac{d+c}{2} = \frac{11}{4} \times \frac{1}{2} = \frac{11}{8}$

2.070 $d^2 + 12d + 11, \; ac = 11 \; (11 \times 1)$
$d^2 + 11d + d + 11$
$d(d + 11) + 1(d + 11)$
$(d + 11)(d + 1)$

2.071 $7h^2 + 12h + 5, \; ac = 35 \; (7 \times 5)$
$7h^2 + 7h + 5h + 5$
$7h(h + 1) + 5(h + 1)$
$(h + 1)(7h + 5)$

2.072 $j^2 - 16j + 39, \; ac = 39 \; (-13 \times -3)$
$j^2 - 13j - 3j + 39$
$j(j - 13) - 3(j - 13)$
$(j - 13)(j - 3)$

2.073 $9k^2 - 15k + 4$, $ac = 36 \, (-12 \times -3)$
$9k^2 - 12k - 3k + 4$
$3k(3k - 4) - 1(3k - 4)$
$(3k - 4)(3k - 1)$

2.074 $r^2 + 48r - 100$, $ac = -100 \, (50 \times -2)$
$r^2 + 50r - 2r - 100$
$r(r + 50) - 2(r + 50)$
$(r + 50)(r - 2)$

2.075 $12s^2 + 5st - 2t^2$, $ac = -24 \, (8 \times -3)$
$12s^2 + 8st - 3st - 2t^2$
$4s(3s + 2t) - t(3s + 2t)$
$(3s + 2t)(4s - t)$

2.076 $u^2 - 23uv - 50v^2$, $ac = -50 \, (-25 \times 2)$
$u^2 - 25uv + 2uv - 50v^2$
$u(u - 25v) + 2v(u - 25v)$
$(u - 25v)(u + 2v)$

2.077 $15w^2 - 2wx - x^2$, $ac = -15 \, (-5 \times 3)$
$15w^2 - 5wx + 3wx - x^2$
$5w(3w - x) + x(3w - x)$
$(3w - x)(5w + x)$

2.078 $\sqrt{n^2 - 9n - 36}$
$\sqrt{n^2 - 12n + 3n - 36}$
$\sqrt{n(n - 12) + 3(n - 12)}$
$\sqrt{(n - 12)(n + 3)}$
$\sqrt{(13 - 12)(13 + 3)}$
$\sqrt{16}$ or ± 4

2.088 $\dfrac{7g + 1 - 4g + 5}{3xc} \Rightarrow \dfrac{3g + 6}{3xc} \Rightarrow \dfrac{3(g + 2)}{3xc} \Rightarrow$
$\dfrac{g + 2}{xc}$

2.089 $\dfrac{p + 3 + 3 - p}{p - q} \Rightarrow \dfrac{6}{p - q}$

2.090 $\dfrac{(g - 2)(g + 2) + 5 \times 1}{5(g + 2)} \Rightarrow \dfrac{g^2 - 4 + 5}{5(g + 2)} \Rightarrow$
$\dfrac{g^2 + 1}{5(g + 2)}$

2.091 $\dfrac{k}{2(9 - k)} \times \dfrac{(9 + k)(9 - k)}{3k^2} \Rightarrow \dfrac{9 + k}{6k}$

2.092 $\dfrac{c + d}{ab} \times \dfrac{ab^2}{c^2 - d^2} \Rightarrow \dfrac{c + d}{1} \times \dfrac{b}{(c + d)(c - d)}$
$\Rightarrow \dfrac{b}{c - d}$

2.093 $\dfrac{3(2t - 3)}{2(7 + s)} \cdot \dfrac{(s + 7)(s - 3)}{5 - 2t - 2} \Rightarrow$
$\dfrac{-3(3 - 2t)}{2} \cdot \dfrac{s - 3}{3 - 2t} \Rightarrow \dfrac{3(3 - s)}{2}$

2.094 $\dfrac{na - ba^2}{a^3} \Rightarrow \dfrac{a(n - ba)}{a^3} \Rightarrow \dfrac{n - ba}{a^2}$

2.095 $\dfrac{\frac{f}{1} - \frac{1}{f}}{\frac{1 + f}{f}} \Rightarrow \dfrac{f^2 - 1}{f} \cdot \dfrac{f}{1 + f} \Rightarrow \dfrac{(f + 1)(f - 1)}{1 + f}$
$\Rightarrow f - 1$

2.096 $\dfrac{\frac{h}{1} - \frac{1}{e}}{\frac{1 - eh}{3e^2}} \Rightarrow \dfrac{eh - 1}{e} \cdot \dfrac{3e^2}{1 - eh} \Rightarrow -3e$

2.097 $\dfrac{\frac{c}{4a} - \frac{a}{c}}{\frac{c}{a} - \frac{2}{1}} \Rightarrow \dfrac{\frac{c^2 - 4a^2}{4ac}}{\frac{c - 2a}{a}} \Rightarrow \dfrac{c^2 - 4a^2}{4ac} \cdot \dfrac{a}{c - 2a}$
$\Rightarrow \dfrac{(c + 2a)(c - 2a)}{4c} \cdot \dfrac{1}{c - 2a} \Rightarrow \dfrac{c + 2a}{4c}$

2.107 $\pm 7acb$

2.108 $\sqrt{2 \times 7^2 d^2 (e^2)^2 (f^2)^2} \Rightarrow 7de^2 f^2 \sqrt{2}$

2.109 $\sqrt[3]{\dfrac{5^3 g^3}{h^3}} \Rightarrow \dfrac{5g}{h}$

2.110 $\dfrac{x\sqrt{3x^2 (y^2)^2 y}}{3\sqrt{y}} \Rightarrow \dfrac{x \cdot xy^2 \sqrt{3} \sqrt{y}}{3\sqrt{y}} \Rightarrow$
$\dfrac{x^2 y^2 \sqrt{3}}{\sqrt{3}\sqrt{3}} \Rightarrow \dfrac{x^2 y^2}{\sqrt{3}}$

2.111 $\dfrac{3\sqrt{7 \times 10 \times 10 \times f}}{\sqrt{3 \times 3 \times 7 \times e^2 f^2 f}} \Rightarrow \dfrac{30\sqrt{7f}}{3ef\sqrt{7f}} \Rightarrow \dfrac{10}{ef}$

2.112 $\dfrac{\sqrt[3]{2^3 \times 2p^3 pqr^3 r}}{\pm 2rp} \Rightarrow \dfrac{2pr \sqrt[3]{2pqr}}{\pm 2rp} \Rightarrow$
$\pm \sqrt[3]{2pqr}$

2.113 $\dfrac{\sqrt[3]{8(-1)}}{\sqrt{-2(-2a^2)}} \Rightarrow \dfrac{\sqrt[3]{-8}}{\sqrt{4a^2}} \Rightarrow \dfrac{-2}{\pm 2a} \Rightarrow \pm \dfrac{1}{a}$

2.114 $\dfrac{\frac{\sqrt{h}}{\sqrt{s}} - \frac{\sqrt{s}}{\sqrt{h}}}{\sqrt{\frac{h^2 - s^2}{sh}}} \Rightarrow \dfrac{\frac{h - s}{\sqrt{sh}}}{\frac{\sqrt{(h + s)(h - s)}}{\sqrt{sh}}} \Rightarrow$
$\dfrac{h - s}{\sqrt{sh}} \cdot \dfrac{\sqrt{sh}}{\sqrt{(h + s)(h - s)}} \Rightarrow$
$\dfrac{\sqrt{h - s}\sqrt{h - s}}{\sqrt{h + s}\sqrt{h - s}} \Rightarrow \sqrt{\dfrac{h - s}{h + s}}$

2.115 $12a\sqrt{3a} - \sqrt{4^2 \times 3a^2 a} - 4\sqrt{2^2 \times 3a} \Rightarrow$
$12a\sqrt{3a} - 4a\sqrt{3a} - 8\sqrt{3a} \Rightarrow$
$8a\sqrt{3a} - 8\sqrt{3a} \Rightarrow 8\sqrt{3a}(a - 1)$

2.131 $-5b = 2 - 12 \Rightarrow -5b = -10 \Rightarrow b = 2$

2.132 $\dfrac{x}{4} = 12 \Rightarrow x = 4 \times 12 \Rightarrow x = 48$

2.133 $\frac{7}{5z} + \frac{7}{z} = 2 - 1 \Rightarrow \frac{7z + 35z}{5z^2} = 1 \Rightarrow$

$\frac{42z}{5z^2} = 1 \Rightarrow \frac{42}{5z} = 1 \Rightarrow 5z = 42 \Rightarrow$

$z = \frac{42}{5}$

2.134 $2a - c - 3a = 8 \Rightarrow -a - c = 8 \Rightarrow$
$-(a + c) = 8 \Rightarrow a + c = -8$

2.135 Equate cross products, and simplify.
$2(p + 4) = 1(11 - 2q) \Rightarrow 2p + 8 =$
$11 - 2q \Rightarrow 2p + 2q = 11 - 8 \Rightarrow$
$2(p + q) = 3 \Rightarrow p + q = \frac{3}{2} \Rightarrow \frac{p+q}{2} = \frac{3}{4}$

2.136 $11f + 4 - 8 = -3 - 8 \Rightarrow 11f - 4 =$
$-11 \Rightarrow (11f - 4)^2 = (-11)^2 = 121$

2.137 $\frac{c}{d} - \frac{a}{b} = -\left(\frac{a}{b} - \frac{c}{d}\right) = -(-12) = 12$

2.138 $\frac{2k}{3m} = \frac{3k}{2m} \times \frac{2}{3} \times \frac{2}{3} = 18 \times \frac{2}{3} \times \frac{2}{3} = 8$

2.139 $3g - 51 = 9 \Rightarrow 3(g - 17) = 9 \Rightarrow$
$g - 17 = 3$

2.140 $\sqrt{\frac{8h+21}{h-3}} = 3 \Rightarrow \frac{8h+21}{h-3} = \frac{9}{1} \Rightarrow$
$8h + 21 = 9(h - 3) \Rightarrow 8h + 21 =$
$9h - 27 \Rightarrow -h = -48 \Rightarrow h = 48$

2.141 $\sqrt{6y + 1} = 10 - 3 \Rightarrow \sqrt{6y + 1} = 7 \Rightarrow$
$6y + 1 = 49 \Rightarrow 6y = 48 \Rightarrow y = 8$

2.146 Subtract the 2nd equation from the 1st.
$g + 2b - (g + b) = 7 - 5 \Rightarrow$
$g + 2b - g - b = 2 \Rightarrow b = 2$
Substitute this in the 1st equation:
$g + 2(2) = 7 \Rightarrow g = 7 - 4 \Rightarrow g = 3$

2.147 Subtract the 2nd equation from the 1st.
$4n + m - (m - 6n) = 39 - (-31) \Rightarrow$
$4n + m - m + 6n = 70 \Rightarrow 10n = 70 \Rightarrow$
$n = 7$. Substitute this in the 1st eqn.:
$4(7) + m = 39 \Rightarrow m = 39 - 28 = 11$

2.148 Add the two equations.
$5y + x + 2x - 5y = 26 + 7 \Rightarrow 3x = 33$
$\Rightarrow x = 11$. Substitute this in the 1st eqn:
$5y + 11 = 26 \Rightarrow 5y = 15 \Rightarrow y = 3$

2.149 Multiply the 1st eqn. by 3, and 2nd by 5.
$15s + 18t = 96$, and $15s + 20t = 110$
Subtract one from the other.
$15s + 18t - (15s + 20t) = 96 - 110$
$\Rightarrow 15s + 18t - 15s - 20t = -14$
$\Rightarrow -2t = -14 \Rightarrow t = 7$
Substitute this in the 2nd eqn.:
$3s + 4(7) = 22 \Rightarrow 3s = -6 \Rightarrow s = -2$

2.152 Substitute the value of u from the 1st eqn. into the 2nd eqn.
$2(6 - 2v) + v = 15 \Rightarrow 12 - 4v + v$
$= 15 \Rightarrow -3v = 3 \Rightarrow v = -1$
Substitute this in the 1st eqn.
$6 - 2(-1) = u \Rightarrow 6 + 2 = u \Rightarrow 8 = u$

2.153 Substitute the value of s from the 2nd eqn. into the 1st eqn.
$3(3t - 7) + 2t = 12 \Rightarrow 9t - 21 + 2t$
$= 12 \Rightarrow 11t = 33 \Rightarrow t = 3$
Substitute this in the 2nd eqn.
$s = 3(3) - 7 \Rightarrow s = 9 - 7 \Rightarrow s = 2$

2.158 $m^2 + 12m + 27 = 0$
$m^2 + 9m + 3m + 27 = 0$
$m(m + 9) + 3(m + 9) = 0$
$(m + 9)(m + 3) = 0$
$m + 9 = 0 \Rightarrow m = -9$
$m + 3 = 0 \Rightarrow m = -3$

2.159 $6u^2 + 5u - 4 = 0$
$6u^2 + 8u - 3u - 4 = 0$
$2u(3u + 4) - 1(3u + 4) = 0$
$(3u + 4)(2u - 1) = 0$
$3u + 4 = 0 \Rightarrow u = -\frac{4}{3}$
$2u - 1 = 0 \Rightarrow u = \frac{1}{2}$

2.160 $f^2 - 7f + 12 = 0$
$f^2 - 4f - 3f + 12 = 0$
$f(f - 4) - 3(f - 4) = 0$
$(f - 4)(f - 3) = 0$.
$f - 4 = 0 \Rightarrow f = 4$
$f - 3 = 0 \Rightarrow f = 3$

2.161 $10z^2 - z - 3 = 0$
$10z^2 - 6z + 5z - 3 = 0$
$2z(5z - 3) + 1(5z - 3) = 0$
$(5z - 3)(2z + 1) = 0$
$5z - 3 = 0 \Rightarrow z = \frac{3}{5}$
$2z + 1 = 0 \Rightarrow z = -\frac{1}{2}$

2.167 $z^2 - 5z - 19 - 17 = 0$
$z^2 - 5z - 36 = 0$
$z^2 - 9z + 4z - 36 = 0$
$z(z - 9) + 4(z - 9) = 0$
$(z - 9)(z + 4) = 0$
$z - 9 = 0 \Rightarrow z = 9$
$z + 4 = 0 \Rightarrow z = -4$

2.168 $2d^2 + 7d^2 = 11 + 25$

$9d^2 = 36$

$d^2 = 4$

$d = 2, \text{not} - 2$ because $d > 0$

2.169 $y^2 + 2y^2 - 3y - 10y + 27 - 15 = 0$

$3y^2 - 13y + 12 = 0$

$3y^2 - 9y - 4y + 12 = 0$

$3y(y - 3) - 4(y - 3) = 0$

$(y - 3)(3y - 4) = 0$

$y - 3 = 0 \Rightarrow y = 3$

$3y - 4 = 0 \Rightarrow y = \dfrac{4}{3}$

Yes, y can have two *different* values.

2.170 $-e^2 - e^2 = -40 - 10$

$-2e^2 = -50$

$e^2 = 25$

$e = \pm 5$

The minimum possible value of e is –5.

2.171 $p - 6 = 0 \Rightarrow p = 6$

$p^3 + 8 = 0 \Rightarrow p^3 = -8 \Rightarrow p = -2$

$p^2 - \dfrac{4}{9} = 0 \Rightarrow p^2 = \dfrac{4}{9} \Rightarrow p = \pm\dfrac{2}{3}$

$\text{Mean} = \dfrac{6 - 2 + \dfrac{2}{3} - \dfrac{2}{3}}{4} = \dfrac{4}{4} = 1$

2.178 Solving for x: $ax^2 - cx^2 = dy - by \Rightarrow$

$x^2(a - c) = y(d - b) \Rightarrow x^2 = \dfrac{y(d - b)}{a - c}$

$\Rightarrow x = \sqrt{\dfrac{y(d - b)}{a - c}}$

Solving for y: $by - dy = cx^2 - ax^2 \Rightarrow$

$y(b - d) = x^2(c - a) \Rightarrow y = \dfrac{x^2(c - a)}{b - d}$

2.179 Solving for a: $b - y^2 + 3y^2 = 4ax \Rightarrow$

$b + 2y^2 = 4ax \Rightarrow \dfrac{b + 2y^2}{4x} = a$

Solving for b: $b = 4ax - 3y^2 + y^2 \Rightarrow$

$b = 4ax - 2y^2$

Solving for x: $b - y^2 + 3y^2 = 4ax \Rightarrow$

$b + 2y^2 = 4ax \Rightarrow \dfrac{b + 2y^2}{4a} = x$

Solving for y: $-y^2 + 3y^2 = 4ax - b \Rightarrow$

$2y^2 = 4ax - b \Rightarrow y^2 = \dfrac{4ax - b}{2} \Rightarrow$

$y = \sqrt{\dfrac{4ax - b}{2}}$

2.180 Solving for p: $p = s(2q - 3r)$

Solving for q: $\dfrac{2q - 3r}{p} = \dfrac{1}{s} \Rightarrow$

$(2q - 3r)s = p \Rightarrow 2qs - 3rs = p$

$\Rightarrow 2qs = p + 3rs \Rightarrow q = \dfrac{p + 3rs}{2s}$

Solving for r: $\dfrac{2q - 3r}{p} = \dfrac{1}{s} \Rightarrow$

$(2q - 3r)s = p \Rightarrow 2qs - 3rs = p$

$\Rightarrow -3rs = p - 2qs \Rightarrow r = \dfrac{2qs - p}{3s}$

Solving for s: It's already solved for s!

2.181 The cx term needs to go.

From the 1st equation: $m - 4f = cx$

From the 2nd equation: $\dfrac{n + 1}{3} = cx$

Equating the two values of cx:

$m - 4f = \dfrac{n + 1}{3}$ $\Rightarrow$ $m = 4f + \dfrac{n + 1}{3}$

2.182 The c term needs to go.

From the 1st eqn.: $h^2 = 81c^2$, or $\dfrac{h^2}{81} = c^2$

From the 2nd eqn.: $4e = 9c^2$, or $\dfrac{4e}{9} = c^2$

Equating the two values of c^2:

$\dfrac{4e}{9} = \dfrac{h^2}{81} \Rightarrow e = \dfrac{h^2}{81} \cdot \dfrac{9}{4} \Rightarrow e = \dfrac{h^2}{36}$

2.183 Express all bases in terms of 5, the smallest base.

$\dfrac{5^3}{(5^2)^{a+b}} = \dfrac{5^y}{(5^3)^{c-d}} \Rightarrow \dfrac{5^3}{5^{2a+2b}} = \dfrac{5^y}{5^{3c-3d}}$

$5^{3-2a-2b} = 5^{y-3c+3d} \Rightarrow$

$3 - 2a - 2b = y - 3c + 3d \Rightarrow$

$y = 3 - 2a - 2b + 3c - 3d$

2.190 $p + 8 = 20 \Rightarrow p = 12$

$-(p + 8) = 20 \Rightarrow -p - 8 = 20 \Rightarrow$

$-p = 28 \Rightarrow p = -28$

$\text{Average} = \dfrac{-28 + 12}{2} = \dfrac{-16}{2} = -8$

2.191 $|8 + 3b| = 9$

$8 + 3b = 9 \Rightarrow 3b = 1 \Rightarrow b = \dfrac{1}{3}$

$-(8 + 3b) = 9 \Rightarrow -8 - 3b = 9 \Rightarrow$

$-3b = 17 \Rightarrow b = -\dfrac{17}{3}$

Since b is positive, it must be $\dfrac{1}{3}$.

2.192 $3u - 14 = 1 \Rightarrow 3u = 15 \Rightarrow u = 5$

$-(3u - 14) = 1 \Rightarrow -3u + 14 = 1 \Rightarrow$

$-3u = -13 \Rightarrow u = \dfrac{13}{3} \Rightarrow u = 4\dfrac{1}{3}$

Their sum $= 5 + 4\dfrac{1}{3} = 9\dfrac{1}{3}$

2.193 $15 - 4k = 17 \Rightarrow -4k = 2 \Rightarrow k = -\dfrac{1}{2}$

$-(15 - 4k) = 17 \Rightarrow -15 + 4k = 17$

$\Rightarrow 4k = 32 \Rightarrow k = 8 =$ integer value.

2.194 $40 - h^2 = 9 \Rightarrow -h^2 = -31 \Rightarrow$

$h^2 = 31 \Rightarrow h = \pm\sqrt{31}$

$-(40 - h^2) = 9 \Rightarrow -40 + h^2 = 9$

$\Rightarrow h^2 = 49 \Rightarrow h = \sqrt{49} \Rightarrow h = \pm 7$

The maximum value of h is 7.

2.195 The least value of a modulus is zero.

$50 - 6g = 0 \Rightarrow -6g = -50 \Rightarrow g = 8\dfrac{1}{3}$

But g is to be an *integer*. Clearly, it would be 8, not 9 because $8\dfrac{1}{3}$ is closer to 8 than it is to 9.

2.207 1st inequality: $3f \geq 25 - 7 \Rightarrow f \geq 6$

2nd inequality: $3f < 23 + 19 \Rightarrow f < 14$

Mean of max. and min. integer values would be $\dfrac{13+6}{2} = \dfrac{19}{2} = 9.5$

2.208 Divide the 1st inequality by c^2 $(c^2 > 0)$

$3 \leq \dfrac{4c}{5} \Rightarrow 3 \times \dfrac{5}{4} \leq c \Rightarrow c \geq \dfrac{15}{4} \Rightarrow c \geq 3\dfrac{3}{4}$

Divide the 2nd inequality by d^2 $(d^2 > 0)$

$d < 5$

The only integer value common to both c and d is 4.

2.209 (i) $20 + 69 = 89$

(ii) $20 - 14 = 6$

(iii) $69 - (-7) = 69 + 7 = 76$

2.210 Simplify the inequality (isolate y).

$-24 < -3y \leq 11 \Rightarrow 8 > y \geq -\dfrac{11}{3}$

$\Rightarrow -3\dfrac{2}{3} \leq y < 8$

y can assume integer values from -3 up to 7, a total of 11 values.

2.211 $w + x + y + z \leq 13 - 9$

$w + x + y + z \leq 4$

$\dfrac{w + x + y + z}{4} \leq 1$

Their maximum possible average is 1.

2.212 The question is about p. So, isolate p.

$2p \geq 15 - q \quad \Rightarrow \quad p \geq \dfrac{15 - q}{2}$

(i) For p to be at its minimum value, we should subtract the most from 15 in the inequality, i.e. we should consider the maximum value of q (i.e. 19). When $q = 19, p \geq \dfrac{15-19}{2} \Rightarrow p \geq -\dfrac{4}{2} \Rightarrow$ $p \geq -2$. Therefore, the minimum integer value of p is -2.

(ii) This time, q can't be 19, but slightly less than it. Also, q does not have to be an *integer*. Meaning, its maximum value could be 18.9999999...... Hence, $p \geq \dfrac{15 - 18.99999\ldots}{2} \Rightarrow p \geq \dfrac{-3.9999999\ldots\ldots}{2} \Rightarrow$ $p \geq -1.99999\ldots$ The minimum integer value of p would now be -1 (not -2).

2.213 Given inequality: $r < q < 0 < p^2$

(i) $p - q < p - r \Rightarrow -q < -r \Rightarrow q > r$

In line with the given inequality. Therefore, statement holds true.

(ii) $p > qr$ We can tell that qr will be positive, because both q and r are negative. p^2 being positive, p may be positive or negative. If negative, then $p < qr$, but if positive, then it may be greater or less than qr, depending upon their actual values. Hence, this statement does not always hold true.

(iii) $\dfrac{q}{r} < 1$ Multiplying both sides by r (negative), we get $q > r$. Matches with what's given. Therefore, holds true.

(iv) $\dfrac{p}{q} < \dfrac{p}{r}$ Multiplying both sides by qr (positive), we get $pr < pq$. We know that p could be positive or negative. If p is positive, then dividing both sides by p, we get $r < q$ which matches up with what's given. But if p is negative, then we get $r > q$ which doesn't match up. Therefore, the statement is not always true.

(v) $\dfrac{p}{r} < \dfrac{q}{r}$ Multiplying both sides by r (negative), we get $p > q$. Once again, p is greater than q if positive, but may be greater or less than q if negative. Therefore, statement not always true.

(vi) $pqr > 0$ qr is positive, but when multiplied by p, may turn out to be positive or negative. Statement not always true.

2.218 $a + 19 \geq 4 \Rightarrow a \geq -15$
$-(a + 19) \geq 4 \Rightarrow -a - 19 \geq 4 \Rightarrow$
$-a \geq 23 \Rightarrow a \leq -23$

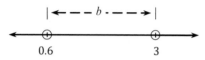

2.219 $9 - 5b < 6 \Rightarrow -5b < -3 \Rightarrow b > 0.6$
$-(9 - 5b) < 6 \Rightarrow -9 + 5b < 6 \Rightarrow$
$5b < 15 \Rightarrow b < 3$

2.220 $4m + 7 > 19 - 2m \Rightarrow 6m > 12$
$\Rightarrow m > 2$
$-(4m + 7) > 19 - 2m \Rightarrow$
$-4m - 7 > 19 - 2m \Rightarrow$
$-2m > 26 \Rightarrow m < -13$

2.221 $13 - 5n \leq 11 + 3n \Rightarrow -8n \leq -2 \Rightarrow$
$n \geq 0.25$
$-(13 - 5n) \leq 11 + 3n \Rightarrow$
$-13 + 5n \leq 11 + 3n \Rightarrow$
$2n \leq 24 \Rightarrow n \leq 12$

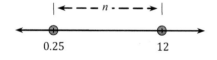

2.226

$-11 \leq (e - 44) \leq 11$
$e - 44 \leq 11$
$e - 44 \geq -11 \Rightarrow -(e - 44) \leq 11$
Together, $|e - 44| \leq 11$

2.227

$-6 < (f + 15) < 6$
$f + 15 < 6$
$f + 15 > -6 \Rightarrow -(f + 15) < 6$
Together, $|f + 15| < 6$

2.228

$u - 3 \geq 11$
$u - 3 \leq -11 \Rightarrow -(u - 3) \geq 11$
Together, $|u - 3| \geq 11$

2.229

$v + 2.5 > 4.5$
$v + 2.5 < -4.5 \Rightarrow -(v + 2.5) > 4.5$
Together, $|v + 2.5| > 4.5$
Alternatively, multiply both sides by 2
to get $|2v + 5| > 9$

CHAPTER 3: WORD PROBLEMS

3.010 Three times a number increased by 8 means $3x + 8$. Four times the number decreased by 3 means $4x - 3$. Equate and solve.
$3x + 8 = 4x - 3 \Rightarrow -x = -11 \Rightarrow x = 11$

3.011 Twice the difference between a number and 4 means $2(x - 4)$. Add this to $2x$ to get 4.
$2x + 2(x - 4) = 4 \Rightarrow 2x + 2x - 8 = 4$
$\Rightarrow 4x = 12 \Rightarrow x = 3$

3.012 12 *less* a number means $12 - x$. Quotient of the number and 12 means $\frac{x}{12}$.
$12 - x = 4 \cdot \frac{x}{12} \Rightarrow 12 - x = \frac{x}{3} \Rightarrow$
$-x - \frac{x}{3} = -12 \Rightarrow -\frac{4x}{3} = -12 \Rightarrow x = 9$

3.013 Assume *smaller* number as x. The larger number would be $6x$. Add the two to get 28.
$x + 6x = 28 \Rightarrow 7x = 28 \Rightarrow x = 4$

3.014 Both digits are to be determined. Conveniently assume the tens digit as x. Units digit would be $2x$. The *value* of this number is $10x + 2x$, or $12x$. After reversing the digits, the units digit is x; the tens digit is $2x$. The value of this new number is $20x + x$, or $21x$. It is 12 less than twice the original number.
$21x = 2(12x) - 12 \Rightarrow 21x = 24x - 12$
$\Rightarrow -3x = -12 \Rightarrow x = 4$ The number=48

3.015 Assume x women. 15 fewer men, means $(x - 15)$ men.
Men + Women = 65 $\Rightarrow (x - 15) + x = 65$
$2x - 15 = 65 \Rightarrow 2x = 80 \Rightarrow x = 40$

3.016 Assume width as x. Length + Width = 15
$\Rightarrow$ Length = 15 – Width $\Rightarrow$ Length = 15 – x
2 × Length = 3 × Width $\Rightarrow 2(15 - x) = 3x$
$\Rightarrow 30 - 2x = 3x \Rightarrow 30 = 5x \Rightarrow 6 = x$

3.017 Assume x number of \$10 bottles sold. Total bottles sold = 200. Therefore, $(200 - x)$ number of \$5 bottles sold.
Revenue from \$5 bottles
+ Revenue from \$10 bottles = Total Revenue
$5(200 - x) + 10x = 1250$
$\Rightarrow 1000 - 5x + 10x = 1250 \Rightarrow 5x = 250$
$\Rightarrow x = 50$

3.022 Doesn't matter which one you assume as x. Let's say the smallest one. The others would be $x + 2$ and $x + 4$.
Product of 1st and 2nd : $x(x + 2)$ or $x^2 + 2x$.
12 more than that: $x^2 + 2x + 12$.
Product of 2nd and 3rd: $(x + 2)(x + 4) \Rightarrow$
$x^2 + 6x + 8$.
Equate the two: $x^2 + 2x + 12 = x^2 + 6x + 8 \Rightarrow$
$-4x = -4 \Rightarrow x = 1$
The numbers are 1, 3, 5. Their sum is 9.

3.023 The largest is x. The others are $x - 1$ and $x - 2$. Their sum would be $3x - 3$.   4 less than 4 times the smallest would be $4(x - 2) - 4$, or $4x - 8 - 4$, or $4x - 12$.   Equate the two: $3x - 3 = 4x - 12 \Rightarrow -x = -9 \Rightarrow x = 9$

3.024 Assume the integers as x, $x + 1$, $x + 2$, $x + 3$ and $x + 4$. Their sum would be $5x + 10$. Equate this to zero: $5x + 10 = 0 \Rightarrow$
$5x = -10 \Rightarrow x = -2$

3.035 4½ years out of 10 years.
$\dfrac{4\frac{1}{2}}{10} \Rightarrow \dfrac{\frac{9}{2}}{10} \Rightarrow \dfrac{9}{2} \times \dfrac{1}{10} \Rightarrow \dfrac{9}{20}$

3.036 $\dfrac{\text{business class seats}}{\text{total seats}} \Rightarrow \dfrac{30}{30+80} \Rightarrow \dfrac{30}{110} \Rightarrow \dfrac{3}{11}$

3.037 Total fleet = 45 helicopters + 60 jets. 35 out of 60 jets are small. So, 25 jets are large. Fleet of aircraft *not* made of small jets would be 45 helicopters + 25 large jets. This, as a fraction of the total fleet is:
$\dfrac{\text{heli.} + \text{large jets}}{\text{total fleet}} = \dfrac{45 + 25}{45 + 60} = \dfrac{70}{105} = \dfrac{2}{3}$

3.038 $\dfrac{4}{15}$ of the trip is done. So, $\dfrac{15}{15} - \dfrac{4}{15} = \dfrac{11}{15}$ of the trip is not done. $\dfrac{11}{15}$ of 450 miles would be:
$\dfrac{11}{15} \times 450 = \dfrac{11 \times \cancel{450}^{\,30}}{\cancel{15}^{\,1}} = 330$ miles.

3.039 P dollars to be split among 6 students. So, each would pay $\dfrac{P}{6}$ dollars. If one student drops out, then each of the remaining 5 would pay $\dfrac{P}{5}$ dollars. Clearly, this amount is greater than the previous amount, the difference being:
$\dfrac{P}{5} - \dfrac{P}{6} = P\left(\dfrac{1}{5} - \dfrac{1}{6}\right) = P\left(\dfrac{6-5}{30}\right) = P\left(\dfrac{1}{30}\right) = \dfrac{P}{30}$

3.040 Assume x applications received. $\dfrac{1}{4}x$ didn't qualify. It means, $\dfrac{3}{4}x$ qualified, and were interviewed. $\dfrac{1}{3}$ of those $\dfrac{3}{4}x$ made it in, i.e. $\dfrac{1}{3} \cdot \dfrac{3}{4}x$, or $\dfrac{1}{4}x$ were selected to fill the 16 vacancies. Thus, $\dfrac{1}{4}x = 16 \Rightarrow x = 16 \times 4 = 64$.

3.041 The least miles would be when I'd put the least fuel (10 gallons), and my car would drive with the worst fuel efficiency (35 mpg). Distance = 10 gallons × 35 mpg = 350 miles. The most miles would be when I'd put the most fuel (15 gallons), and my car would drive with the best fuel efficiency (40 mpg). Distance = 15 gallons × 40 mpg = 600 miles.

3.042 $\dfrac{2}{3}$ of $x = 10$.   Notice that $\dfrac{1}{3}$ is simply $\dfrac{1}{2}$ of $\dfrac{2}{3}$. Meaning, $\dfrac{1}{3}x = \dfrac{1}{2} \cdot \dfrac{2}{3}x = \dfrac{1}{2} \cdot 10 = 5$

3.043 List A: 2, 4, 6.   List B: 1, 3, 5.
The following divisions result:
$\dfrac{2}{1}, \dfrac{2}{3}, \dfrac{2}{5};$ $\dfrac{4}{1}, \dfrac{4}{3}, \dfrac{4}{5};$ $\dfrac{6}{1}, \dfrac{6}{3}, \dfrac{6}{5}$
Reducing and rewriting, we get:
$2, \dfrac{2}{3}, \dfrac{2}{5};$ $4, \dfrac{4}{3}, \dfrac{4}{5};$ $6, 2, \dfrac{6}{5}$
The *unique* integers are only 2, 4 and 6.

3.044 There's no need to list every number between 50 and 80. Simply look at the numbers dividing them: 5, 10, 15, 20. Only the multiples of 5 *between* 50 and 80 (means, exclude 50 and 80) would need to be considered. All others when divided by 5, 10, 15 or 20 will result in fractions. The list is now: 55, 60, 65, 70, 75. A quick look will tell us that 55, 65, 70 and 75 cannot be divided by 20. Only 60 remains which, when divided by each of 5, 10, 15 and 20 results in an integer.

3.055 $\frac{12.1}{100} \times 400 = x \Rightarrow 48.4 = x$

3.056 $35 = \frac{x}{100} \times 175 \Rightarrow \frac{35 \times 100}{175} = x \Rightarrow$

$\frac{35^5 \times 100^4}{175^1} = x \Rightarrow 20 = x$

3.057 $19p = \frac{25}{100} \times x \Rightarrow 19p = \frac{25^1}{100^4} \times x$

$19p \times 4 = x \Rightarrow 76p = x$

3.058 Assume it sold x tickets. 5% of x generally get cancelled. 95 tickets were cancelled. Equate the two:

$\frac{5}{100} x = 95 \Rightarrow x = \frac{95 \times 100^{20}}{5^1} = 1900$

3.059 Down payment + monthly payments: (20% of 20,000) + (300 × 60) $\Rightarrow$

$\frac{20}{100} \times 20{,}000 + 18{,}000 \Rightarrow 4{,}000 + 18{,}000 \Rightarrow$ $22,000. That's $2,000 more than the price tag.

3.060 Assume her annual salary as x.

20% of x = 45,000 $\Rightarrow \frac{20^1}{100^5} x = 45{,}000 \Rightarrow$

$x = 45{,}000 \times 5 \Rightarrow x = \$225{,}000$

3.061 Both of their ages are in comparison to my age. So, assume my age as x. Think convenience. My sister's age would then be $\frac{4}{5}x$, and my mom's age would be $2x$. My sister's age, as a percentage of my mom's age would be:

$\frac{\text{sister's age}}{\text{mom's age}} \times 100\% \Rightarrow \frac{\frac{4}{5}x}{2x} \times 100\%$

$\Rightarrow \frac{4}{5} \times \frac{1}{2} \times 100\% \Rightarrow 40\%$

3.062 Given information:

	Sharks	Dolphins	TOTAL
Male	25% of 12		12
Female			
TOTAL	30% of 30	70% of 30	30

Calculate the numbers from the percentages:

	Sharks	Dolphins	TOTAL
Male	3		12
Female			
TOTAL	9	21	30

Fill in the rest through subtraction:

	Sharks	Dolphins	TOTAL
Male	3	9	12
Female	6	13	18
TOTAL	9	21	30

Male sharks : Female dolphins = 3 : 13

3.063 Number of wins it already has:

80% of first 15 = $\frac{80^4}{100^5} \times 15 = 12$

Total number of wins it needs:

75% of total 20 = $\frac{75^3}{100^4} \times 20 = 15$

So, it needs 3 more wins to get to 15. This 3, as a percentage of 5 is: $\frac{3}{5} \times 100\% = 60\%$

3.064 Assume x dollars invested in each fund.

Interest from fund A: 13% of $x = \frac{13}{100}x$

Interest from fund B: 12% of $x = \frac{12}{100}x$

Total interest: $\frac{13}{100}x + \frac{12}{100}x = \frac{25}{100}x = \frac{1}{4}x$

Interest from A, as percentage of total interest:

$\frac{\frac{13}{100}x}{\frac{1}{4}x} \times 100\% = \frac{13}{100} \times \frac{4}{1} \times 100\% = 52\%$

3.073 72 number of employees, after hiring 12 new ones. Meaning, there must have been 72 − 12 = 60 employees originally.

Percentage increase = $\frac{\text{Change}}{\text{Original value}} \times 100\%$

$= \frac{12}{60} \times 100\% = \frac{1}{5} \times 100\% = 20\%$

3.074 $25 loss, after selling the TV for $100. Meaning, I must have bought it for $125.

Percentage loss = $\frac{\text{Loss}}{\text{Original value}} \times 100\%$

$= \frac{25}{125} \times 100\% = \frac{1}{5} \times 100\% = 20\%$

3.075 Value of my house after 1 year: 100,000 − (10% of 100,000) = 100,000 − 10,000 = $90,000   After 2 years: 90,000 − (10% of 90,000) = 90,000 − 9,000 = $81,000

After 3 years: $81{,}000 - (10\% \text{ of } 81{,}000) = 81{,}000 - 8{,}100 = \$72{,}900$

3.076 Discounted price $= 200 - (10\% \text{ of } 200) = 200 - 20 = \$180$. Sales tax $= 10\%$ of $180 = \$18$ Discounted price $+$ Sales tax $= 180 + 18 = \$198$. This is what the TV will cost on Saturdays.

3.077 In 1970, 10 gallons would go 120 miles. Today, only 60% of that fuel would go the same 120 miles. Meaning, 6 gallons would go 120 miles. Every gallon would go $\dfrac{120}{6}$ miles, or 20 miles. 10 gallons would go $10 \times 20 = 200$ miles.

3.078 I've got to make a 40% profit on the $1,000 I spent. Meaning, I need to make $1,400 from the sale. Assume the list price as x. Then, 30% discounted off this price should get me my $1,400. $x - (30\% \text{ of } x) = 1400 \Rightarrow x - 0.3x = 1400 \Rightarrow 0.7x = 1400 \Rightarrow$

$x = \dfrac{1400}{0.7} = \dfrac{1400}{0.7} \times \dfrac{10}{10} = \dfrac{14{,}000}{7} = \$2{,}000$

3.079 Assume x as the original worth of the house. After losing 30%, it will be worth $x - (30\% \text{ of } x) = x - 0.3x = 0.7x$ If this new value of $0.7x$ is to increase to x, then:

Percentage increase $= \dfrac{\text{Change}}{\text{Original value}} \times 100\%$

$= \dfrac{x - 0.7x}{0.7x} \times 100\% = \dfrac{0.3x}{0.7x} \times 100\%$

$= \dfrac{3}{7} \times 100\% = \dfrac{300}{7}\% = 42.86\%$

3.080 Assume the original price as x dollars. Price of blue chairs $= x + (10\% \text{ of } x) = x + 0.1x = 1.1x$ Price of yellow chairs $= x - (20\% \text{ of } x) = x - 0.2x = 0.8x$ Their difference $= 1.1x - 0.8x = 0.3x$ This difference is given as $12. Therefore, $0.3x = 12$

$\Rightarrow x = \dfrac{12}{0.3} = \dfrac{12}{0.3} \times \dfrac{10}{10} = \dfrac{120}{3} = \$40$

3.089 dollar : quarter : 2 dimes $= 100 : 25 : 20$ Reduce it by 5 to get $20 : 5 : 4$.

3.090 Number of seconds in a day on planet X is $60 \times 24 \times 24$.   Number of seconds in a day on Earth is $24 \times 60 \times 60$.

$\dfrac{\text{Seconds on planet } X}{\text{Seconds on Earth}} = \dfrac{60 \times 24 \times 24}{24 \times 60 \times 60} = \dfrac{24}{60} = \dfrac{2}{5}$

Therefore, the ratio is $2 : 5$.

3.091 The ratio of # of meat-eaters to the total # of people is $4 : (13 + 4)$, or $4 : 17$. In terms of a fraction, it would be $\dfrac{4}{17}$.

3.092 econ. : eng. : lib. arts $= 5 : 2 : 3$ Engineering would comprise 2 parts out of a total of $5 + 2 + 3 = 10$ parts. Total 60 students comprise the total 10 parts. Meaning, each part is worth 6 students. Engineering comprises 2 parts, or $2(6) = 12$ students.

3.093 Assume it lays off x men and x women. Ratio of men to women after layoff would be $(40 - x) : (50 - x)$. This ratio is given as $5 : 7$. Equate the two. $(40 - x) : (50 - x) = 5 : 7$

$\dfrac{40 - x}{50 - x} = \dfrac{5}{7} \Rightarrow 7(40 - x) = 5(50 - x) \Rightarrow$

$280 - 7x = 250 - 5x \Rightarrow -2x = -30$

$\Rightarrow x = 15$

3.094 First, 1 inch $= 2.5$ cm $= 2.5(75)$ miles Distance between the two towns would be: 2 inches $= 2 \times 2.5(75) = 5(75) = 375$ miles

3.095 In 1 min, 300 words. Therefore, in 60 min, $60(300) = 18{,}000$ words, i.e. in 1 hour, 18,000 words.

1 hr : 18,000 words $= x$ hrs : 27,000 words

$1 : 18{,}000 = x : 27{,}000 \Rightarrow \dfrac{1}{18{,}000} = \dfrac{x}{27{,}000}$

$\Rightarrow \dfrac{27{,}000}{18{,}000} = x \Rightarrow \dfrac{27}{18} = x \Rightarrow x = 1.5$ hrs

3.096 1 hr $= 60 \times 60 = 3600$ seconds

Therefore, 1 second $= \dfrac{1}{3600}$ hrs

S seconds $= \dfrac{S}{3600}$ hrs

P pages : $\dfrac{S}{3600}$ hrs $= x$ pages : $\dfrac{1}{2}$ hr

$P : \dfrac{S}{3600} = x : \dfrac{1}{2} \Rightarrow \dfrac{P}{\frac{S}{3600}} = \dfrac{x}{\frac{1}{2}} \Rightarrow$

$P \times \dfrac{3600}{S} = x \times \dfrac{2}{1} \Rightarrow \dfrac{3600P}{S} = 2x \Rightarrow$

$x = \dfrac{1800P}{S}$ pages

3.103 Grades   Compared to avg. 85
         85, 79, 80, 92 $0 - 6 - 5 + 7 = -4$
I was 4 points short of 85. So I must have made up for it in the fifth course by scoring 89.

3.104 $\text{avg} = \dfrac{\text{sum}}{17}$ Average, as a fraction of the sum would be $\dfrac{\text{avg}}{\text{sum}}$, i.e. $\dfrac{1}{17}$.

3.105 avg. gain in speed $= \dfrac{\text{gain in speed}}{\text{time elapsed}} = \dfrac{160-0}{5} = 32$ mph per minute.

3.106 $\dfrac{\text{sum}}{3} = \text{avg} \Rightarrow \dfrac{\text{sum}}{3} = 120 \Rightarrow$ sum $= 3 \times 120 = 360$

If one increases by 12, second drops by 3, and third drops by 6, the sum increases by 12 – 3 – 6 = 3. The new sum is 360 + 3 = 363. The new average would be $\dfrac{363}{3} = 121$ mph.

3.107 Total maximum score = 6 × 100 = 600. Abigail needs a total of 6 × 90 = 540.
She has scored 79 + 88 + 86 = 253.
To see just how low she could score on the 6th test, let's assume she scores a full 100 on each of the 4th and 5th tests. Then her score would reach 253 + 100 + 100 = 453.
She would need to score 540 – 453 = 87 or more on the 6th test.

3.108 Assume x families relocated into 500 sq. ft. homes. Then $(15 - x)$ families must have been relocated into 800 sq. ft. homes.
avg. area of apts. $= \dfrac{\text{total area of apts.}}{\text{\# of apts.}} \Rightarrow$
$660 = \dfrac{500x + 800(15 - x)}{15} \Rightarrow$
$660 \times 15 = 500x + 12000 - 800x \Rightarrow$
$9900 = 12000 - 300x \Rightarrow -2100 = -300x$
$\dfrac{-2100}{-300} = x \Rightarrow 7 = x$

3.112 Total speed of 12 electrons = 200 × 12 = 2400 km/s;   Total speed of 3 protons = 300 × 3 = 900 km/s;
Total speed of 5 neutrons = 700 × 5 = 3500 km/s.
Avg. speed of all $= \dfrac{\text{sum of speeds of all}}{\text{total \# of particles}} =$
$\dfrac{2400 + 900 + 3500}{12 + 3 + 5} = \dfrac{6800}{20} = 340$ km/s

3.113 Total value of cars = 18,000 × 30 = $540,000; Total value of vans = 26,000 × 25 = $650,000;   Total value of trucks = 14,000 × 15 = $210,000.

Avg. value of all vehicles $= \dfrac{\text{Total value of all}}{\text{Total \# of vehicles}}$
$= \dfrac{540,000 + 650,000 + 210,000}{30 + 25 + 15} = \dfrac{1,400,000}{70}$
$= \$20,000$

3.114 Total UFO sightings in the south = 200 × 7 = 1400;
Total UFO sightings in the northeast = 150 × 6 = 900;   Total UFO sightings in the west = $x \times 5 = 5x$
Avg. of all sightings $= \dfrac{\text{Total \# of sightings}}{\text{\# of states where seen}} \Rightarrow$
$300 = \dfrac{1400 + 900 + 5x}{18} \Rightarrow$
$300 \times 18 = 2300 + 5x \Rightarrow 5400 = 2300 + 5x$
$3100 = 5x \Rightarrow 620$

3.119 Value of mixture = Value of peanut oil + Value of olive oil = $(35 \times 4) + (3 \times 45) = 140 + 135 = \$275$

3.120 Assume x kg of brown rice. Then, the mixture would comprise $(x + 60)$ kg of rice.
Value of mixture =
Value of brown rice + Value of basmati rice $\Rightarrow$
$150x + 180(60) = 168(x + 60) \Rightarrow$
$150x + 10,800 = 168x + 10,080 \Rightarrow$
$-18x = -720 \Rightarrow x = \dfrac{-720}{-18} = 40$

3.121 Assume x pounds of pistachios. The mixture has 15 pounds in all. So there must be $(15-x)$ pounds of cashews. Value of mixture = Value of cashews + Value of pistachios $\Rightarrow$
$3.20(15) = 2.50(15 - x) + 6x \Rightarrow$
$32(15) = 25(15 - x) + 60x \Rightarrow$
$480 = 375 - 25x + 60x \Rightarrow 105 = 35x$
$\Rightarrow x = 3$

3.122 Vegetable oil = 75% of 20 gallons = $0.75 \times 20 = 15$ gallons
Diesel = the remaining 5 gallons
(i) Add x gallons vegetable oil.   Now, there's $(x + 15)$ gallons of vegetable oil. Diesel is the same at 5 gallons.
Veg. oil : diesel $= (x + 15):5 = 90\% : 10\%$
$\dfrac{x + 15}{5} = \dfrac{9}{1} \Rightarrow x + 15 = 45 \Rightarrow x = 30$
(ii) Add y gallons of diesel. Diesel's now $(y + 5)$ gallons. Veg. oil is the same at 15 gallons.
Veg. oil : Diesel $= 15:(y + 5) = 25\%:75\%$

$$\frac{15}{y+5}=\frac{25}{75} \quad \Rightarrow \quad \frac{15}{y+5}=\frac{1}{3} \quad \Rightarrow$$

$$15(3)=y+5 \quad \Rightarrow \quad 45=y+5 \quad \Rightarrow \quad y=40$$

3.127

	Rebecca	Peter
Today	x	$x+6$
18 yrs ago	$x-18$	$x+6-18=x-12$

$$x-12=2(x-18) \quad \Rightarrow \quad x-12=2x-36$$
$$\Rightarrow \quad 24=x$$

3.128

	Liz	Coleen
Today	x	$2x$
5 yrs later	$x+5$	$2x+5$

$$2(2x+5)=3(x+5) \quad \Rightarrow \quad 4x+10=3x+15$$
$$\Rightarrow \quad x=5 \qquad \text{Coleen's age} = 2x=2(5)=10$$

3.129

	Ulrich	Ulster
Today	$2x$	x
9 yrs later	$2x+9$	$x+9$

$$(2x+9)+(x+9)=30 \quad \Rightarrow \quad 3x+18=30$$
$$\Rightarrow \quad 3x=12 \quad \Rightarrow \quad x=4$$

3.130

	Sister	Me
Today	x	$x-3$
6 yrs ago	$x-6$	$x-3-6=x-9$
6 yrs later	$x+6$	$x-3+6=x+3$

$$4(x-9)=x+6 \quad \Rightarrow \quad 4x-36=x+6$$
$$\Rightarrow \quad 3x=42 \quad \Rightarrow \quad x=14 \qquad \text{My age} = 11$$

3.133 Temp. $= 96 \times \left(\frac{1}{2}\right)^6 = 96 \div 64 = 1.5°F$

3.134 $1280 = 20 \times 4^t \quad \Rightarrow \quad \dfrac{1280}{20} = 4^t \quad \Rightarrow$

$$4^t = 64 \quad \Rightarrow \quad 4^t = 4^3 \quad \Rightarrow \quad t = 3 \text{ months}$$

3.140 From 3:45 p.m. to 7:30 p.m. is $3\frac{3}{4}$ hours, or $\frac{15}{4}$ hours. Distance = 225 miles.

$$s = \frac{d}{t} = \frac{225}{\frac{15}{4}} = 225 \times \frac{4}{15} = 60 \text{ mph}$$

3.141 m minutes means $\frac{m}{60}$ hours. First, use the rate of *walking* to determine the distance. $d = st \quad \Rightarrow \quad d = 2 \times \frac{m}{60} \quad \Rightarrow \quad d = \frac{m}{30}$ miles. Now, use this distance to determine the rate of *jogging*. Once again, keep in mind that $m + 15$ minutes means $\frac{m+15}{60}$ hours. Also, the distance is twice of d (house to school, and back to house).

$$s = \frac{2d}{t} = \frac{2 \times \frac{m}{30}}{\frac{m+15}{60}} = \frac{2m}{30} \times \frac{60}{m+15} = \frac{4m}{m+15}$$

3.142

5 miles	3 miles
⟵——————⟶	⟵——————⟶
30 mph	36 mph

Time for the 1st part $= t = \dfrac{d}{s} = \dfrac{5}{30} = \dfrac{1}{6}$ hour

Time for the 2nd part $= \dfrac{3}{36} = \dfrac{1}{12}$ hour

Total time $= \dfrac{1}{6} + \dfrac{1}{12} = \dfrac{12+6}{72} = \dfrac{18}{72} = \dfrac{1}{4}$ hour

Avg. speed for the entire trip $= \dfrac{\text{total distance}}{\text{total time}}$

$$= \frac{5+3}{\frac{1}{4}} = 8 \times 4 = 32 \text{ mph}$$

3.143

375 miles	175 miles
⟵——————⟶	⟵——————⟶
7.5 hours	x mph

Total distance is 550 miles, and the average speed for that distance is to be 55 mph.

Hence, total time available $= \dfrac{d}{s} = \dfrac{550}{55} = 10$ hrs

Out of those 10 hours, 7.5 hours are already used up in the first part of the trip. 2.5 more hours remain to travel 175 miles.

$$s = \frac{d}{t} = \frac{175}{2.5} = \frac{175}{2.5} \times \frac{10}{10} = \frac{1750}{25} = 70 \text{ mph}$$

3.144

x miles	x miles
⟵——————⟶	⟵——————⟶
64 mph	48 mph
Home	Office	Home

Time from home to office $= t = \dfrac{d}{s} = \dfrac{x}{64}$ hours

Time from office to home $= t = \dfrac{d}{s} = \dfrac{x}{48}$ hours

Total time = Home to office + Office to home

$$1\frac{3}{4} = \frac{x}{64} + \frac{x}{48} \quad \Rightarrow \quad \frac{7}{4} = x\left(\frac{1}{64} + \frac{1}{48}\right) \quad \Rightarrow$$

$$7 = x\left(\frac{1}{16} + \frac{1}{12}\right) \quad \Rightarrow \quad 7 = x\left(\frac{12+16}{16 \times 12}\right) \quad \Rightarrow$$

$$7 = \frac{28x}{16 \times 12} \quad \Rightarrow \quad 28x = 7 \times 16 \times 12 \quad \Rightarrow$$

$$x = \frac{7 \times 16 \times 12}{28} = 48$$

3.148 The two buses are going away from each other at $50 + 45 = 95$ mph.

$$t = \frac{d}{s} = \frac{380}{95} = 4 \text{ hours}$$

3.149 The two trains are 429 miles away at noon. By 2 p.m., the first train has already covered $57 \times 2 = 114$ miles. So, at 2 p.m., the two trains are $429 - 114 = 315$ miles apart. Now, they start closing in on each other at $57 + 48 = 105$ mph. The time it will take them to reach each other $= \frac{d}{s} = \frac{315}{105} = 3$ hours. They will cross each other 3 hours after 2 p.m., i.e. at 5 p.m.

3.150 In 15 minutes (¼ hour) from 1 p.m. to 1:15 p.m., the starving plane will have gone $300 \times \frac{1}{4} = 75$ miles away. The tanker plane will have 75 miles to cover, closing in on the starving plane at $450 - 300 = 150$ mph.
$t = \frac{d}{s} = \frac{75}{150} = \frac{1}{2}$ hour after $1:15$ p. m.,
i.e. at 1:45 p.m.

3.155 In one hour John can plaster $\frac{1}{8}$ the wall; Fred can plaster $\frac{1}{12}$ the wall; Chris can plaster $\frac{1}{6}$ the wall. Together, in one hour, they can plaster: $\frac{1}{8} + \frac{1}{12} + \frac{1}{6} = \frac{12+8}{96} + \frac{1}{6} = \frac{20}{96} + \frac{1}{6} = \frac{5}{24} + \frac{1}{6} = \frac{30+24}{144} = \frac{54}{144} = \frac{3}{8}$ the wall. If they work for x hours, and finish the entire wall, then: $x \times \frac{3}{8} = 1 \quad \Rightarrow \quad x = \frac{8}{3} = 2\frac{2}{3}$ hours $= 2$ hours, 40 minutes. Notice that the size of the wall (8' × 25') is not important.

3.156 If Aurash can mow the lawn in 20 minutes, then his father can do it in 10 minutes. In one minute, Aurash can mow $\frac{1}{20}$ the lawn; his father can mow $\frac{1}{10}$ the lawn. Together, in one minute, they would mow $\frac{1}{20} + \frac{1}{10} = \frac{10+20}{200} = \frac{30}{200} = \frac{3}{20}$ the lawn. In 5 minutes, they would mow $5 \times \frac{3}{20} = \frac{15}{20} = \frac{3}{4}$ the lawn. The portion that would *remain* to be mowed would be $1 - \frac{3}{4} = \frac{4}{4} - \frac{3}{4} = \frac{1}{4}$ the lawn, i.e. 25% of the lawn.

3.157 In one hour, the large pipe can fill $\frac{1}{3}$ the tank; the small pipe can fill $\frac{1}{12}$ the tank; the two pipes together can fill $\frac{1}{3} + \frac{1}{12} = \frac{12+3}{36} = \frac{15}{36} = \frac{5}{12}$ the tank. In 2 hours, the large pipe will have filled $2 \times \frac{1}{3} = \frac{2}{3}$ the tank. So, the portion of the tank still to be filled would be $1 - \frac{2}{3} = \frac{3}{3} - \frac{2}{3} = \frac{1}{3}$. If it takes x more hours to fill this portion by both the pipes working together, then, $x \times \frac{5}{12} = \frac{1}{3} \quad \Rightarrow \quad x = \frac{1}{3} \times \frac{12}{5} = \frac{4}{5}$ hour $= 48$ minutes.

3.158 Assume it takes her husband x minutes to wash the car by himself. Sophia can wash the car in h hours, or $60h$ minutes. In one minute, Sophia can wash $\frac{1}{60h}$ the car; her husband can wash $\frac{1}{x}$ the car. Together, in one minute, they can wash $\frac{1}{60h} + \frac{1}{x} = \frac{x+60h}{60hx}$ the car. Working together for m minutes, they finish washing the entire car. Therefore,
$m \times \frac{x+60h}{60hx} = 1 \quad \Rightarrow \quad \frac{m(x+60h)}{60hx} = 1 \quad \Rightarrow$
$mx + 60mh = 60hx \quad \Rightarrow \quad 60mh = 60hx - mx$
$60mh = x(60h - m) \quad \Rightarrow \quad x = \frac{60mh}{60h - m}$

3.163 Amount of work $=$ (# of tractors) $\times$ (time) $= 3 \times 8 = 24$. Assume x more tractors. New amount of hours $= 8 - 2 = 6$.
Same amount of work $=$
(new # of tractors) $\times$ (new time needed)
$24 = (x + 3) \times 6 \quad \Rightarrow \quad 24 = 6x + 18 \quad \Rightarrow$
$6 = 6x \quad \Rightarrow \quad x = 1$

3.164 Amount of food $=$ (# of cows) $\times$ (# of days) $= 2 \times 3 = 6$. Assume x fewer days. New # of days $= (3 - x)$.
Same amount of food $=$
(new # of cows) $\times$ (new # of days)
$6 = 6 \times (3 - x) \quad \Rightarrow \quad 6 = 18 - 6x \quad \Rightarrow$
$-12 = -6x \quad \Rightarrow \quad 2 = x$

3.165 Amount of water $=$ (# of hoses) $\times$ (time) $= h \times m = hm$
Assume x more hoses. New time $= (m - 15)$.
Same amount of water $=$
(new # of hoses) $\times$ (new time needed)
$hm = (h + x)(m - 15) \quad \Rightarrow$
$hm = hm - 15h + xm - 15x \quad \Rightarrow$
$xm - 15x = 15h \quad \Rightarrow \quad x(m - 15) = 15h$
$\Rightarrow \quad x = \frac{15h}{m - 15}$

3.166 Find the individual rate of manufacturing for each machine: the # of wheels each machine makes per day. That would be the # of wheels divided by the # of machines divided by the # of days (not weeks).

$$\text{Individual rate of mfg.} = \frac{\frac{2100}{5}}{7} = \frac{2100}{5} \times \frac{1}{7}$$

$= 60$ wheels/machine/day.

Assume x days for 8 machines to make 4,800 wheels. The individual rate of mfg. would be:

$$\frac{\frac{4800}{8}}{x} = \frac{600}{x} \text{ wheels/machine/day}$$

Equate the two individual rates of mfg.:

$$60 = \frac{600}{x} \quad \Rightarrow \quad 60x = 600 \quad \Rightarrow \quad x = 10$$

3.171 2 choices of doors, 5 choices of ext. colors, 3 choices of int. color, 3 choices of engine, 2 choices of transmission.

Total variations $= 2 \times 5 \times 3 \times 3 \times 2 = 180$

3.172 x types of fruit (assume), 3 types of yogurt, 4 types of sweeteners. Total 84 unique smoothies.

$$x \times 3 \times 4 = 84 \quad \Rightarrow \quad 12x = 84 \quad \Rightarrow \quad x = 7$$

3.179 Only odd integers in the place of each digit. Meaning, each digit can be one of five numbers: 1, 3, 5, 7, 9.

5 5 5 5

⊠ ⊠ ⊠ ⊠ Total numbers=5×5×5×5=625

3.180 (i) '4-digit odd numbers' means the units digit has to be odd; the others may be odd or even. The units digit can only be one of 3 or 9, a total of two numbers. The other digits may be one of 2, 3, 4, 6, or 9, a total of five numbers.

5 5 5 2

⊠ ⊠ ⊠ ⊠ Total numbers=5×5×5×2=250

(ii) Once the units digit assumes *one* of 3 or 9, the first digit from the left can still be one of the remaining *four* numbers. Once the first digit assumes a number, the second digit can be one of three numbers. Likewise, the third digit can be one of the remaining two.

4 3 2 2

⊠ ⊠ ⊠ ⊠ Total numbers=4×3×2×2=48

3.181 First, 6-digit odd numbers have their units digit odd; other digits may be odd or even. Also, the first digit from the left can't be 0 for it to be a 6-digit number. For example, 051743 is actually 51743, not a 6-digit number. The configuration would be:

9 10 10 10 10 5

⊠ ⊠ ⊠ ⊠ ⊠ ⊠

Now, the odd-numbered digits are to be even. Meaning, the first, third, and fifth digits can each be one of 0, 2, 4, 6, or 8, a total of five numbers. But the first digit can't be 0; it can only be 2, 4, 6, or 8, a total of four numbers.

4 10 5 10 5 5

⊠ ⊠ ⊠ ⊠ ⊠ ⊠

Total numbers=4×10×5×10×5×5=50,000

3.182 If every 5-digit number is to begin with 9, then it's really a 4-digit question. Since any number can't be repeated on a license plate, and the number 9 is already going to be on it, there are only 0, 1, 2,......,8 a total of nine numbers available for each of the remaining four digits. But the license plate number can't end with 4. So, the units digit can only be one of eight numbers: 0, 1, 2, 3, 5, 6, 7, or 8.

                          8

⊠ ⊠ ⊠ ⊠

Once the units digit has assumed one of those numbers, say 2, the first digit from the left can be one of *eight* numbers. That's because while the units digit can't be 4, the first digit from the left can. So, 4 simply replaces the number unavailable to occupy the first digit (here, 2). Once the first digit from the left assumes a number, the second digit can be one of seven remaining numbers. Likewise, the third digit can be one of six.

8 7 6 8

⊠ ⊠ ⊠ ⊠

Total arrangements $= 8 \times 7 \times 6 \times 8 = 2,688$

3.183 Group the *T*s together and call them only *T*, not *TT*. Similarly, group the *O*s into one *O*. Clearly, it's only a 4-slot question: *C O T N*

While rearranging these four letters, the first slot from the left can assume one of the four letters: *C*, *O*, *T*, or *N*. Once it assumes one letter, say *T*, the second slot from the left can only hold one of the remaining three letters. Likewise, the third slot can only assume one of the remaining two letters; the fourth slot can only hold the last remaining letter.

4 3 2 1

⊠ ⊠ ⊠ ⊠

Total arrangements $= 4 \times 3 \times 2 \times 1 = 24$

3.184 First, there are only 6 unique letters: K, A, T, R, I and N. The question is about 3-letter displays, i.e. 3-slots. The first slot is always K. The middle slot can never be K, because it is to be different from the other two slots, and the first slot is always K. So, the middle slot can assume one of A, T, R, I or N, a total of five different letters. Once the middle slot assumes a letter, say N, the right-most slot can hold one of *five* letters out of the six, because the letter held by the middle slot (here, N) is no longer available. Keep in mind that the letter K is allowed in the right-most slot.

1	5	5
☒	☒	☒

Total arrangements = $1 \times 5 \times 5 = 25$

3.193 Imagine that the 6 students are standing in the room, and the 4 chairs rise up in the air, as if by magic, and float on top of the heads of the students. In how many ways can 4 chairs float on top of 6 students? This is the same as: In how ways can 6 students be seated in 4 chairs? After all, it's a question of pairing up students with chairs. If two chairs exchange their places, then it's a new configuration. Meaning, this is permutation.

$$_6P_4 = \frac{6!}{(6-4)!} = \frac{6!}{2!} = \frac{6\times5\times4\times3\times2!}{2!} = 360$$

3.194 Each of the 11 courses is unique, just like individual persons. So, we can make teams of 5 courses out of 11 courses in $_{11}C_5$ ways.

$$_{11}C_6 = \frac{11!}{5!(11-5)!} = \frac{11!}{5!6!} = \frac{11\times10\times9\times8\times7\times6!}{5\times4\times3\times2\times6!}$$
$$= 462$$

3.195 12 names being called in 12 different ranks is the same as 12 people sitting in 12 chairs. They can be moved around in those 12 chairs, and each configuration would be different. This is permutation, or $_{12}P_{12}$.

$$_{12}P_{12} = \frac{12!}{(12-12)!} = \frac{12!}{0!} = \frac{12!}{1} = 12!$$

Alternative method: The 1st name to be called out can be one of 12 names; the 2nd name can be one of 11; the 3rd can be one of 10; and so on. The 12th name would be the last one remaining. In essence, this is a 12-slot question.

12	11	10	9		3	2	1
☒	☒	☒	☒	....	☒	☒	☒

Total sequences = $12 \times 11 \times 10 \times \dots\dots\dots \times 3 \times 2 \times 1 = 12!$

3.196 Butter knives: B_1, B_2, B_3, B_4, B_5
Steak knives: S_1, S_2, S_3, S_4, S_5, S_6, S_7, S_8
Carving knives: C_1, C_2, C_3, C_4
Of all the combinations, the ones with 3 of each type of knives would have 3 of 5 butter knives (drawn in $_5C_3$ ways), 3 of 8 steak knives (drawn in $_8C_3$ ways), and 3 of 4 carving knives (drawn in $_4C_3$ ways). When combining each of these sequences, we would take their product:

$$_5C_3 \times {}_8C_3 \times {}_4C_3 = \frac{5!}{3!(5-3)!} \times \frac{8!}{3!(8-3)!} \times$$
$$\frac{4!}{3!(4-3)!} = \frac{5!}{3!2!} \times \frac{8!}{3!5!} \times \frac{4!}{3!1!} =$$
$$\frac{5\times4\times3!}{3!\times2} \times \frac{8\times7\times6\times5!}{3\times2\times5!} \times \frac{4\times3!}{3!} =$$
$$10 \times 56 \times 4 = 2240$$

3.197 When the person starts walking downwards from point A, he has 5 different paths to walk through the first section. Then, he has 4 paths to walk through the second section. Finally, 3 paths to go through the third section. Each of the first 5 paths can be considered in combination with each of the second 4 paths, which can be combined with each of the 3 paths. The total number of paths from A to B would be their product $5 \times 4 \times 3 = 60$.

3.198 13 entrants. Make 2-member teams. That could be in $_{13}C_2$ different ways.

$$_{13}C_2 = \frac{13!}{2!(13-2)!} = \frac{13!}{2!11!} = \frac{13\times12\times11!}{2\times11!} = 78$$

3.199 From *List A*, 2 letters can be teamed out of 7 letters in $_7C_2$ ways. From *List B*, 2 letters can be teamed out of 5 letters in $_5C_2$ ways. Their product would give the number of ways in which 4 letters (2 from each list) can be *selected* from the two lists.

$$_7C_2 \times {}_5C_2 = \frac{7!}{2!(7-2)!} \times \frac{5!}{2!(5-2)!} = \frac{7!}{2!5!} \times \frac{5!}{2!3!} =$$
$$\frac{7\times6\times5!}{2\times5!} \times \frac{5\times4\times3!}{2\times3!} = \frac{7\times6\times5\times4}{2\times2} = 210$$

Now, each of the 4-letters in a team selected (for example *CMXY*) can be rearranged within four places:

4	3	2	1
☒	☒	☒	☒

Rearrangements = $4! = 24$

Number of teams × Number of rearrangements
= 210 × 24 = 5040
This is total the number of 4-letter arrangements possible.

3.200 *$30 at the most* means $30 or less. That can be obtained by pulling out 2 bills in the following sequences:

5, then 10   10, then 5
5, then 20   20, then 5
10, then 20   20, then 10
10, then 10   5, then 5

Total number of ways = 8

3.201 *At least 4* means 4 or 5.
Four candies can be selected from 5 candies in

$$_5C_4 = \frac{5!}{4!(5-4)!} = \frac{5!}{4!1!} = \frac{5 \times 4!}{4!} = 5 \text{ ways.}$$

Five candies can be selected from 5 candies in

$$_5C_5 = \frac{5!}{5!(5-5)!} = \frac{5!}{5!1!} = 1 \text{ way.}$$

In total, at least 4 candies can be selected out of 5 candies in 5 + 1 = 6 ways.

3.206

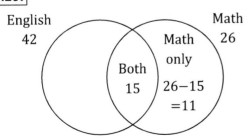

Adding the two crescents, the intersection region, and x should total 90, the total number of students.

12 + 41 + 17 + x = 90   ⇒
70 + x = 90   ⇒   x = 20

3.207

English   Math
42   Math   26

Both   Math
15   only
            26−15
            =11

Percentage of students majoring in math only

$$= \frac{\text{math only}}{\text{total students}} \times 100\% = \frac{11}{55} \times 100\% = 20\%$$

3.208

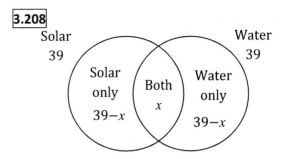

Imagine the two circles pulling away from each other. The intersection region (x) starts diminishing. If the two circles completely separate, then $x = 0$. Then 39 + 39 would have to be 70, the total number of homes in the township. But 39 + 39 = 78. Clearly, those extra 8 homes that overlap are the ones that have both, a solar panel and a water well. $x = 8$. The overlap can always be greater than 8, but not less, i.e. $x \geq 8$.

3.209 $X \equiv \{50, 51, 52, \ldots\ldots\ldots, 60\}$
            $Y \equiv \{\ldots\ldots -4, -2, 0, 2, 4, \ldots\ldots\}$
            $Z \equiv \{\ldots\ldots -6, -3, 0, 3, 6, 9, \ldots\ldots\}$

p is the # of integers common to X and Y. Those integers are: 50, 52, 54, 56, 58, 60. $p = 6$
q is the # of integers common to X and Z. Those integers are: 51, 54, 57, 60. $q = 4$
$(p + q)(p - q) = (6 + 4)(6 - 4) = 20$

3.224   (i) From 7 a.m. to 10 p.m. is a 15-hour period. There are 24 hours in a day. The probability that it's between 7 a.m. and 10 p.m. right now is the probability that we are in that 15-hour window right now:

$$P = \frac{15}{24} = \frac{5}{8}$$

(ii) Please note the words **7 o'clock** and **10 o'clock**. There's no mention of a.m. or p.m. The 7 o'clock to 10 o'clock period occurs twice a day: once in the morning, and once in the evening, i.e. a total of 6 hours. The time outside of that period is 24 – 6 = 18 hours. The probability that we're NOT in the 7 o'clock to 10 o'clock window is the probability that we're in the remaining 18-hour window right now:

$$P = \frac{18}{24} = \frac{3}{4}$$

3.225 Let's consider the possible genders of each of the three.   M = male, and F = female

Principal	Parent	Child	
F	F	F	F = 3
M	F	F	F = 2
F	M	F	F = 2
F	F	M	F = 2
M	M	F	F = 1
M	F	M	F = 1
F	M	M	F = 1
M	M	M	F = 0

Total possibilities = 8

At least one female means F ≥ 1. That's a total of 7 possibilities. Therefore the probability is 7 in 8, or $\frac{7}{8}$.

3.226 The outcomes of the two dice are:

1,1	1,2	1,3	1,4	1,5	1,6
2,1	2,2	2,3	2,4	2,5	2,6
3,1	3,2	3,3	3,4	3,5	3,6
4,1	4,2	4,3	4,4	4,5	4,6
5,1	5,2	5,3	5,4	5,5	5,6
6,1	6,2	6,3	6,4	6,5	6,6

(i) Odd + Odd = Even   Even + Even = Even
The highlighted pairs in the *above table* will have even sums.

$$P(\text{sum} = \text{even}) = \frac{18}{36} = \frac{1}{2}$$

1,1	1,2	1,3	1,4	1,5	1,6
2,1	2,2	2,3	2,4	2,5	2,6
3,1	3,2	3,3	3,4	3,5	3,6
4,1	4,2	4,3	4,4	4,5	4,6
5,1	5,2	5,3	5,4	5,5	5,6
6,1	6,2	6,3	6,4	6,5	6,6

(ii) The highlighted pairs in the above table will have their sums at least 7 (sum ≥ 7).

$$P(\text{sum} \geq 7) = \frac{21}{36} = \frac{7}{12}$$

1,1	1,2	1,3	1,4	1,5	1,6
2,1	2,2	2,3	2,4	2,5	2,6
3,1	3,2	3,3	3,4	3,5	3,6
4,1	4,2	4,3	4,4	4,5	4,6
5,1	5,2	5,3	5,4	5,5	5,6
6,1	6,2	6,3	6,4	6,5	6,6

(iii)   The highlighted pairs above will have their sums equal to 6.

$$P(\text{sum} = 6) = \frac{5}{36}$$

1,1	1,2	1,3	1,4	1,5	1,6
2,1	2,2	2,3	2,4	2,5	2,6
3,1	3,2	3,3	3,4	3,5	3,6
4,1	4,2	4,3	4,4	4,5	4,6
5,1	5,2	5,3	5,4	5,5	5,6
6,1	6,2	6,3	6,4	6,5	6,6

(iv) The highlighted pairs above will have their sums at most 5 (sum ≤ 5).

$$P(\text{sum} \leq 5) = \frac{10}{36} = \frac{5}{18}$$

3.227 *B, A, T,* and *S* can be reshuffled in 4 slots in the following way:

4   3   2   1
☒   ☒   ☒   ☒

Total arrangements = 4 × 3 × 2 × 1 = 24
TABS and *STAB* are 2 arrangements out of 24.

$$P = \frac{2}{24} = \frac{1}{12}$$

3.228 Five members are standing side by side. That sounds like a 5-slot example.

5   4   3   2   1
☒   ☒   ☒   ☒   ☒

Total arrangements = 5 × 4 × 3 × 2 × 1 = 120
Out of the 120 arrangements, the ones in which the oldest member is standing to the far left, and the youngest to the far right, would be the ones in which only the remaining 3 members are shuffled in the middle 3 slots.

	3	2	1	
oldest	☒	☒	☒	youngest

Total arrangements = 3 × 2 × 1 = 6

$$P = \frac{6}{120} = \frac{1}{20}$$

3.229 A probability of 20% is $\frac{20}{100}$, or $\frac{1}{5}$.
Probability is, after all, the proportion in which that type of object is present in a group of objects.

Black	Green	Red
$\frac{5}{8}$	$\frac{1}{5}$	x

Equalize the denominators of the first two fractions: *Raise* the first fraction by 5, and the second by 8.

Black	Green	Red
$\frac{5 \times 5}{8 \times 5}$	$\frac{1 \times 8}{5 \times 8}$	x

Black	Green	Red
$\frac{25}{40}$	$\frac{8}{40}$	x

Adding the three fractions, we should get 100%, or 1.

$$\frac{25}{40} + \frac{8}{40} + x = 1 \quad \Rightarrow \quad x = 1 - \frac{25}{40} - \frac{8}{40}$$

$$\Rightarrow \quad x = \frac{40}{40} - \frac{25}{40} - \frac{8}{40} = \frac{40 - 25 - 8}{40} = \frac{7}{40}$$

Black Green Red
$\frac{25}{40}$ $\frac{8}{40}$ $\frac{7}{40}$

These fractions are already in their lowest terms. Meaning, there are at least 7 red markers, 8 green markers, and 25 black markers in the box.

3.230 $M =$ me, $B =$ my brother

B M

| 1st | 2 | 3 | ... | ... | ... | 26 | 27th | ... | ... | ... | 70 |

The number of ranks between my rank (27th) and my brother's rank (1st) is: 26 – 1 = 25.

Total possible ranks (students) = 70.

The probability that my cousin ranked between me and my brother is: 25 ranks out of a total of 70 ranks, or $\frac{25}{70}$ or $\frac{5}{14}$.

3.231 Let's find all the unique products by picking each number from *List A*, and multiplying it by each number from *List B*.

1 × 2	1 × 5	1 × 6
3 × 2	3 × 5	3 × 6
5 × 2	5 × 5	5 × 6

2	5	6
6	15	18
10	25	30

The unique products are: 2, 5, 6, 10, 15, 18, 25 and 30, a total of 8 different numbers. Among these, the odd numbers are 5, 15 and 25, a total of 3 different numbers.

Probability of the product being odd is 3 out of 8, or $\frac{3}{8}$.

3.241 (i) There's a $\frac{1}{7}$ individual chance that each of them was born on a Thursday. The probability that they were *all* born on a Thursday is their joint probability, i.e. their product: $\frac{1}{7} \times \frac{1}{7} \times \frac{1}{7} = \frac{1}{343}$

(ii) There's a $\frac{31}{365}$ individual chance that each of the students was born in October. The probability that the tutor was born on a Monday, and both the students were born in October is: $\frac{1}{7} \times \frac{31}{365} \times \frac{31}{365} = \frac{31^2}{7 \times 365^2}$

3.242 The probability that the first adopted animal is a cat is: $\frac{\text{\# of cats}}{\text{total \# of animals}} = \frac{12}{25}$

Presuming the first animal adopted is a cat, the probability that the second adopted animal is a cat is: $\frac{\text{\# of remaining cats}}{\text{total \# of remaining animals}} = \frac{11}{24}$

Presuming the second animal adopted is a cat, the probability that the third adopted animal is a cat is: $\frac{\text{\# of remaining cats}}{\text{total \# of remaining animals}} = \frac{10}{23}$

Their joint probability is their product: $\frac{12}{25} \times \frac{11}{24} \times \frac{10}{23} = \frac{11}{115}$

3.243 # of face cards = 12

total # of cards = 52

(i) The # of face cards and the total # of cards remains constant. The joint probability of all 3 cards being face cards is: $\frac{12}{52} \times \frac{12}{52} \times \frac{12}{52} = \frac{3}{13} \times \frac{3}{13} \times \frac{3}{13} = \frac{27}{2197}$

(ii) Presuming the first and second cards drawn are face cards, the face cards will decline from 12 to 11 to 10. The total number of cards will decline from 52 to 51 to 50. The joint probability of all 3 cards being face cards is:

$$\frac{12}{52} \times \frac{11}{51} \times \frac{10}{50} = \frac{11}{(13 \times 17 \times 5)} = \frac{11}{1105}$$

Note: The above calculations are within reason on the GRE/GMAT. Be sure to *reduce* first before multiplying.

3.244 Going part-time to college, and studying liberal arts are two independent events.

Meaning, they do not influence each other, but can happen at the same time. A student can go to college part time, but does not have to study liberal arts, although he/she can if he/she wishes to.

$P(PT) = 45\%$　　　　　　$P(LA) = 40\%$

$P(PT \text{ and } LA) = P(PT) \cdot P(LA) =$

$\dfrac{45}{100} \times \dfrac{40}{100} = \dfrac{9}{20} \times \dfrac{2}{5} = \dfrac{18}{100} = 18\%$

$P(PT \text{ or } LA) = P(PT) + P(LA) - P(PT \text{ and } LA) =$
$45\% + 40\% - 18\% = 67\%$

3.245　Total # of items = 14

6 items can be selected out of 14 in $_{14}C_6$ ways.

$_{14}C_6 = \dfrac{14!}{6!(14-6)!} = \dfrac{14!}{6!8!} =$

$\dfrac{14 \times 13 \times 12 \times 11 \times 10 \times 9 \times 8!}{6 \times 5 \times 4 \times 3 \times 2 \times 8!} =$

$7 \times 13 \times 11 \times 3 = 77 \times 39 = 3003$

Out of these 3003 combinations only some will have 2 of each. 2 ball pens can be selected out of

5 in $_5C_2$ ways $= \dfrac{5!}{2!(5-2)!} = \dfrac{5!}{2!3!} = \dfrac{5 \times 4 \times 3!}{2!3!} =$

10 ways; 2 pencils out of 3 in $_3C_2 = \dfrac{3!}{2!(3-2)!} =$

$\dfrac{3!}{2!1!} = 3$ ways; 2 sketch pens out of 6 in $_6C_2$

$= \dfrac{6!}{2!(6-2)!} = \dfrac{6!}{2!4!} = \dfrac{6 \times 5 \times 4!}{2!4!} = 15$ ways. 2 of

each of the above three items can be combined in $_5C_2 \times {}_3C_2 \times {}_6C_2$ ways $= 10 \times 3 \times 15 = 450$ ways.

The probability that 2 of each type will be selected is:

$\dfrac{_5C_2 \times {}_3C_2 \times {}_6C_2}{_{14}C_6} = \dfrac{450}{3003} = \dfrac{150}{1001}$

3.246　Drawing a diamond, and drawing a black card are mutually exclusive events; they cannot happen at the same time, because diamonds are red cards, not black.

$P(D) = \dfrac{13}{52} = \dfrac{1}{4}$　　　　$P(BC) = \dfrac{26}{52} = \dfrac{1}{2}$

$P(D \text{ or } BC) = P(D) + P(BC) - P(D \text{ and } BC) =$
¼ + ½ - 0 = ¾

3.247　$P(C) = 40\%$　　$P(F) = 25\%$

$P(C \text{ and } F) = 8\%$

(i)　$P(\text{at least one}) = P(\text{one or both}) = P(C \text{ or } F)$
　　$= P(C) + P(F) - P(C \text{ and } F)$
　　$= 40\% + 25\% - 8\% = 57\%$

(ii)　$P(\text{exactly one}) = P(\text{one or both}) - P(\text{both})$
　　$= P(C \text{ or } F) - P(C \text{ and } F) = 57\% - 8\% = 49\%$

CHAPTER 4: GEOMETRY

4.009　The identical markings on line segments mean $BP = PE = EC$. Assume it as x. Since $\overleftrightarrow{m}$ bisects DC, $DE = EC = x$. Since $\overleftrightarrow{m}$ also bisects OB, $BP = PO = x$. Since $\overleftrightarrow{l}$ bisects AB, $BO = OA$. But $BO = BP + PO = x + x = 2x$. Hence, $OA = 2x$. $AP = PO + OA = x + 2x = 3x$.

Finally, $DE : AP = x : 3x = 1 : 3$

4.010

Extend the two line segments as shown. A segment appears at right angles to both of them. Meaning, they run parallel to each other. Also, $\overleftrightarrow{l}$ is a transversal to them. The angle to the left of x, above $\overleftrightarrow{l}$ will be 41°, because it and the 41° angle are corresponding angles of two parallel lines. (They look like the letter F).

Finally, $x = 180 - 41 = 139°$

4.011　a is greater than b by 80° means that $a = b + 80$. Also, from the figure, the sum of all the angles is 360°.　　$3a + 3b = 360$　⇒
$3(a + b) = 360$　⇒　$a + b = 120$
Substitute $a = b + 80$ in this.
$(b + 80) + b = 120$　⇒　$2b + 80 = 120$　⇒
$2b = 40$　⇒　$b = 20$

Therefore, $a = b + 80 = 20 + 80 = 100$

Finally,　　　$a^2 - b^2 = (a + b)(a - b) =$
$(100 + 20)(100 - 20) = 120 \times 80 = 9600$

4.012　Notice that y and the 112° angle add up to 180°. Therefore, $y = 180 - 112 = 68°$.

Also, y is vertically opposite to the two x's taken together. Meaning, $y = 2x$

$y = 2x \quad \Rightarrow \quad x = \dfrac{y}{2} = \dfrac{68}{2} = 34°$

Finally, $y - x = 68 - 34 = 34°$

4.013 (i) $\overleftrightarrow{l}$ and $\overleftrightarrow{m}$ would be parallel if either $\overleftrightarrow{p}$ or $\overleftrightarrow{q}$ (or both) would be perpendicular to **both** $\overleftrightarrow{l}$ and $\overleftrightarrow{m}$. But that's not obvious from the figure. Therefore, $\overleftrightarrow{l}$ and $\overleftrightarrow{m}$ are not necessarily parallel. So the statement is not true.

(ii) Same explanation as (i) above, only in regard to $\overleftrightarrow{m}$ and $\overleftrightarrow{n}$. Hence, statement untrue.

(iii) We see that $\overleftrightarrow{p} \perp \overleftrightarrow{l}$ and $\overleftrightarrow{p} \perp \overleftrightarrow{n}$. Therefore, $\overleftrightarrow{l}$ and $\overleftrightarrow{n}$ must be parallel. The statement is therefore true.

4.014

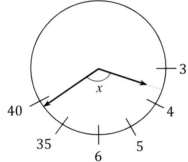

Every minute, the minute hand advances 6°. At 3:39, it has traveled 4 minutes beyond the 35 minute marker, i.e. a total of 4 × 6° = 24° beyond the 35 minute marker.

Every minute, the hour hand advances ½°. At 3:39, it has traveled for 39 minutes after 3 o'clock, i.e. a total of 39 × ½° = 19.5° beyond the 3 o'clock marker. The angle it still needs to travel to reach the 4 o'clock marker is $30° - 19.5° = 10.5°$.

Therefore, $x = 24° + 90° + 10.5° = 124.5°$

4.015 Notice that $2a - 70$ is vertically opposite to a. Therefore, they are both equal in measure.

$2a - 70 = a \quad \Rightarrow \quad a = 70°$

Also, the angle to the right of b below the transversal is equal to $2a - 70$, because they are both corresponding angles of parallel lines. (They look like the letter F). The measure of this angle is $2a - 70 = a = 70°$. This angle and b add up to 180°.

$70 + b = 180 \quad \Rightarrow \quad b = 110°$

Finally, $b - a = 110 - 70 = 40°$

Notice that in the figure, angle a is obtuse, and angle b is acute. But our calculations conclude the opposite. It doesn't matter because the figure is *not drawn to scale*. Such questions do appear on the GRE/GMAT.

4.016 From the figure, angles p and q add up to 180°. Same with angles s and r.

$p + q = 180°$ and $s + r = 180°$

If $p = s$, then we would end up with:

$s + q = 180°$ and $s + r = 180°$

Therefore, $s + q = s + r$

Subtracting s from both sides, we get $q = r$

4.025 The triangle in the left portion is isosceles; the angles facing the two marked sides measure x each. Redraw the figure:

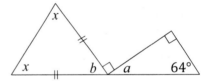

In the triangle in the right portion,

$64° + 90° + a = 180° \quad \Rightarrow \quad a = 26°$

The three adjacent angles between the two triangles add up to 180°, i.e. $a + 90° + b = 180°$

$\Rightarrow \quad 26° + 90° + b = 180° \quad \Rightarrow \quad b = 64°$

In the triangle on the left, $x + x + b = 180° \quad \Rightarrow$

$2x + 64° = 180° \Rightarrow 2x = 116° \Rightarrow x = 58°$

4.026 (i) From the proportion, we can say that there are a total of $1 + 2 + 3 = 6$ parts. The angles would be: $x = \dfrac{1}{6} \times 180° = 30°$,

$y = \dfrac{2}{6} \times 180° = 60°$, and $z = \dfrac{3}{6} \times 180° = 90°$.

Therefore, $\triangle ABC$ is a right triangle.

(ii) $\triangle ABC$ is not isosceles because the three angles have different measures.

(iii) $\dfrac{1}{BC} = \dfrac{2}{AB} \quad \Rightarrow \quad AB = 2 \cdot BC$ We see that $\triangle ABC$ is a 30–60–90 triangle in which, we must have the hypotenuse twice as long as the smallest side. Therefore, statement is true.

4.027 Imagine $\triangle PQR$ resting on side PR. Clearly, all the triangles have the same height (say h), but different bases a, b and c. Given the proportion, we can conveniently write $a = 2x$, $b = 3x$ and $c = 5x$.

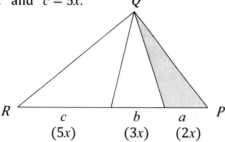

$$R \quad \underset{(5x)}{c} \quad \underset{(3x)}{b} \quad \underset{(2x)}{a} \quad P$$

Base of $\triangle PQR = 5x + 3x + 2x = 10x$

Area of $\triangle PQR = \dfrac{10x \cdot h}{2} \quad \Rightarrow \quad 5xh$

$\Rightarrow \quad xh = 14$

Area of shaded triangle $= \dfrac{(2x) \cdot h}{2} = xh = 14$

4.028 $\angle ABC$ and the 30° angle at C are alternate interior angles (they look like the letter Z), and are therefore equal in measure. It means $\triangle ABC$ is a 30–60–90 triangle. Hypotenuse $AB = 8$. Therefore, smallest side $AC = 4$. Draw altitude $CD \perp AB$.

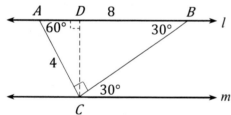

Then, $\triangle ADC$ is a 30–60–90 triangle in which $AD = \frac{1}{2} \cdot AC = \frac{1}{2}(4) = 2$

Using Pythagorean Theorem in $\triangle ADC$,

$(AC)^2 = (AD)^2 + (DC)^2 \quad \Rightarrow \quad 4^2 = 2^2 + (DC)^2$

$\Rightarrow \quad 16 = 4 + (DC)^2 \quad \Rightarrow \quad 12 = (DC)^2 \quad \Rightarrow$

$DC = \sqrt{12} = 2\sqrt{3} =$ dist. between $\overleftrightarrow{l}$ and $\overleftrightarrow{m}$.

4.029 In $\triangle PTR$, $PT \perp TR$. Also, hypotenuse PR is twice as long as leg TR. Meaning, $\triangle PTR$ is a 30–60–90 triangle. $\triangle PQR$ is also a 30–60–90 triangle because $\angle Q = 30°$, $\angle R = 60°$, and therefore $\angle QPR = 90°$. $\triangle QSP$ is isosceles as marked on segments QS and SP. Therefore, $\angle QPS = 30°$. $\angle SPT = \angle QPR - (\angle QPS + \angle TPR)$

$= 90° - (30° + 30°) = 30°$. It means that $\triangle PST$ and $\triangle PRT$ are mirror images, i.e. $a = 4$ and $b = SP = PR = 8$　　Finally, $a : b = 4 : 8 = 1 : 2$

4.030 Assume the smallest angle as x. Then, the other non-right angle would be $x + 24$. The sum of all three angles is 180°. Therefore,

$x + (x + 24) + 90 = 180 \quad \Rightarrow$

$2x + 114 = 180 \quad \Rightarrow \quad 2x = 66 \quad \Rightarrow \quad x = 33°$

4.031 In the 4 hours from 2 p.m. to 6 p.m., the bus traveling due north will have covered $4 \times 45 = 180$ miles. The bus traveling due west will have covered $4 \times 60 = 240$ miles.

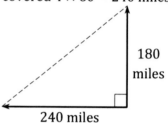

Should we use the Pythagorean Theorem to find the dotted distance between them? 180 and 240 are the 3rd and 4th multiples of 60. Meaning, the above triangle is a 3–4–5 triangle. The dotted hypotenuse must be the 5th multiple of 60, i.e. $5 \times 60 = 300$ miles.

4.032 $\triangle BDC$ is a 30–60–90 triangle in which hypotenuse BD is twice as long as the smallest side BC, i.e. $BD = 8$. $\triangle AED$ and $\triangle AEB$ are also 30–60–90 triangles mirrored about AE. Therefore, $DE = EB = \frac{1}{2} \cdot BD = \frac{1}{2}(8) = 4$

Hypotenuse AD is twice as long as the smallest side DE, i.e. $AD = 2(4) = 8$. Using Pythagorean Theorem in $\triangle AED$, $(AE)^2 + (DE)^2 = (AD)^2 \quad \Rightarrow$

$(AE)^2 + 4^2 = 8^2 \quad \Rightarrow \quad (AE)^2 + 16 = 64 \quad \Rightarrow$

$(AE)^2 = 48 \quad \Rightarrow \quad AE = \sqrt{48} = \sqrt{16 \times 3} = 4\sqrt{3}$

In $\triangle AED$, DE can be considered the base, and AE can be considered the height.

Area of $\triangle AED = \dfrac{DE \times AE}{2} = \dfrac{4 \times 4\sqrt{3}}{2} = 8\sqrt{3}$

4.044 Assume the polygon has n sides.

$180(n - 2) = 720 \quad \Rightarrow \quad 180n - 360 = 720 \quad \Rightarrow \quad 180n = 1080 \quad \Rightarrow \quad n = 6$

# of diagonals $= {}_nC_2 - n = {}_6C_2 - 6 =$

$\dfrac{6!}{2!\,(6-2)!} - 6 = \dfrac{6!}{2!\,4!} - 6 = \dfrac{6 \times 5 \times 4!}{2 \times 4!} - 6 =$

$15 - 6 = 9$

4.045 $\triangle ABE$ is a 30-60-90 triangle in which,

$AE = 2 \times BE = 2 \times 4 = 8$

$\angle AEB + \angle AED + \angle DEC = 180° \Rightarrow$

$60 + 90 + \angle DEC = 180 \Rightarrow \angle DEC = 30°$

$\angle ADE = \angle DEC = 30°$ because they are both alternate interior angles of parallel line segments (they look like the letter Z).

$\triangle AED$ is a 30-60-90 triangle in which,

$AD = 2 \times AE = 2 \times 8 = 16 = BC$

Also, $EC = BC - BE = 16 - 4 = 12$

$(AB)^2 + (BE)^2 = (AE)^2 \Rightarrow (AB)^2 + 4^2 = 8^2$

$\Rightarrow (AB)^2 = 8^2 - 4^2 = 64 - 16 = 48$

$\Rightarrow AB = \sqrt{48} = \sqrt{16 \times 3} = 4\sqrt{3} = DC$

Area of $\triangle DEC = \dfrac{EC \times DC}{2} = \dfrac{12 \times 4\sqrt{3}}{2} = 24\sqrt{3}$

4.046 The shaded region is a trapezoid. The perpendicular distance between $\overleftrightarrow{l}$ and $\overleftrightarrow{m}$ would be the height (say, h) of that trapezoid.

Area of trapezoid $= \dfrac{\text{(sum of parallel sides)} \times \text{ht}}{2}$

$75 = \dfrac{(6 + 9) \times h}{2} \Rightarrow 150 = 15h \Rightarrow h = 10$

4.047

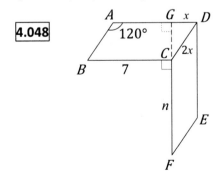

Add 8' to the inner dimensions to get the outer dimensions of the pool: 58' × 33'.

Area of the marble pavement = (area of outer rectangle) – (area of inner rectangle) = $(58 \times 33) - (50 \times 25) = 1914 - 1250 = 664$ ft²

4.048

Redraw the figure with altitude $GC \perp AD$.

$\angle DCB = \angle DAB = 120°$ out of which, $\angle GCB = 90°$

Therefore, $\angle GCD = 30°$, i.e. $\triangle GCD$ is a 30-60-90 triangle. x and $2x$ are marked accordingly.

Area of parallelogram $ABCD =$ base × height

$35\sqrt{3} = 7 \times GC \Rightarrow GC = 5\sqrt{3}$

Using the Pythagorean Theorem in $\triangle GCD$,

$(GC)^2 + x^2 = (2x)^2 \Rightarrow \left(5\sqrt{3}\right)^2 + x^2 = 4x^2$

$75 = 3x^2 \Rightarrow 25 = x^2 \Rightarrow x = 5$

Area of parallelogram $CDEF = CF \times GD = n \times x$

$= n \times 5 = 5n$ This has to be greater than $35\sqrt{3}$.

$5n > 35\sqrt{3} \Rightarrow n > 7\sqrt{3} \Rightarrow \dfrac{n}{\sqrt{3}} > 7$

Hence, the least integer value of $\dfrac{n}{\sqrt{3}}$ would be the very next integer after 7, which is 8.

4.049 $\triangle PQR \sim \triangle AQB$, and $QR = PS = 10$

$\dfrac{QA}{QB} = \dfrac{QP}{QR} \Rightarrow \dfrac{6}{QB} = \dfrac{8}{10} \Rightarrow QB = \dfrac{60}{8} = 7.5$

Area of lower shaded region = area of $\triangle PQR$ –

area of $\triangle AQB = \dfrac{10 \times 8}{2} - \dfrac{7.5 \times 6}{2} = 40 - 22.5 = $ **17.5**

Next, $\triangle DSC \sim \triangle PSR$, and $SR = PQ = 8$

$\dfrac{SC}{SD} = \dfrac{SR}{SP} \Rightarrow \dfrac{SC}{7} = \dfrac{8}{10} \Rightarrow SC = \dfrac{56}{10} = 5.6$

Area of upper shaded region = area of $\triangle PSR$ –

area of $\triangle DSC = \dfrac{10 \times 8}{2} - \dfrac{7 \times 5.6}{2} = 40 - 19.6 = $ **20.4**

Total shaded area = 17.5 + 20.4 = 37.9

4.050 Assume length as l and width as w. It's given that three times l is five times w:

$3l = 5w \Rightarrow w = \dfrac{3}{5}l$

Both l and w are to be odd integers. First of all, if w is to be an integer, then the denominator 5 on the right side would have to go. Meaning, l would have to be a multiple of 5, i.e. 5, 10, 15 etc. But to yield the least value of w, l would have to be at its least value as well, which is 5.

$w = \dfrac{3}{5}l = \dfrac{3}{5}(5) = 3$

Therefore, $w = 3$ and $l = 5$ which are both at their least possible odd integer values.

$$w = \frac{3}{5}l = \frac{3}{5}(5) = 3$$

Therefore, $w = 3$ and $l = 5$ which are both at their least possible odd integer values.

4.051 (i) Imagine that the rectangle is *really* thin and flat. Its small sides have infinitesimally small lengths; its large sides have lengths slightly less than 50 each. The perimeter is 100. But the area is almost nothing. Now imagine the rectangle starts to stand up, i.e. inflates and gets thicker and thicker. Its area increases, reaching its maximum value when the rectangle is a square. At that point, each side will have a length of 25. The maximum area will be $25 \times 25 = 625$. The area will never reach 700.

(ii) If the rectangle is really thin and flat, then its length **can be** 700 times its width, or 7000 times its width, or 7,000,000,000 times......

(iii) The diagonals will be of least length when the rectangle has completely inflated into a square of sides 25, the length of diagonals being $\sqrt{25^2 + 25^2} = \sqrt{625 + 625} = \sqrt{625 \times 2} = 25\sqrt{2}$

This is greater than 25. Therefore, the length of its diagonals will never fall down to 25.

4.052 The three sides with identical lengths of 4 are only meant to mislead you. Knowing those lengths won't help you in any way! The key lies in recognizing that the sum of all vertical lengths on the right side will add up to 12. Likewise, the sum of all horizontal lengths above the base will add up to 16. Therefore, the perimeter is simply $12 + 16 + 12 + 16 = 56$.

4.065 The non-shaded sector covers 65% of the circle. x as a fraction of 360° would equal 65% (65 as a fraction of 100).

$$\frac{x}{360} = \frac{65}{100} \quad \Rightarrow \quad x = \frac{360 \times 65}{100} = 234°$$

4.066

$$\frac{\text{area of shaded sector}}{\text{area of nonshaded sector}} \times 100\% =$$
$$\frac{20\% \times \pi r^2}{80\% \times \pi r^2} \times 100\% = \frac{20}{80} \times 100\% =$$
$$\frac{1}{4} \times 100\% = 25\%$$

4.067 $EC = CF = 12$, and $CD = 16$

12 and 16 are the 3rd and 4th multiples of 4. Hence, ED must be the 5th multiple of 4, i.e. 20. $\triangle ECD$ is a 3–4–5 triangle.

In $\triangle ECD$ and $\triangle BCF$, $\angle B = \angle D$, and $\angle C = 90°$. Therefore, $\angle E = \angle F$. Moreover, $EC = CF$. Meaning, the two triangles have identical dimensions. Thus, $BC = CD = 16 = AB$.

Radius of the semi-circle = ½(AB) = ½$(16) = 8$

ED : radius = 20 : 8 = 5 : 2 (after reducing by 4)

4.068 The horizontal side of the triangle is the diameter of the semi-circle. The diameter subtends a right angle at the circumference (Thales' Theorem). Therefore, the triangle is a right triangle, with the right angle at the circumference. One of the sides is 6, whereas the hypotenuse is the horizontal side, i.e. diameter whose length is twice the radius, or twice of 5, or 10. 6 and 10 are the 3rd and 5th multiples of 2. Therefore, the third side must be the 4th multiple of 2, i.e. 8.

Area of the right triangle = $\dfrac{\text{base} \times \text{height}}{2}$

Imagine the triangle resting on the side with length 6. Then, its height would be the side with length 8.

Area = $\dfrac{6 \times 8}{2} = 24$

Area of shaded region = area of semi-circle – area of triangle =

$$\frac{\pi 5^2}{2} - 24 = \frac{25\pi}{2} - 24 = 12.5\pi - 24$$

4.069 Find the two areas and equate them. Then solve for d. Radius of semi-circle = $\dfrac{d}{2}$.

Area of semi-circle = $\dfrac{\pi\left(\frac{d}{2}\right)^2}{2} = \dfrac{\pi\left(\frac{d^2}{4}\right)}{2} = \dfrac{\pi d^2}{8}$

Area of triangle = $\dfrac{dh}{2}$

Equating the two areas, we get:

$$\frac{\pi d^2}{8} = \frac{dh}{2} \quad \Rightarrow \quad 2\pi d^2 = 8dh \quad \Rightarrow$$

$$\frac{d^2}{d} = \frac{8h}{2\pi} \quad \Rightarrow \quad d = \frac{4h}{\pi}$$

4.070 Four quarter circles together make a full circle, the area of which is $\pi(11)^2 = 121\pi$. The square will have its side equal to twice the length of the radius of each quarter circle, i.e. $2 \times 11 = 22$. Area of the square $= 22 \times 22 = 484$. Area of the shaded region is simply the area of the square, minus area of the full circle, i.e. $484 - 121\pi$

4.071 Assume the smaller circle has a radius r, and the larger has a radius R.

Circumference of the smaller circle $= 2\pi r$

This is the radius of the larger circle: $R = 2\pi r$

Area of the smaller circle $= \pi r^2$. Call this a.

Area of the larger circle $= \pi R^2 = \pi(2\pi r)^2 = \pi(4\pi^2 r^2) = 4\pi^3 r^2$. Call this A.

$$\frac{a}{A} = \frac{\pi r^2}{4\pi^3 r^2} = \frac{1}{4\pi^2} \quad \Rightarrow \quad a : A = 1 : 4\pi^2$$

4.072 A tangent is perpendicular to the radius drawn at the point of tangency. Therefore, $\vec{l} \perp \overline{OA}$, and $\vec{m} \perp \overline{OC}$. Meaning, $\angle OAB = \angle OCB = 90°$. The sum of the interior angles of quadrilateral $AOCB = 180(n-2) = 180(4-2) = 180(2) = 360°$. This should equal $\angle OAB + \angle ABC + \angle OCB + \angle AOC$.

$90 + 37 + 90 + \angle AOC = 360 \quad \Rightarrow \quad \angle AOC = 360 - (217) = 143°$

4.073

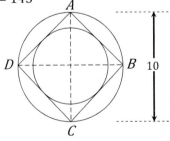

Figure shows a circle inscribed in square $ABCD$, with the square itself inscribed in a larger circle of diameter 10. Let $AB = BC = CD = DA = x$

Diameter of outer circle $= 10 =$ diagonal AC

In right $\triangle ACB$, $(AC)^2 = (AB)^2 + (BC)^2 \quad \Rightarrow$

$10^2 = x^2 + x^2 \quad \Rightarrow \quad 100 = 2x^2 \quad \Rightarrow$

$50 = x^2 \quad \Rightarrow \quad x = \sqrt{50} = \sqrt{25 \times 2} = 5\sqrt{2}$

This ($5\sqrt{2}$) is also the diameter of the inner circle, because each *side* of the square is as long as the diameter of the circle inscribed within it. Therefore, area of the inner circle is:

$$\pi\left(\frac{5\sqrt{2}}{2}\right)^2 = \pi\left(\frac{25 \times 2}{4}\right) = 12.5\pi$$

4.074

In $\triangle AEC$, let $\angle ACE = x$. Then, $\angle CAE = 90 - x$.

Diameter BC subtends a right angle at A (Thales' Theorem).

So, $\angle BAE = \angle BAC - \angle EAC = 90 - (90 - x) = x$.

Then, $\angle ABE = 90 - x$.

$\triangle BEA$ and $\triangle AEC$ have their three angles with the same set of measures, i.e. $\triangle BEA \sim \triangle AEC$

Their sides must be in proportion:

$$\frac{AE}{BE} = \frac{CE}{AE} \quad \Rightarrow \quad \frac{6}{4} = \frac{CE}{6} \quad \Rightarrow \quad CE = \frac{6 \times 6}{4} = 9$$

Diameter $= BC = CE + BE = 9 + 4 = 13$

Radius $= \frac{1}{2}$(diameter) $= \frac{1}{2}(13) = 6.5$

4.075 Chords BD and EA intersect at C. Therefore, the products of their segments must be equal:

$(BC)(CD) = (EC)(CA) \quad \Rightarrow \quad 12 \times 10 = 8 \times CA$

$\Rightarrow \quad \dfrac{12 \times 10}{8} = CA \quad \Rightarrow \quad CA = 15$

In $\triangle ABC$, BC and CA are the 4th and 5th multiples of 3. AB must be the 3rd multiple of 3, i.e. 9.

Area of $\triangle ABC = \dfrac{(AB)(BC)}{2} = \dfrac{9 \times 12}{2} = 54$

4.084 Volume of each cone $= \dfrac{1}{3}\pi r^2 h = \dfrac{1}{3}\pi(1)^2(4) = \dfrac{4\pi}{3}$.

Volume of cylindrical pot $= \pi r^2 h = \pi(3)^2(6) = 54\pi$.

The number of cones that can be filled with the available ice cream would be the total volume of ice cream (i.e. volume of the pot), divided by the volume of each cone:

$$\frac{54\pi}{\frac{4\pi}{3}} = 54\pi \times \frac{3}{4\pi} = \frac{162}{4} = 40.5$$

The least number of cones would be 41, because there's no such thing as 40.5 cones.

4.085 Let r_c = radius of cone, and r_s = radius of sphere. Then, height of cone = diameter of sphere = $2r_s$.

Volume of cone = $\frac{1}{3}\pi r_c^2 (2r_s) = \frac{2\pi r_c^2 r_s}{3}$

Volume of sphere = $\frac{4}{3}\pi r_s^3 = \frac{4\pi r_s^3}{3}$

It's given that the two volumes are equal:

$$\frac{2\pi r_c^2 r_s}{3} = \frac{4\pi r_s^3}{3} \quad\Rightarrow\quad 2\pi r_c^2 r_s = 4\pi r_s^3 \quad\Rightarrow$$

$$r_c^2 r_s = 2r_s^3 \quad\Rightarrow\quad \frac{1}{2} = \frac{r_s^3}{r_c^2 r_s} \quad\Rightarrow\quad \frac{1}{2} = \frac{r_s^2}{r_c^2}$$

$$\sqrt{\frac{1}{2}} = \frac{r_s}{r_c} \quad\Rightarrow\quad \frac{1}{\sqrt{2}} = \frac{r_s}{r_c} \quad\Rightarrow\quad r_s : r_c = 1 : \sqrt{2}$$

4.086 There are several ways the bricks can be oriented. But to use the *least* number of bricks, the face with the greatest area (12"×6") would have to be along the face of the wall.

The wall is 19' wide, i.e. there would need to be 19 bricks along the horizontal direction.

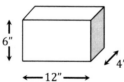

The wall is 14' high, i.e. there would need to be 28 bricks in the vertical direction. Hence, the total number of bricks would be 19 × 28 = 532.

4.087 The rise in the water level will correspond to the volume of the four plates released into it. That volume is 4×(11"×7"×2") = 616 cubic inches. Let d = diameter of drum. Then, volume of water in the 4" level rise = volume of the four plates:

$$\pi\left(\frac{d}{2}\right)^2 (4) = 616 \quad\Rightarrow\quad \pi\left(\frac{d^2}{4}\right)(4) = 616 \quad\Rightarrow$$

$$\pi d^2 = 616 \quad\Rightarrow\quad d^2 = \frac{616}{\pi} = \frac{616}{\frac{22}{7}} = 616 \times \frac{7}{22}$$

$$= 28 \times 7 \quad\Rightarrow\quad d = \sqrt{28 \times 7} = \sqrt{(4 \times 7) \times 7}$$

$$= \sqrt{2 \times 2 \times 7 \times 7} = 2 \times 7 = 14''$$

4.088 Let e = edge of cube = diameter of sphere.

Surface area of cube, $A = 6e^2 \quad\Rightarrow\quad e^2 = \frac{A}{6}$

Surface area of sphere, $a = 4\pi r^2 = 4\pi\left(\frac{e}{2}\right)^2 =$

$$\frac{4\pi e^2}{4} = \pi e^2 \quad\Rightarrow\quad a = \pi e^2 \quad\Rightarrow\quad e^2 = \frac{a}{\pi}$$

Equate the above two values of e^2:

$$\frac{A}{6} = \frac{a}{\pi} \quad\Rightarrow\quad A = \frac{6a}{\pi}$$

4.089

$$P(\text{tile} = \text{defective}) = \frac{\text{\# of defective tiles}}{\text{total \# of tiles}}$$

Along the 12' length of the floor, there would be 24 tiles, and along the 10' width of the floor, there would be 20 tiles. Therefore, the total # of tiles would be 24 × 20 = 480

$$P(\text{tile} = \text{defective}) = \frac{3}{480} = \frac{1}{160}$$

4.090 Please do not count the blocks one by one! Instead, count them *layer by layer*.

The top layer has   2 × 1 = 2 blocks
The second layer has   3 × 2 = 6 blocks
The third layer has   4 × 3 = 12 blocks
The bottom layer has 5 × 4 = 20 blocks
Total # of blocks = 2 + 6 + 12 + 20 = 40.

4.099 If we consider the two endpoints of the diameter, then the center of the circle $(-4, -1)$ is their midpoint. One endpoint is $(1, 2)$. Let the other be (x, y). Using the midpoint formula:

$$\frac{x + 1}{2} = -4 \quad\Rightarrow\quad x + 1 = -8 \quad\Rightarrow\quad x = -9$$

$$\frac{y + 2}{2} = -1 \quad\Rightarrow\quad y + 2 = -2 \quad\Rightarrow\quad y = -4$$

4.100 This example is easier than it seems. $OP = 9$ means the coordinates of P are $(0, 9)$. $OQ = 14.6$ means the coordinates of Q are $(14.6, 0)$. Also, $\overline{OP} \perp \overline{OQ}$. If arc POQ is a part of a circle, then *P* and *Q* can only be the endpoints of a diameter of that circle (Thales' Theorem). Only a diameter (*PQ*) subtends a right angle at a point (*O*) on the circumference.

Meaning, the center of the circle is simply the midpoint of diameter *PQ*. Using the midpoint formula, its coordinates are:

$$\left(\frac{0 + 14.6}{2}, \frac{9 + 0}{2}\right) \quad \text{i.e.} \quad (7.3, 4.5)$$

4.101 Slope of $\overleftrightarrow{l}$ = slope of $\overline{OB} = \frac{4}{3}$

$$\frac{8 - 0}{d - 0} = \frac{4}{3} \Rightarrow \frac{8}{d} = \frac{4}{3} \Rightarrow 4d = 24 \Rightarrow d = 6$$

Slope of $\overleftrightarrow{m}$ = slope of $\overline{OA} = -2$

$$\frac{c - 0}{-5 - 0} = -2 \Rightarrow \frac{c}{-5} = -2 \Rightarrow c = 10$$

Use the two-point form for equation of $\overline{AB}$:

$$\frac{y - y_1}{y_1 - y_2} = \frac{x - x_1}{x_1 - x_2} \Rightarrow \frac{y - 10}{10 - 8} = \frac{x - (-5)}{-5 - 6} \Rightarrow$$

$$\frac{y - 10}{2} = \frac{x + 5}{-11} \Rightarrow -11(y - 10) = 2(x + 5)$$

$$\Rightarrow -11y + 110 = 2x + 10 \Rightarrow$$

$$-11y = 2x - 100 \Rightarrow y = -\frac{2}{11}x + \frac{100}{11}$$

4.102 Notice that for $\overleftrightarrow{l}$, the *y*-intercept is −4. We need to find the slope of $\overleftrightarrow{l}$, but we don't have enough information for that. So let's find the slope of $\overleftrightarrow{m}$, and then use the *product of slopes* property for $\overleftrightarrow{l} \perp \overleftrightarrow{m}$.

Slope of $\overleftrightarrow{m} = \frac{y_2 - y_1}{x_2 - x_1} = \frac{18 - 0}{6 - 0} = \frac{18}{6} = 3$

(Slope of $\overleftrightarrow{l}$) × (slope of $\overleftrightarrow{m}$) = −1.

Slope of $\overleftrightarrow{l} = \frac{-1}{\text{slope of } \overleftrightarrow{m}} = \frac{-1}{3} = -\frac{1}{3}$

Slope-intercept equation of $\overleftrightarrow{l}$ is:

$$y = -\frac{1}{3}x - 4$$

4.103 (i) (−6, 3) would lie on the circle if its distance from the center (−3, 7) is equal to the radius 5. That distance is:

$$\sqrt{(7 - 3)^2 + \left(-3 - (-6)\right)^2} = \sqrt{4^2 + 3^2} =$$

$$\sqrt{16 + 9} = \sqrt{25} = 5 = \text{radius}$$

Therefore, (−6, 3) lies on the circle.

(ii) (1, 6) would lie outside the circle if its distance from the center (−3, 7) is greater than the radius 5. That distance is:

$$\sqrt{(7 - 6)^2 + (-3 - 1)^2} = \sqrt{1^2 + 4^2} =$$

$$\sqrt{1 + 16} = \sqrt{17} < \text{radius (i.e. } \sqrt{25})$$

So (1, 6) lies inside, not outside, the circle.

(iii) The center of the circle (−3, 7) lies at a distance of 7 above the *x*-axis. The radius is only 5. Meaning, the lowermost point on the circle will not reach the *x*-axis. Therefore, the circle will not cross over into the fourth quadrant.

4.104 The coordinates of *C* mean that it is at a horizontal distance of 24 from the *y*-axis, and a vertical distance of 32 from the *x*-axis.

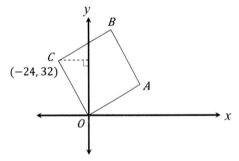

OC is the hypotenuse of the right triangle (dotted and below) whose legs are 24 and 32, i.e. 3rd and 4th multiples of 8. *OC* = (5th multiple of 8) = 40, the triangle being a 3–4–5 triangle. In the square, the diagonal *AC* would be the hypotenuse of the isosceles right triangle *OAC*.

$(AC)^2 = (OA)^2 + (OC)^2 = 40^2 + 40^2 =$

$1600 + 1600 = 1600 \times 2$

$AC = \sqrt{1600 \times 2} = 40\sqrt{2}$

4.105

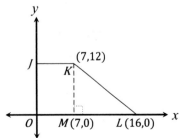

OJKL is a trapezoid means that $\overline{JK} \parallel \overline{OL}$. Drop an altitude *KM*. Then, *ML* = 16 − 7 = 9

Also, $KM = 12 = JO$. ML and KM are the 3rd and 4th multiples of 3. $KL = $ (5th multiple of 3) = 15 Perimeter of $OJKL = JO + JK + KL + OL = 12 + 7 + 15 + 16 = 50$.

4.106 The two shaded regions in Quadrants II and IV have a total area equal to that of a semicircle with radius 3. That area (say A_1) is:

$$A_1 = \frac{\pi r^2}{2} = \frac{\pi (3)^2}{2} = \frac{9\pi}{2}$$

Each shaded region in Quadrants I and III has an area equal to area of a square of side 3, minus area of a quarter circle of radius 3. Together, their area (call it A_2) is:

$$A_2 = 2\left(s^2 - \frac{\pi r^2}{4}\right) = 2\left(3^2 - \frac{\pi (3)^2}{4}\right) =$$

$$2\left(9 - \frac{9\pi}{4}\right) = 18 - \frac{9\pi}{2}$$

Total area of the shaded region $= A_1 + A_2 = \frac{9\pi}{2} + \left(18 - \frac{9\pi}{2}\right) = 18$

CHAPTER 5: ODD TYPES OF QUESTIONS
5.004

```
    2  B  4  A
    A  9  B  C
 +  5  3  6  8
 ----------------
 A  C  C  B  9
```

Three individual digits when added, will result in $9 + 9 + 9 = 27$ at the most. Meaning, when three 1-digit numbers are added, there will at the most be a 2 as a carried-over digit. In the highlighted numbers, $2 + 5 = 7$.   7 plus a maximum of 9 (for A) will result in at the most 18 even if there's a carried-over 2 above the highlighted 2. Therefore, at the bottom, $A = 1$. A is not zero because the bottom row is a 5-digit number. Replace all As with 1, and rewrite.

```
    2  B  4  1
    1  9  B  C
 +  5  3  6  8
 ----------------
 1  C  C  B  9
```

In the highlighted column, C must be **0**. Replace all Cs with 0, and rewrite.

```
    2  B  4  1
    1  9  B  0
 +  5  3  6  8
 ----------------
 1  0  0  B  9
```

For the shaded numbers, there must be a carried-over 2 above the shaded 2. That's the only way they will add up to 10.

Also, there must be a carried-over 1 above B, from the addition of $4 + B + 6$, because $4 + 6 = 10$ and B can at the most be 9, i.e. the result would be between 10 and 19 inclusive. Mark the carry-overs, and rewrite.

```
       2     1
    2  B  4  1
    1  9  B  0
 +  5  3  6  8
 ----------------
 1  0  0  B  9
```

The shaded numbers should add up to 20,   i.e. $B + 9 + 3 + 1 = 20 \Rightarrow B = 7$

5.005

```
    X  2  4
    Y  9  7
 +  Z  5  5
 ----------------
 1  0  7  6
```

The shaded and solid numbers will result in carry-overs. Mark them, and rewrite.

```
    1     1
    X  2  4
    Y  9  7
 +  Z  5  5
 ----------------
 1  0  7  6
```

The shaded column should add up to 10. $X + Y + Z + 1 = 10 \Rightarrow Z = 9 - (X + Y)$ For Z to be at its maximum value, X and Y would have to be at their minimum values. But neither X nor Y can be 0, because the addition involves 3-digit numbers. Also, $X \neq Y \neq Z$. So, X and Y would be the next possible values greater than 0 which would be 1 and 2. It doesn't matter which one is which. $X + Y = 1 + 2 = 3$ Maximum value of $Z = 9 - 3 = 6$.

5.006

$$\begin{array}{r} 1 \ A \ 3 \\ \times \quad 9 \ A \\ \hline 5 \ B \ C \\ 1 \ C \ 8 \ B \ + \\ \hline 1 \ 3 \ A \ A \ 2 \end{array}$$

The shaded portion means $C = 2$. Rewrite.

$$\begin{array}{r} 1 \ A \ 3 \\ \times \quad 9 \ A \\ \hline {}_1 \ 5 \ B \ 2 \\ 1 \ 2 \ 8 \ B \ + \\ \hline 1 \ 3 \ A \ A \ 2 \end{array}$$

Notice the carried-over 1. In the shaded region, $3 \times 4 = 12$ is the only viable option. Meaning, $A = 4$. In the solid region, since there is no carry-over, $B + B = A = 4$, i.e. $B = 2$.

5.008 Let's find $(5 \ \blacktriangledown \ 1)$ and $(10 \ \blacktriangle \ 5)$ separately, and then divide the first by the second:

$5 \blacktriangledown 1 = 5^2(1) - 1 = 25 - 1 = 24$

$10 \blacktriangle 5 = 2(10) - 3(5) = 20 - 15 = 5$

$$\frac{5 \blacktriangledown 1}{10 \blacktriangle 5} = \frac{24}{5} = 4\frac{4}{5} = 4.8$$

5.010 Any speed that's divisible by 2, 3 and 4 is also divisible by the LCM of 2, 3 and 4, the LCM being 12. So the speeds are simply the multiples of 12 between 0 and 50:

12, 24, 36 and 48 mph

Also, 0 itself is divisible by 2, 3, and 4, or for that sake, by any number.

So, the "preferred speeds" are: 0, 12, 24, 36, 48. To find their average, please don't perform the actual calculations. Notice that the numbers are equally spaced out, and are in ascending order. Their average is simply the midpoint of the list, which is 24 mph.

5.014 1st term: 1

2nd term: $3(1) + 1 = 3 + 1 = \underline{4}$

3rd term: $3(4) + 1 = 12 + 1 = 1\underline{3}$

4th term: $3(13) + 1 = 39 + 1 = 4\underline{0}$

5th term: $3(40) + 1 = 120 + 1 = 12\underline{1}$

6th term: $3(121) + 1 = 363 + 1 = 36\underline{4}$

Notice the pattern in the units digit of the terms: 1, 4, 3, 0, ... The arithmetic mean of the units digits of every four consecutive terms is:

$\frac{1+4+3+0}{4} = \frac{8}{4} = 2$ The first 100 terms comprise 25 groups, each having 4 consecutive terms. Since the average units digit of any 4 consecutive terms is 2, the average units digit of all 100 terms will remain the same, i.e. 2.

5.015 1st term: 3   **Odd term**

2nd term: $3 - 7 = -4$ **Even term**

3rd term: $|-4| = 4$ **Odd term**

4th term: $4 - 7 = -3$ **Even term**

5th term: $|-3| = 3$ **Odd term**

6th term: $3 - 7 = -4$ **Even term**

Notice the pattern: 3, −4, 4, −3, ...

Every 4th term is −3. So, the 28th term will be −3. Following the sequence, the 30th term will be −4.

5.016 Consider the first two terms: 2, 0

0 is q more than p times 2:

$0 = q + 2p \ \Rightarrow \ 2p + q = 0$

Yes, it can be this easy!

5.018 Let's find $f(g(h(-1)))$ by evaluating the functions from the inside out.

$h(-1) = 2(-1) - 1 = -2 - 1 = -3$

$g(h(-1)) = g(-3) = (-3)^2 - 6 = 9 - 6 = 3$

$f(g(h(-1))) = f(3) = 3^2 + 1 = 9 + 1 = 10$

Let's find $|g(h(0))|$ in a similar way.

$h(0) = 2(0) - 1 = -1$

$g(h(0)) = g(-1) = (-1)^2 - 6 = 1 - 6 = -5$

$|g(h(0))| = |-5| = 5$

$f(g(h(-1))) \cdot |g(h(0))| = (10) \cdot (5) = 50$

5.020 First evaluate $h(4)$, i.e. the y-value on the line $h(x)$ when $x = 4$.   From the graph, $h(4) = 0$. Next, find $g(h(4))$, i.e. $g(0)$. When $x = 0$, $g(x) = g(0) = 1$. Therefore, $g(h(4)) = 1$. Finally, locate $f(g(h(4)))$, i.e. $f(1)$. This is the point on $f(x)$ when $x = 1$. It appears to be 2. Therefore, $f(g(h(4))) = 2$.

CHAPTER 6: VISUAL DATA ANALYSIS

6.004 Upon hospitalization the first time, Judy would have received $2,000. Upon the two subsequent hospitalizations, she would have received $1,000 each time, i.e. $2,000 more. The total she would have received over three hospitalizations would have been $2,000 + $2,000 = $4,000. The answer is (C).

6.005 The question is about her monthly premium. So, look up the monthly premium for an individual in the 18 – 50 age bracket. That's $80. When she got married to Donald they were both still in the 18 – 50 age bracket, but were now under the two parents category. Remember, this category is for all couples with or without children. Their new monthly premium would be $120 for the policy, and an additional $5 + $15 = $20 for riders A and C. The total monthly premium would be $140. If it is equally split between Stacy and Donald, then Stacy's premium would be half of $140, i.e. $70. Her original monthly premium was $80. It means, her new monthly premium decreased by $10. The answer is (D).

6.009 Basically, 25% of 40% (from the second chart) of 19% (from the first chart) of $101 billion is what we need to find out.

$$\frac{25}{100} \times \frac{40}{100} \times \frac{19}{100} \times 101$$

This is where approximation helps. 19% is close to 20%, and $101 bn. is close to $100 bn.

$$\frac{25}{100} \times \frac{40}{100} \times \frac{20}{100} \times 100 = \frac{1}{4} \times \frac{2}{5} \times \frac{1}{5} \times 100 = 2$$

The approximate value is 2, i.e. $2 billion. The answer is (A).

6.010 (i) This statement is about the **number of vehicles** sold. The two graphs are about percentages of $101 billion. To find the number of vehicles sold of any kind, we would need to know the unit price of those vehicles. That information is not available, and therefore this statement cannot be inferred.

(ii) Interior body parts are not any of the specifically listed types of parts in the second chart. Meaning, interior body parts would be a part of *other parts*, their greatest share being 15%. So, if the other parts comprised fully of interior body parts alone, then that would at the most be 15% of the total parts. This statement can be inferred.

(iii) Once again, there's no need to calculate the actual values; simply compare the totals of percentages. From the first chart, Trucks + SUVs = 17% + 11% = 28% of $101 billion.
From the first and second charts,
Vans + Powertrain Parts = 22% + (40% of 19%) = 22% + 7.6% = 29.6% of $101 billion, which is greater than Trucks + SUVs. This statement can be inferred from the data.
The answer is (E).

6.014

$$\text{percent increase} = \frac{\text{actual increase}}{\text{original value}} \times 100$$

In the 5-year period from 1970 to 1975, the service sector grew from $5 billion to $10 billion, a total increase of 100%. During none of the other listed 5-year periods did the service sector double. The answer is (A). Please do not merely look at how steep the incline is from one year to another. You would need to look at the incline if the question were about the greatest *actual* increase, not the greatest percent increase.

6.015 The figure is drawn to scale; believe in what you see! The service sector (right hand scale) grew from $20 billion to about $26 billion, a total of about $6 billion. The manufacturing sector (left hand scale) declined from about $38 billion to $35 billion, a total of about $3 billion. The ratio of growth in service sector to decline in manufacturing sector would be $6 bn. : $3 bn., or 2 : 1. The answer is (E).

6.019 The graph is about the number of homes. But to find the sales taxes paid we would also need to know the average prices of the new and existing homes sold. That information is not available. Therefore, we cannot calculate the taxes. The answer is (E).

6.020 Number of existing homes sold in 2006 = 4 million. Number of existing homes sold in 2009 = 4.5 million.

$$\text{percent increase} = \frac{\text{actual increase}}{\text{original value}} \times 100\% =$$

$$\frac{4.5 - 4}{4} \times 100\% = \frac{0.5}{4} \times 100\% = \frac{50}{4} = 12.5\%$$

The $100,000 home I bought in 2006 as an existing home was worth 12.5% more in 2009, i.e. $12,500 more. Its new worth in 2009 would be $112,500. The answer is (D).

6.024 In 2004, the *increase* in domestic oil consumption was about 5 million barrels over its previous year's consumption of 15 million barrels, i.e. 5 out of 15, or 1 out of 3, or $33.\overline{3}\%$. In 2005, the increase was 10 million barrels over its previous year's consumption of 20 million barrels, i.e. 10 out of 20, or 50%. There was no increase in 2006, and the increase in 2007 or 2008 certainly doesn't look anywhere close to 50%. The answer is (B).

6.025 (i) From 2007 to 2008, the *amount* by which its domestic oil consumption increased was 5 million barrels. Also, the *percent* by which its oil production dropped was:

$$\frac{60 - 50}{60} \times 100 = \frac{10}{60} \times 100 = \frac{100}{6} \approx 16\%$$

In 2009, the domestic oil demand must have been 40 + 5 = **45 million barrels**. The oil production must have been about 16% less than 50 million barrels, or 84% of 50 =

$$\frac{84}{100} \times 50 = \frac{84}{2} \approx \textbf{42 million barrels}.$$

Therefore, this statement is true.

(ii) We just calculated that the domestic oil demand in 2009 was 45 million barrels, not more than it. Hence, this statement is not true.

(iii) It looks like Zigzagistan didn't even produce enough oil to meet its domestic

demand. So, maybe it didn't export any oil at all. Or maybe it exported 60% less oil at a certain time of the year, but ended up importing much more oil at other times of the year. Maybe its oil fields are frozen and inoperable in winter. Too many variables are at work. Therefore, this statement cannot be verified. The answer is (A).

6.029 A quick look at the labor costs and total costs per mile will reveal the following fractions for the listed highways:

H-1: $\dfrac{1}{3}$

H-2: $\dfrac{0.7}{2.5} = \dfrac{7}{25} \approx \dfrac{1}{3.5}$

H-4: $\dfrac{1.7}{5.0} = \dfrac{17}{50} \approx \dfrac{1}{3}$

H-6: $\dfrac{2}{8} = \dfrac{1}{4}$

H-8: $\dfrac{0.5}{3.0} = \dfrac{5}{30} = \dfrac{1}{6}$

Clearly, H-8 had the smallest fraction of the total cost as the labor cost. The answer is (E).

6.030 The length of highway H-5 is 400 miles. Half of that is 200 miles due for repairs. The cost of materials is 50% less, while labor cost and overheads remain the same. Therefore, the total cost per mile would be 1 + 0.5 + 0.5 = $2 million. Total cost of repair of the due section of highway H-5 = 2 × 200 = $400 million.

CHAPTER 7: QUANTITATIVE COMPARISON

7.021 Draw approximate sketches of the clock at 3:30 and 5:45.

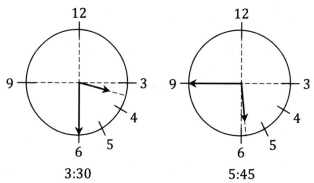

3:30   5:45

Simply by looking at the sketches you can tell that the minor angle between the two hands of the clock at 3:30 is less than 90°, whereas that at 5:45 is greater than 90°. No calculations are required. The answer is (B).

7.022 $\frac{27}{100} \times 3800 = \frac{27 \times 3800}{100} = 27 \times 38$

This is the same as column B. The answer is (C).

7.023 If $0 \leq x \leq 1$, then $x^2 < 1$ if x is a fraction. But if x is 1, then $x^2 = 1$. Therefore, $x^2 \leq 1$. Similarly, if x is a fraction, then:

$\frac{1}{x} > 1$ i.e. $\frac{1}{x^2} = \left(\frac{1}{x}\right)^2 > 1$ also.

But if x is 1, then $\frac{1}{x^2} = \frac{1}{1^2} = 1$. Hence, $\frac{1}{x^2} \geq 1$.

The value $x = 1$ is common to both. Therefore, it is not possible to generally say which column will always be greater. The answer is (D).

7.024 Initially, let x and y both be 90°. Imagine a scenario in which $\angle D$ remains the same, and $\angle C$ increases considerably than $\angle D$, i.e. $x > y$.

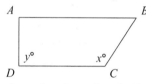

Notice that the length of diagonal AC would remain the same, whereas B goes farther away from D, i.e. the length of diagonal BD would considerably increase, and be greater than AC.

Now imagine a scenario in which $\angle C$ remains the same, and $\angle D$ drops considerably, i.e. $y < x$.

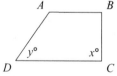

Notice that A gets relatively closer to C, i.e. the length of diagonal AC decreases, whereas the length of diagonal BD remains the same. AC is less than BD.

In either case, $BD > AC$. If both the scenarios happen simultaneously, i.e. if $\angle C$ increases *and* $\angle D$ decreases, then the effect will be even more pronounced, i.e. $BD > AC$ even more. The answer is therefore (B).

7.025 $\frac{a}{c} = \frac{-1}{\pi}$ = a negative constant. Meaning, either a is negative or c is negative, but neither a nor c is zero. Therefore, product ac must be negative, i.e. $ac < 0$. If $abc = 0$, then b must be zero, because neither a nor c is zero, i.e. $b = 0$. Column B is greater; the answer is (B).

7.026 When a rectangle is inscribed in a circle, the diagonals of the rectangle are the diameters of the circle.

3 × diagonal = 3 × diameter = 3 × 2 × radius

Circumference of the circle = 2 × π × radius = 2 × 3.14 × radius.   Clearly, circumference is slightly greater than 3 times the diameter (or the diagonal). The answer is (B).

7.027 Notice the relationship between the corresponding terms in the two columns:

$\frac{1}{3} > \frac{1}{4}$; $\frac{1}{5} > \frac{1}{6}$;   and $\frac{1}{7} > \frac{1}{8}$

Meaning, in column A, all the greater terms are being added, and in column B, all the smaller terms are being added. The answer is (A).

7.028 In the triangle, the length of the side facing angle x is equal to the length of the rectangle, minus the width of the rectangle. The length of the side facing angle y is equal to the length of the rectangle. Meaning, x faces a smaller side, and y faces a larger side. In a triangle, the smaller the angle, the smaller the side facing it. Hence, $y > x$. The answer is (B).

7.029 For column A, it would be the ratio of women to men, i.e. 7 : 3, i.e. $\frac{7}{3}$. For column B, it would be the ratio of children to women, i.e. 16 : 7, i.e. $\frac{16}{7}$. Now, compare the fractions:

Column A is greater; the answer is (A).

7.030

$$\frac{27 \times 26!}{25!} = \frac{27 \times 26 \times 25!}{25!} = 27 \times 26$$

$$\frac{27!}{25 \times 24!} = \frac{27 \times 26 \times 25 \times 24!}{25 \times 24!} = 27 \times 26$$

Both columns are equal. The answer is (C).

7.031 Height (h) = diameter = $2 \times$ radius (r)

i.e. $h = 2r$ The cap is *hemispherical*; its outer (curved) surface area is:

$$\frac{4\pi r^2}{2} = 2\pi r^2$$

The outer (curved) surface area of the can is:

$2\pi rh = 2\pi r(2r) = 4\pi r^2$

Column B is greater; the answer is (B).

7.032 If you run the positive integer values of k through your mind, $k = 1, 2, 3, \ldots\ldots$ you will see that at $k = 2$, $|3k - 7| = |3(2) - 7| = |-1| = 1$. This is the minimum value of $|3k - 7|$.

Similarly, the minimum value of $|5m - 9|$ is $|5(2) - 9| = |1| = 1$.

They're both equal; the answer is (C).

7.033 x is at least 0.5. When rounding off x to the nearest units place, the choice would be between $x = 0$ and $x = 1$. Since x is ½ or greater, it would be rounded off to 1.

y is at the most 0.9. When rounding off y to the nearest tens place, the choice would be between 0 and 10. Since y is closer to 0 than to 10, it would be rounded off to 0.

Column A is greater; the answer is (A).

7.034 On the slanted line shown, the x and y coordinates are equal. Below the slanted line in the 1st quadrant, x coordinate is greater than y coordinate. Meaning, $p > q$.

Above the slanted line in the 3rd quadrant, $|x| >$ $|y|$. But since both x and y coordinates are negative, y coordinate is greater than x coordinate. Meaning, $s > r$.

In column A, both the greater quantities are added. In column B, the smaller ones are added. Hence, column A is greater; the answer is (A).

7.035 To clear any confusion, always first express one variable in terms of the other(s).

$2m > 3n \quad \Rightarrow \quad m > \frac{3}{2}n \quad \Rightarrow \quad m > 1.5n$

$2t > 3h \quad \Rightarrow \quad 3h < 2t \quad \Rightarrow \quad h < 0.\overline{6}t$

Column B = $\frac{n}{t}$

Column A = $\frac{m}{h} = \frac{(> 1.5n)}{(< 0.\overline{6}t)} > \frac{n}{t}$

The numerator is greater than n, whereas the denominator is less than t. Therefore, the result is greater than $\frac{n}{t}$. Column A is greater; the answer is (A).

7.036 $35 = 7 \times 5$

$36 = 6 \times 6 = 3 \times 2 \times 3 \times 2$

Sum of unique prime factors of 35 is $7 + 5 = 12$

Sum of unique prime factors of 36 is $3 + 2 = 5$

Column A is greater; the answer is (A).

7.037 Drop an altitude from $(c, 9)$ onto the dotted diagonal of the rectangle. By subtracting the y coordinates, we can tell that the length of the altitude is 4.

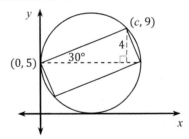

The diagonals of a rectangle inscribed in a circle are the diameters of that circle. Also, since the circle is tangential to the x and y axes, the point $(0, 5)$ is the endpoint of the dotted diameter (diagonal). The radius is the vertical distance of the dotted diameter from the x axis, i.e. 5. Meaning, the diameter is 2 times the radius, or $2 \times 5 = 10$. The dotted diagonal divides the rectangle into two equal right triangles. Each has an area of $\frac{\text{diagonal} \times \text{altitude}}{2} = \frac{10 \times 4}{2} =$ 20.   Area of the rectangle is twice that, i.e. 40. Both columns are equal; the answer is (C).

7.038 Let p = Polly's age, and s = Sarah's age. They are both teenagers, i.e. $p \geq 13$, and $s \geq 15$.

Column A = $\dfrac{p}{s}$, and Column B = $\dfrac{p-2}{s-2}$

The fact that $p < s$ means column A is a proper fraction. If the same quantity is added to the numerator and the denominator of a proper fraction, the value of the fraction increases. By the same token, if the same quantity is subtracted from them both, then the value of the fraction *decreases*. Therefore, column B is less than column A; the answer is (A). Notice that knowing their actual ages was not important.

7.039 If $0 < x < 1$, then $x^2 < x < 1$.

Also, $\dfrac{1}{x} > 1$. Therefore, $\left(\dfrac{1}{x}\right)^2 > 1$, i.e. $\dfrac{1}{x^2} > 1$

Meaning, $\dfrac{1}{x^2} > x$

0.638 is between 0 and 1, i.e. $0 < 0.638 < 1$.

Therefore, $\dfrac{1}{(0.638)^2} > (0.638)^2$

The answer is (B).

7.040 In column A, each of the four digits XXXX can take on the values 1, 3, 5, 7, 9. Meaning, 5 possible values for the 1st digit, 5 for the 2nd digit, 5 for the 3rd, and 5 for the 4th. The total number of 4-digit numbers that can be formed would be $5 \times 5 \times 5 \times 5$.

In column B, the values available are 0, 2, 4, 6, 8. For the number to be a 4-digit number, the 1st digit can only take on the values 2, 4, 6, or 8 (not 0), i.e. a total of 4 values. The 2nd, 3rd and 4th digits can be any of 0, 2, 4, 6, 8, i.e. a total of 5 values each. The total number of 4-digit numbers that can be formed would be $4 \times 5 \times 5 \times 5$.

Column B is lesser; the answer is (A).

7.041 For the product ab to be negative, one of them would have to be negative, whereas the other would have to be positive. If b is the negative number (say –6), then $|a - b|$ is the same as $|a - (-6)|$, i.e. $|a + 6|$, where a is already positive. Meaning, $|a + b|$ would be the same as $|a| + |b|$. That's greater than $|a| - |b|$. The answer is (A). Remember: We're trying to find the *greatest* value of $|a - b|$. Therefore, a would have to be positive, and b negative. The actual value of b (–6 or any other value) doesn't matter, as long as it is negative.

7.042 If you consider $\overline{AB}$ as the base of $\triangle ADB$, then its height will be maximum when $\angle A = \angle B = 45°$. $\angle D$ will always be 90° (Thales' Theorem). The more $\angle B$ deviates from 45°, the shorter the triangle will be, and the lesser its area will be. We see that 67° deviates from 45° by $67 - 45 = 22°$. Using similar logic in $\triangle ACB$, 37° deviates from 45° by $45 - 37 = 8°$ only. Therefore, $\triangle ADB$ is the shorter of the two triangles, and will have lesser area. The answer is (B).

7.043 Odd numbered years are not leap years. Meaning, February 1975 had only 28 days, not 29 days. The probability that I was born in February 1975 is 28 days out of 365 days, i.e. $\dfrac{28}{365}$. The probability that I was born in January 1975 is $\dfrac{31}{365}$. The answer is (A). Please note that it's not correct to calculate the two probabilities as $\dfrac{1}{12}$ each (i.e. 1 month out of 12 months). You'd need to be specific about the *number of days* in each month.

7.044 $12 \times 52 = 1300$, and $26 \times 51 = 1326$. The answer is (B).

7.045

$\left(\dfrac{u+v}{-2}\right)^2 < \left(\dfrac{u-v}{-2}\right)^2 \Rightarrow \dfrac{(u+v)^2}{4} < \dfrac{(u-v)^2}{4}$

$\Rightarrow (u+v)^2 < (u-v)^2 \Rightarrow$

$u^2 + 2uv + v^2 < u^2 - 2uv + v^2 \Rightarrow$

$2uv < -2uv \Rightarrow 4uv < 0 \Rightarrow uv < 0$

The answer is (A).

7.046 Assume x as the total number of employees. Then, the number of male employees is greater than or equal to 25% of x, i.e. greater than or equal to $\frac{1}{4}x$. Meaning, the number of female employees is less than or equal to $\frac{3}{4}x$. Of them, at the most $\frac{2}{3}$ are married, i.e. less than or equal to $\frac{2}{3}$ of $\frac{3}{4}x$ are married. That would be less than or equal to:

$$\frac{2}{3} \times \frac{3}{4}x = \frac{1}{2}x = 50\% \ of \ x$$

Meaning, ($\geq 25\%$ of x) is being compared with ($\leq 50\%$ of x). The answer is indeterminate (D).

7.047

$$p^2 = -64q \quad \Rightarrow \quad q = -\frac{p^2}{64}$$

Regardless of the value of p (positive or negative), p^2 will always be positive. Therefore, q will be negative.

In $r^3 = 64s$, if s is positive, then r will be positive. If s is negative, then r will be negative. Meaning, the product rs will be positive either way. Therefore, $rs > q$. The answer is (B).

7.048

If $\frac{a}{b} = \frac{c}{d}$, then $\frac{a-b}{a+b} = \frac{c-d}{c+d}$. See page 26.

Let $\frac{a-b}{a+b} = \frac{c-d}{c+d} = x.$

$$\frac{2a-2b}{3a+3b} = \frac{2(a-b)}{3(a+b)} = \frac{2}{3}\left(\frac{a-b}{a+b}\right) = \frac{2}{3}x$$

$$\frac{4c-4d}{5c+5d} = \frac{4(c-d)}{5(c+d)} = \frac{4}{5}\left(\frac{c-d}{c+d}\right) = \frac{4}{5}x$$

So, the comparison is between $\frac{2}{3}x$ and $\frac{4}{5}x$, or simply between $\frac{2}{3}$ and $\frac{4}{5}$. $\frac{4}{5}$ is obtained when the same integer (2) is added to the numerator as well as the denominator of the proper fraction $\frac{2}{3}$. Therefore, its value is greater than that of $\frac{2}{3}$. The answer is (B).

7.049 If $k < c$, then $\frac{k}{c}$ is a proper fraction. Adding 1.22 to both the numerator and the denominator will result in a greater fraction.

But if $k > c$, then $\frac{k}{c}$ is an improper fraction. Adding the same number 1.22 to both the numerator and the denominator will result in a smaller fraction. Therefore, the relationship is indeterminate. The answer is (D).

7.050 Area remaining fixed, there are infinitely many configurations possible for a rectangle. Its perimeter would be the least when it is a square. For rectangle X (square X) of area 36, its side would be $\sqrt{36} = 6$. Its least perimeter would be $6 + 6 + 6 + 6 = 24$.

Perimeter remaining fixed, there are infinitely many configurations possible for a rectangle. Its area would be maximum when it is a square. For rectangle Y (square Y) of perimeter 36, its side would be $36 \div 4 = 9$. Its maximum area would be $9 \times 9 = 81$.

Column B is greater; the answer is (B).

7.051 If the rate of working is uniform, then simply consider that in column A, 2 fences are to be built by 4 men. Think to yourself how many men would 20 fences take? 40. How many men for 30 fences? 60. Meaning, to finish one extra fence in a certain amount of time, it takes two extra men. How many men for 7 fences (from column B)? 14. So, the amount of time taken would be the same, as long as two extra men are hired for every extra fence to be built. The answer is (C).

7.052 $\sqrt[3]{x} = x^{\frac{1}{3}}$, $\sqrt[7]{x} = x^{\frac{1}{7}}$, $\sqrt[3]{\sqrt[7]{x}} = \left(x^{\frac{1}{7}}\right)^{\frac{1}{3}} = x^{\frac{1}{21}} = \sqrt[21]{x}$

$\sqrt[5]{x} = x^{\frac{1}{5}}$, $\sqrt[9]{x} = x^{\frac{1}{9}}$, $\sqrt[5]{\sqrt[9]{x}} = \left(x^{\frac{1}{9}}\right)^{\frac{1}{5}} = x^{\frac{1}{45}} = \sqrt[45]{x}$

When x is between 0 and 1 (i.e. a proper fraction), its square root is larger than itself. The cube root, the 4th root, 5th root etc. will get even larger. Therefore, column B is greater; the answer is (B).

7.053 If C bisects minor arc AB, and D bisects major arc AB, then DC must be the diameter passing through the center O.

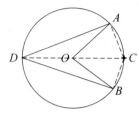

Draw line segments AC and BC. ΔDOA and ΔAOC have the same base length (radius) and the same height. Therefore, they have equal areas. But there's some extra area between the side AC and the minor arc AC. That area, along with area of ΔAOC will be greater than area of ΔDOA. Similarly, the total area of the loop $OCBO$ will be greater than area of ΔDOB. Together, the area of loop $AOBCA$ will be greater than the area of loop $AOBDA$. The answer is (A).

7.054 In column A, proper fractions are added, resulting in a greater fraction. In column B, proper fractions are multiplied, resulting in a smaller fraction. The answer is (A).

7.055 The diagonal of the rectangular box is $\sqrt{3^2 + 4^2 + 5^2}$. The diagonal of the cube is $\sqrt{4^2 + 4^2 + 4^2}$. So, the real comparison is between $3^2 + 5^2 = 9 + 25 = 34$, and $4^2 + 4^2 = 16 + 16 = 32$. Column A is greater; the answer is (A).

7.056 Clearly, you want to buy as many cheaper $3 items as possible. In column A, you're looking for the greatest multiple of $3 which, when subtracted from $20, will leave behind $5. That would be 5, i.e. $3 × 5 = $15. So, buy 5 items for $3 each, and one $5 item for the remaining $5, the total being 5 + 1 = 6 items.

In column B, you're looking for the greatest multiple of $3 which, when subtracted from $21, will leave behind $5. That would be 2, i.e. $3 × 2 = $6. No other multiple makes sense. So, buy 2 items for $3 each, and three $5 items for the remaining $15, the total being 2 + 3 = 5 items. The answer is (A).

7.057

$$100 \times 10^{-13} = 10^2 \times 10^{-13} = 10^{-11} = \frac{1}{10^{11}}$$

$$0.1^2 \times 0.1^5 \times 0.1^3 = 0.1^{2+5+3} = 0.1^{10} = \frac{1}{10^{10}}$$

The denominator is greater in column A. Therefore, column A is lesser; the answer is (B).

7.058 In the slope intercept equation:

$$p = \text{slope} = \frac{y_2 - y_1}{x_2 - x_1} = \frac{a - 0}{0 - b} = \frac{a}{-b} = \frac{-a}{3a} = \frac{-1}{3}$$

$|a|^b = 1$ could mean one of two things:
$a = \pm 1$, and $b = $ any real number $\neq 0$, or
$a = $ any real number $\neq 0$, and $b = 0$.
But we're given that $b = 3a$. Therefore, $b \neq 0$, because if $b = 0$, then $a = 0$, and $|a|^b = 0^0 \neq 1$. Meaning, a must be ± 1. Given the positions of the two points, it's clear that $a = -1$. Therefore, $q = y$-intercept $= a = -1$.
Therefore, $p > q$. The answer is (A).

7.059

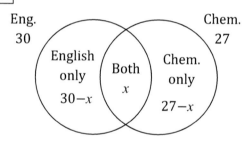

We don't need to know x (>0) to be able to tell which crescent – *English only* or *Chemistry only* – is greater. Clearly, *English only* is greater because $(30 - x)$ is greater than $(27 - x)$; the answer is (B).

7.060 With the exception of 2 and 5, prime numbers always end with 1, 3, 7 or 9. So, if x is a prime number, then $2x$ could have 0, 2, 4, 6 or 8 as its units digit. That's five digits total.

Once again, with the exception of 2 and 5, primes always end with 1, 3, 7 or 9. So, x^2 could have 1, 4, 5 or 9 as its units digit. That's four digits total. The answer is (A).

CHAPTER 8: DATA SUFFICIENCY

8.021 Statement (1): It doesn't say anything about q, and is not sufficient to answer the question.

Statement (2): Rewrite $\frac{p}{qr}$ as the following:

$$\frac{p}{qr} = \frac{\left(\frac{p}{r}\right)}{q}$$

If the numerator $\left(\frac{p}{r}\right)$ is an even multiple of q, then the end result of the given fraction is an integer. This statement is sufficient to answer the question.

The answer is (B).

8.022 Statement (1): Adding $16 to the price at the start of the month would give us the price at the end of the month. But the starting price is unknown. This statement is not sufficient to answer the question.

Statement (2): An increment of 12.78% means nothing unless there's a value of which it is expressed as a percentage. This statement is not sufficient to answer the question.

Let's consider the two statements together. Assume $x as the starting price of the stock. Then, 12.78% of x would equal $16. From this, x can be found out. Adding $16 to $x would give the price of the stock at the end of the month. Statements (1) and (2) are together sufficient to answer the question, although neither of them is independently sufficient. The answer is (C).

8.023 Statement (1): If $x < y$, then $\frac{x}{y}$ is a proper fraction. And if $a = b$, then the same positive number is added to the numerator as well as to the denominator. Meaning, the value of the original fraction increases. This statement is sufficient to answer the question.

Statement (2): If a positive integer b is added to the numerator as well as to the denominator of a proper fraction, then the value of the fraction increases. If a larger integer a is added to the numerator, and a smaller integer b is added to the denominator, then the value of the fraction increases even further. This statement is sufficient to answer the question.

Each statement is independently sufficient to answer the question. The answer is (D).

8.024 Assume the largest integer is x. Statement (1): If the average of the three *smaller* integers is 42, then:

$$\frac{(42 \times 3) + x}{4} = 45\frac{1}{2}$$

Solve to find the value of x. This statement is sufficient to answer the question.

Statement (2): If the four integers are consecutive, then there's only one way their average could be 45½, i.e. if the numbers are 44, 45, 46 and 47. This statement is sufficient to answer the question.

Each statement is independently sufficient to answer the question. The answer is (D).

8.025 Statement (1): The perimeter of a semicircle is equal to half the circumference, plus the diameter (or twice the radius). Meaning, $32 = \frac{1}{2}(2\pi r) + 2r$. Solve to find the value of r. This statement is sufficient to answer the question.

Statement (2): The length of the semicircular arc is equal to half the circumference. Meaning, $\frac{1}{2}(2\pi r) = 20$. Solve to find the value of r. This statement is sufficient to answer the question.

Each statement is independently sufficient to answer the question. The answer is (D).

8.026 Statement (1): $A = 12 = 2B$ means that $A = 12$ and $B = 6$. With that, we can find the value of $A \text{ ⌘ } B$. But if you look closely at the definition: $x \text{ ▣ } y = 2 - 3x$, you will notice that knowing the value of y is not important at all. Meaning, the value of $(A \text{ ⌘ } B) \text{ ▣ } C$ can be calculated without knowing the value of C. This statement is sufficient to answer the question.

Statement (2): This statement does not give the value of either A or B or C. Therefore, this statement is not sufficient to answer the question.

The answer is (A).

8.027 If I were born in 1976, then I would turn 34 sometime in 2010. On December 31, 2009, there would be a 0% chance of my being 34 years old. On December 31, 2010, there would be 100% chance that I am 34. The probability would increase from 0% to 100% as each day in 2010 would pass by.

Statement (1): If today's date is known, then the probability that I'm 34 is:

$$\frac{\text{\# of days in 2010 elapsed until today}}{\text{\# of days in 2010}}$$

This statement is sufficient to answer the question.

Statement (2): You would compare my birthday with today's date. If my birthday in 2010 has already passed, or is today, then the probability that I'm 34 would be 100%. If not, then it would be 0%. This statement does not tell today's date, and is therefore, not sufficient to answer the question. (Please note that the probability that today is June 15 would be $\frac{1}{365}$ or $\frac{1}{366}$ depending upon whether this year is a leap year. Simply knowing my birthday won't help; today's date would still need to be known.)

The answer is (A).

8.028 Points M and N lie on a line segment passing through the origin and (12, 12). It means that every point on the line segment has equal x and y coordinates.

Statement (1): If $d = 4$, then it also means that $c = 4$. But to find the length of segment MN, we would need to know the coordinates of M as well. This statement does not help find either a or b. Therefore, this statement is not sufficient to answer the question.

Statement (2): This statement expresses a relationship between b and c. But that, in

conjunction with the diagram, doesn't help find either c or d. This statement is not sufficient to answer the question.

If we use the statements together, then from statement (1), we can find the value of c, because $c = d$. From statement (2), we can find the value of b, and therefore the value of a, because $a = b$.

Statements (1) and (2) are together sufficient to answer the question, although neither of them is independently sufficient. The answer is (C).

8.029 Assume x as the number of gallons of gasoline added.

Statement (1): The original mixture had 70% of $15 = 10.5$ gallons of alcohol, and 30% of $15 = 4.5$ gallons of gasoline. After the addition of x gallons of gasoline, the amount of gasoline became $(x + 4.5)$ gallons. The alcohol remained steady at 10.5 gallons. This new ratio of alcohol to gasoline is given as 2 : 3.

$$10.5 : (x + 4.5) \quad \Rightarrow \quad \frac{10.5}{x + 4.5} = \frac{2}{3}$$

From this equation, x can be found out. Therefore, this statement is sufficient to answer the question.

Statement (2): The final mixture of 25 gallons can be split in the ratio 2 : 3 to find the amount of alcohol (10 gallons) and gasoline (15 gallons) in it. If x gallons of gasoline is taken out of it, then the original mixture in the ratio 70% : 30% (or 7 : 3) would be obtained:

$$\frac{10}{15 - x} = \frac{7}{3}$$

From this equation, x can be found out. Therefore, this statement is sufficient to answer the question.

Each statement is independently sufficient to answer the question. The answer is (D).

8.030 Statement (1): If $q = \frac{r}{p}$, then $pq = r$.

Therefore, $|pq - r| = |pq - pq| = |0| = 0$. This statement is sufficient to answer the question.

Statement (2): If $p \cdot |q| = r$, then either $p \cdot q = r$, or $p \cdot (-q) = r$, i.e. $-pq = r$. If $pq = r$, then $|pq - r| = 0$. But if $-pq = r$, then $|pq - (-pq)| = |2pq|$. Without knowing the actual values of p and q, it is not possible to estimate the value of $|2pq|$. This statement is not sufficient to answer the question.

The answer is (A).

8.031 Neither of the two statements says anything about the exact value(s) of x. But the two statements together would mean that x could only be 14, 15 or 16. Statements (1) and (2) are together sufficient to answer the question, although neither of them is independently sufficient. The answer is (C).

8.032 The following diagram shows the approximate positions of their heights. Ken could be either taller or shorter than Lisa. All we know is that Ken is taller than Amy, and that Amy is shorter than Lisa.

$$\text{Ken} \quad -----------$$
$$\text{Lisa} \quad \underline{\hspace{3cm}}$$
$$\text{Ken} \quad -----------$$
$$\text{Amy} \quad \underline{\hspace{3cm}}$$

Statement (1): Mike is taller than Ken. So, Mike could be in several places, either shorter or taller than Lisa. Dixon, being shorter than Mike, could be shorter or taller than Lisa. There's nothing that establishes a firm connection between their heights. This statement is not sufficient to answer the question.

Statement (2): This statement introduces Mike, but does not establish anything between Mike's height and Dixon's height that would firmly tell us who is taller between Lisa and Dixon. This statement is not sufficient to answer the question.

Consider the statements together. Statement (1) says that Dixon is shorter than Mike. Statement (2) says that Lisa is taller than Mike. It means that Lisa must be taller than Dixon.

Statements (1) and (2) are together sufficient to answer the question, although neither of them is independently sufficient. The answer is (C).

8.033 Statement (1): This statement means that $PQ \perp QR$. Meaning, $\angle Q = 90°$. In that case, there's no way $\angle P$ or $\angle R$ could be obtuse or right, i.e. $\angle R$ cannot be 106°. This statement is sufficient to answer the question.

Statement (2): $\triangle PQR$ could be isosceles in three different ways. First, $\angle P = \angle R$ while $\angle Q = 37°$. In that case, $\angle P$ or $\angle R$ cannot be obtuse, i.e. $\angle R$ cannot be 106°. Second, $\angle Q = \angle P = 37°$ in which case, $\angle R = 180 - (37 + 37) = 106°$. Third, $\angle Q = \angle R = 37°$ in which case, $\angle R \neq 106°$. This statement is not sufficient to answer the question.

The answer is (A).

8.034 The given statement can be rewritten as
$$3^{ax-b} = 3^{-c} \quad \Rightarrow \quad ax - b = -c \quad \Rightarrow$$
$$ax = b - c \quad \Rightarrow \quad a = \frac{b-c}{x}$$

Statement (1): If $b = c$, then $a = \frac{0}{x} = 0$. This statement is sufficient to answer the question.

Statement (2): If $b^c = 1$, then either $b = 1$, and $c = $ any integer, or $b = -1$, and $c = $ any even integer, or $b = $ any non-zero number and $c = 0$. In any case, the value of a cannot be affirmed. This statement is not sufficient to answer the question.

The answer is (A).

8.035 Assume the trip is x miles long.
$$\text{Speed} = \frac{\text{distance}}{\text{time}} \quad \text{i.e.} \quad s = \frac{d}{t}$$

Statement (1):
$$55 = \frac{d}{4} \quad \Rightarrow \quad d = 55 \times 4 = 220 \text{ miles}$$

This distance is $\frac{2}{3}$ the total distance. Therefore,

$220 = \frac{2}{3}x$ Solve for x to find the total distance. This statement is sufficient to answer the question.

Statement (2): He can finish the trip in a total of 7 hours out of which, he has already used up 4 hours. In the remaining 3 hours, if he drives at an average speed of 60 mph, he will cover 180 more miles. Remember, he has already completed $\frac{2}{3}$ the trip. So, $\frac{1}{3}$ the trip remains to be completed. 180 miles would be the equivalent of $\frac{1}{3}$ the trip. Therefore, $180 = \frac{1}{3}x$. Solving this equation will give the value of x. This statement is sufficient to answer the question.

Each statement is independently sufficient to answer the question. The answer is (D).

8.036 Basically, the question is: Is $\frac{ab}{2}$ an even integer? Length $a \neq 0$, and length $b \neq 0$.

Statement (1): If a and b are consecutive integers, then one of them is odd and the other is even. Their product ab will be even. But when divided by 2, the result could be odd or even. For example:

$4 \times 5 = 20$, $20 \div 2 = 10 = $ even
$5 \times 6 = 30$, $30 \div 2 = 15 = $ odd

This statement is not sufficient to answer the question.

Statement (2): If $a + b > 5$, then individually, a and b could be odd or even. Their sum could be odd or even. The fact that the sum is greater than 5 does not help. This statement is not sufficient to answer the question.

Statements (1) and (2) are together not sufficient to answer the question. The answer is (E).

8.037 Statement (1): If k is odd, then $2k$ is even. If k is even, then too $2k$ is even. k cannot be a fraction, because if it were, then $2k$ would also be a fraction with the exception of ½. ½ × 2 = 1, an integer. This statement only means that k is an integer, but that is not sufficient to answer the question.

Statement (2): If $\frac{k}{2}$ is not a whole number, then it simply means k is not divisible by 2. So k could be odd, or k could be a fraction. This statement is not sufficient to answer the question.

Taken together, the two statements mean that k is an integer not divisible by 2; k must be odd. Statements (1) and (2) are together sufficient to answer the question, although neither of them is independently sufficient. The answer is (C).

8.038 Statement (1): Out of a total of 17 apples a week, he eats 8 on Wednesdays. That leaves 9 apples to be eaten on the remaining 6 days. But he must eat 1 or more each day. Let's say he eats apples in the following sequence:

Monday	?
Tuesday	1
Wednesday	8
Thursday	1
Friday	1
Saturday	1
Sunday	1

All the apples add up to 13. That leaves 17 – 13 = 4 apples to be eaten on Monday. There's no way he can eat 5 apples on Monday. If he did, he would have to eat 0 apples on one of the other days, which is not possible. This statement is sufficient to answer the question.

Statement (2): This statement does not stipulate any condition that he eats at least one apple a day. Meaning, he could eat 8 apples on Wednesdays, **5 apples on Mondays**, and the remaining 4 apples on any day in any combination. This statement is sufficient to answer the question.

Each statement is independently sufficient to answer the question. The answer is (D).

8.039

1997	1998	1999	2000
$27M	$19M	$24M	$30M

Assume $xM as the revenue for Year X.

Statement (1): This statement means that $27 \le x \le 30$.   So, x could be as low as 27, or as high as 30. But it's not possible to tell whether it was definitely *greater than* 27. This statement is not sufficient to answer the question.

Statement (2): If the average of $27M, $30M, and $xM is $27M, then x would have to be less than 27 to balance out the extra $3M from the year 2000. This statement is sufficient to answer the question.

The answer is (B).

8.040 $\overleftrightarrow{l}$ passes through (–5, –5) and the origin (0, 0). Every point on $\overleftrightarrow{l}$ has the same x and y coordinates. For any random point in the coordinate plane, as long as its x coordinate is greater than its y coordinate, the point will lie vertically below the line. It doesn't matter which quadrant the point is in. On the other hand, if its y coordinate is greater than its x coordinate, then it will lie vertically above $\overleftrightarrow{l}$.

Statement (1): Only the x coordinate is 5. There's no information about the y coordinate. This statement is not sufficient to answer the question.

Statement (2): $q < p$ means that (y coordinate) < (x coordinate)   In that case, the point would lie vertically below $\overleftrightarrow{l}$.

This statement is sufficient to answer the question.

The answer is (B).

8.041 Statement (1): Assume there are n number of blocks. Then, n blocks can be arranged side by side in a line in $n!$ unique sequences. It's given that they can be arranged in a straight line in 20! Unique sequences. Meaning, there must be 20 blocks. This statement is sufficient to answer the question.

Statement (2): This statement means that 45% blocks are either red or blue, i.e. 55% of the blocks must be yellow. This only gives the proportion in which the blocks are present. The total number of blocks cannot be found out using this statement. Therefore, this statement is not sufficient to answer the question.

The answer is (A).

8.042 Statement (1): If $AC = BD$, then the quadrilateral could be a rectangle or an isosceles trapezoid, the diagonals of which are of equal lengths. But they will be perpendicular to each other only if the rectangle is a square. This statement is not sufficient to answer the question.

Statement (2): If $BC = AD$, then too the quadrilateral is a rectangle or an isosceles trapezoid. The diagonals will be perpendicular to each other if the rectangle is a square. This statement is not sufficient to answer the question.

Considered together, the statements still only mean that the quadrilateral is a rectangle or an isosceles trapezoid. Statements (1) and (2) are together not sufficient to answer the question.

The answer is (E).

8.043 Assume Sue's age as x, and her son's age as y.

Statement (1): Sue's age will be $x + 6$, and her son's age will be $y + 6$. The equation would be:

$$x + 6 = 2\frac{1}{2}(y + 6)$$

x cannot be found out by solving this equation. This statement is not sufficient to answer the question.

Statement (2): Sue's age was $x - 4$, and her son's age was $y - 4$. The equation would be:

$$x - 4 = 10(y - 4)$$

x cannot be found out by solving this equation. This statement is not sufficient to answer the question.

The statements together give us two linear equations in two variables. Solving them simultaneously would yield the values of x and y. Statements (1) and (2) are together sufficient to answer the question, although neither of them is independently sufficient. The answer is (C).

8.044

$$\frac{\text{length of minor arc}}{\text{circumference}} = \frac{x}{360} = \frac{\text{shaded sector area}}{\text{area of circle}}$$

To calculate the length of minor arc, we would need the values of x and radius.

Statement (1): The radius cannot be found out using this statement because the angle (x) swept by the shaded sector is not known. This statement is not sufficient to answer the question.

Statement (2): Merely knowing the angle swept by the shaded sector is not enough to find the radius. This statement is not sufficient to answer the question.

Taken together, the two statements can be used to find the radius (r) of the circle. Here's how:

$$\frac{x}{360} = \frac{\text{shaded sector area}}{\text{area of circle}} \Rightarrow \frac{51}{360} = \frac{496}{\pi r^2}$$

Once r is found out, the length of minor arc AC can be determined as:

$$\frac{\text{length of minor arc}}{\text{circumference}} = \frac{x}{360} \Rightarrow \frac{AC}{2\pi r} = \frac{51}{360}$$

Statements (1) and (2) are together sufficient to answer the question, although neither of them is independently sufficient. The answer is (C).

8.045 Statement (1): If n is an integer, then mn could be an integer or a fraction depending upon the value of m. For example, if $m = 3$ and $n = \frac{1}{4}$, then $mn = \frac{3}{4}$ = a fraction. But if $n = 5$, then $mn = 15$ = an integer. This statement is not sufficient to answer the question.

Statement (2): This statement only means that m is a multiple of n. m and n could both be integers, or both fractions. $\frac{m}{n}$ would still be an integer. For example, if $m = 4$ and $n = 2$, then $\frac{m}{n} = \frac{4}{2} = 2$ = integer. On the other hand, if $m =$

$\frac{1}{2}$ and $n = \frac{1}{4}$, then too $\frac{m}{n} = \frac{\frac{1}{2}}{\frac{1}{4}} = \frac{1}{2} \times \frac{4}{1} = 2 =$

integer. Also, m could be 0 while n could be an integer or a fraction. In that case, $\frac{m}{n}$ would be 0, an integer. This statement is not sufficient to answer the question.

Consider the two statements together. If n is an integer, then for m to be divisible by n (i.e. for $\frac{m}{n}$ to be an integer), m would also have to be an integer. In that case, the product mn would definitely be an integer. Statements (1) and (2) are together sufficient to answer the question, although neither of them is independently sufficient. The answer is (C).

8.046 Statement (1):

$$x + y = 2000 \quad \Rightarrow \quad y = 2000 - x$$

The percent of my total annual interest that comes from Bank X would be:

$$\frac{\text{interest from Bank X}}{\text{total interest}} \times 100$$

$$= \frac{3\% \text{ of } x}{3\% \text{ of } x + 4\% \text{ of } y} \times 100$$

$$= \frac{\frac{3}{100}x}{\frac{3}{100}x + \frac{4}{100}y} \times 100$$

$$= \frac{3x}{3x + 4y} \times 100 = \frac{3x}{3x + 4(2000 - x)} \times 100$$

$$= \frac{3x}{3x + 8000 - 4x} \times 100 = \frac{3x}{8000 - x} \times 100$$

x cannot be factored out and cancelled. Therefore, we would need to know the value of x to proceed. This statement is not sufficient to answer the question.

Statement (2):

$$2x : 3y = 7 : 4 \quad \Rightarrow \quad \frac{2x}{3y} = \frac{7}{4} \quad \Rightarrow \quad y = \frac{8}{21}x$$

The percent of my total annual interest that comes from Bank X would be:

$$\frac{\text{interest from Bank X}}{\text{total interest}} \times 100$$

$$= \frac{3\% \text{ of } x}{3\% \text{ of } x + 4\% \text{ of } y} \times 100$$

$$= \frac{\frac{3}{100}x}{\frac{3}{100}x + \frac{4}{100}y} \times 100$$

$$= \frac{3x}{3x + 4y} \times 100 = \frac{3x}{3x + 4\left(\frac{8}{21}x\right)} \times 100$$

$$= \frac{3}{3 + \frac{32}{21}} \times 100$$

x got factored out and cancelled. The answer would be a pure number. This statement is sufficient to answer the question.

The answer is (B).

8.047 $\triangle BDC$ is isosceles. $\angle B = \angle C$. Clearly, $\angle B = \angle C = 60°$, which means that $\triangle BDC$ is an equilateral triangle with side length x. $\triangle ABC$ is a right triangle. $\angle B = 90°$. $\angle EBD = 90 - 60 = 30°$. Therefore, $\triangle EBD$ is a 30–60–90 triangle.

Statement (1): In the 30-60-90 $\triangle EBD$, $ED = \frac{1}{2}(DB) = \frac{1}{2}(x) = \frac{1}{2}(10) = 5$. In the shaded right triangle AED, $ED = 5$ and $AD = 10$. The length of AE can be determined by using the Pythagorean Theorem. Area of $\triangle AED = \frac{AE \times ED}{2}$.

This statement is sufficient to answer the question.

Statement (2): In the 30-60-90 $\triangle EBD$, assume $ED = w$. Then $DB = 2w$.

Using the Pythagorean Theorem:

$$ED^2 + EB^2 = DB^2 \quad \Rightarrow \quad w^2 + \left(5\sqrt{3}\right)^2 = (2w)^2$$

From this, w can be determined, i.e. ED can be found. In shaded $\triangle AED$, $AE = EB = 5\sqrt{3}$, and $ED = w$ = now known.

Area of $\triangle AED = \frac{AE \times ED}{2}$.

This statement is sufficient to answer the question.

Each statement is independently sufficient to answer the question. The answer is (D).

8.048 Statement (1): $ab^2 < c$ If we multiply both sides by b, we would get the given statement, which is $ab^3 < cb$. The inequality sign does not reverse, which means b must be

positive, i.e. $b > 0$. But that does not necessarily mean that $b > 1$. This statement is not sufficient to answer the question.

Statement (2): This statement only means that b can be either greater than 1, or less than −1. b just cannot be a proper fraction. This statement is not sufficient to answer the question.

Taken together, statement (2) means that b can be greater than 1, or less than −1, and statement (1) means that b can only be positive. Therefore, b has to be greater than 1, i.e. $b > 1$. Statements (1) and (2) are together sufficient to answer the question, although neither of them is independently sufficient. The answer is (C).

8.049 Statement (1): Assume the population was x in 1900. After 25 years, the population became 120% of what it was, i.e. 1.2 times its original number. In 25 more years, it became 120% of the new number, i.e. $1.2 \times 1.2 = 1.2^2$ times the *original* number. Meaning, the population increased by a factor of 1.2 every 25 years. At the end of t 25-year periods after 1900, the population would be: $x \cdot 1.2^t$. If we break the 25-year period into 25 years, then at the end of n years, the population would be: $x \cdot (1.2)^{\frac{n}{25}}$ By 1970, 70 years had elapsed since 1900, i.e. $n = 70$. But the initial population (x) is not known. Therefore, this statement is not sufficient to answer the question.

Statement (2): A reference population for any particular year (like 1983) won't help us determine the population in 1970, because it doesn't tell us the rate at which the population was growing or declining. This statement is not sufficient to answer the question.

Consider the statements together. By 1983, 83 years had elapsed since 1900, i.e. $n = 83$, and the population was 27 million.

$27{,}000{,}000 = x \cdot (1.2)^{\frac{83}{25}}$ From this, x can be found out. Then, the population in 1970 (for $n = 70$) would be: $x \cdot (1.2)^{\frac{70}{25}}$

Statements (1) and (2) are together sufficient to answer the question, although neither of them is independently sufficient. The answer is (C).

8.050 Statement (1):

$$\frac{x}{100} - 0.02 < 5 \quad \Rightarrow \quad \frac{x}{100} < 5 + 0.02 \quad \Rightarrow$$

$$\frac{x}{100} < 5.02 \quad \Rightarrow \quad x < 502$$

It means x could be any number less than 502. This statement is not sufficient to answer the question.

Statement (2):

$$\frac{x}{100} - 0.002 \geq 5 \quad \Rightarrow \quad \frac{x}{100} \geq 5 + 0.002 \quad \Rightarrow$$

$$\frac{x}{100} \geq 5.002 \quad \Rightarrow \quad x \geq 500.2$$

x could be any number greater than or equal to 500.2. This statement is not sufficient to answer the question.

The two statements together mean that $500.2 \leq x < 502$.   There's only one integer value possible for x, which is 501, an odd integer. But that still does not answer the question because we don't know if x is an *integer*. Statements (1) and (2) are together not sufficient to answer the question. The answer is (E).

8.051 Statement (1):

$$\frac{1}{x} = \frac{1}{z} \quad \Rightarrow \quad x = z$$

Two angles within the triangle are of equal measure. Therefore, the triangle must be isosceles. This statement is sufficient to answer the question.

Statement (2):

$$\frac{2}{x} = \frac{3}{y} = \frac{2}{z} \quad \Rightarrow \quad \frac{2}{x} = \frac{2}{z} \quad \Rightarrow \quad x = z$$

Two angles are of the same measure. Thus, the triangle is isosceles. This statement is sufficient to answer the question.

Each statement is independently sufficient to answer the question. The answer is (D).

8.052 Statement (1): The units digit would have to be 4 times the tens digit. That way, their reversal would result in a greater number. Meaning, the number could only be either 14 or 28. This statement is not sufficient to answer the question.

Statement (2): Assume the units digit as x and tens digit as y. The value of the original number would be $x + 10y$, and that of the new number would be $y + 10x$, which would be greater than the original number by 54. Thus:

$$y + 10x = 54 + (x + 10y)$$

But this equation has two variables (x and y). It cannot be solved to find each variable. This statement is not sufficient to answer the question.

Considering the two statements together, notice that 28, when reversed, results in 82 which is 54 greater than 28.

Statements (1) and (2) are together sufficient to answer the question, although neither of them is independently sufficient. The answer is (C).

8.053 Recall that $a^2 - b^2 = (a + b)(a - b)$.

Statement (1): This statement means:

$$a = 3 + b \quad \Rightarrow \quad a - b = 3$$

It is not possible to determine $a^2 - b^2$ from this equation. This statement is not sufficient to answer the question.

Statement (2): This statement means:

Either $a + b = 7$

or $-(a + b) = 7 \quad \Rightarrow \quad a + b = -7$

It is not possible to determine $a^2 - b^2$ from either equation. This statement is not sufficient to answer the question.

Taken together, we would again have two different values:

$a^2 - b^2 = (a + b)(a - b) = (7)(3)$ or
$a^2 - b^2 = (a + b)(a - b) = (-7)(3)$

Statements (1) and (2) are together not sufficient to answer the question. The answer is (E).

8.054 Statement (1): If x is a non-zero integer, then only x^2 will be positive. The other terms $(x, \sqrt[3]{x}, \frac{1}{x})$ could be positive or negative. Therefore, it is not possible to establish the inequality without knowing the actual value of x. This statement is not sufficient to answer the question.

Statement (2): If $0 < x < 1$, then x is a proper fraction. In that case, notice that in the given inequality $x^2 + x > \sqrt[3]{x} + \frac{1}{x}$, we would have $x^2 < \sqrt[3]{x}$ and $x < \frac{1}{x}$. Meaning, the two smaller quantities are added on the left side, whereas the two larger quantities are added on the right side. So, the left side is less than the right side. This statement is sufficient to answer the question.

The answer is (B).

8.055 There are 40 business class seats. Let's say there are E economy class seats. Also, suppose there are M men in business class.

Statement (1):
$$\frac{M}{40} \times 100\% = \frac{E}{40 + E} \times 100\%$$

To find E, we would need to know M. This statement does not give us the value of M, and is therefore not sufficient to answer the question.

Statement (2): This statement makes no mention of economy class, and is therefore not sufficient to answer the question.

Consider the statements together. 16 women being in business class does not necessarily mean $40 - 16 = 24$ men being in business class; maybe some seats in business class are empty. So we can't find M this way either.

Statements (1) and (2) are together not sufficient to answer the question. The answer is (E).

8.056 Statement (1): The sum of the interior angles of a polygon with n sides is $180(n - 2)$. Meaning, $360 = 180(n - 2)$. Solving this equation would yield n, the number of sides. This statement is sufficient to answer the question.

Statement (2): The sum of the exterior angles of a polygon with *any* number of sides is $360°$. This statement is not sufficient to answer the question.

The answer is (A).

8.057 $x(2x + 7)(x - 3) = 0$ means three cases: $x = 0$ or

$2x + 7 = 0 \implies 2x = -7 \implies x = -\frac{7}{2}$ or

$x - 3 = 0 \implies x = 3$

Statement (1): It means that x could be 0 or 3. This statement is not sufficient to answer the question.

Statement (2): It means that x could only be 3. This statement is sufficient to answer the question.

The answer is (B).

8.058 The two triangles have a common side DE. Imagine the two triangles resting on DE as the base. Then, each triangle will have its greatest height (perpendicular to DE) when its angle x (or angle y) is 90°. The more x (or y) deviates from 90°, the lesser the height of the triangle.

Also, $x + y = 360 - \angle ADC = 360 - 90 = 270°$

Statement (1):
$x = 115 \implies y = 270 - 115 = 155$

x is only 25° away from 90°, whereas y is 65° away from 90°. Clearly, $\triangle ADE$ is taller than $\triangle CDE$, and will have a greater area. This statement is sufficient to answer the question.

Statement (2):
$y = 155 \implies x = 270 - 155 = 115$

x is only 25° away from 90°, whereas y is 65° away from 90°. Hence, $\triangle ADE$ is taller than $\triangle CDE$, and will have a greater area. This statement is sufficient to answer the question.
Each statement is independently sufficient to answer the question. The answer is (D).

8.059 Statement (1): It means a^2 and c are both integers, or both fractions, or a is zero and c is any non-zero number, or a^2 is an integer and c is a fraction. This statement doesn't mention b, and is therefore not sufficient to answer the question.

Statement (2): It means b could be an integer (5, 41, −29 etc.) or a radical ($\sqrt{2}, \sqrt{11}$ etc.) because an integer or a radical squared would give an integer. But this statement says nothing about a or c, and is therefore not sufficient to answer the question.

Consider the statements together. b could be an integer or a radical. $\dfrac{a^2}{c}$ is an integer, but $\dfrac{a}{c}$ could be an integer or a fraction or a radical.

Statements (1) and (2) are together not sufficient to answer the question. The answer is (E).

4.060 Standard deviation is given by:

$$\sigma = \sqrt{\frac{\Sigma(\mu - x)^2}{n}}$$

x is each of the numbers deviating from the mean μ. If we open the parenthesis, and multiply, we will get: $\mu^2 - 2\mu x + x^2$ for each value of x. Adding up these terms for n different values of x, we would get:

$n\mu^2 - 2\mu(x_1 + x_2 + x_3 ... + x_n)$
$\qquad + (x_1^2 + x_2^2 + x_3^2 ... + x_n^2)$

Statement (1): The median has no place in the formula. This statement is not sufficient to answer the question.

Statement (2): $x_1^2 + x_2^2 + x_3^2 ... + x_n^2 = 476$. Also, if their average (μ) is 11, then their sum must be $4 \times 11 = 44$, i.e. $x_1 + x_2 + x_3 ... + x_n = 44$.

That much information is enough to use the formula. This statement is sufficient to answer the question.
The answer is (B).

APPENDIX B

MULTIPLICATION TABLES

	TABLE OF :							
	2	**3**	**4**	**5**	**6**	**7**	**8**	**9**
1	2	3	4	5	6	7	8	9
2	4	6	8	10	12	14	16	18
3	6	9	12	15	18	21	24	27
4	8	12	16	20	24	28	32	36
5	10	15	20	25	30	35	40	45
6	12	18	24	30	36	42	48	54
7	14	21	28	35	42	49	56	63
8	16	24	32	40	48	56	64	72
9	18	27	36	45	54	63	72	81
10	20	30	40	50	60	70	80	90

	TABLE OF :							
	12	**13**	**14**	**15**	**16**	**17**	**18**	**19**
1	12	13	14	15	16	17	18	19
2	24	26	28	30	32	34	36	38
3	36	39	42	45	48	51	54	57
4	48	52	56	60	64	68	72	76
5	60	65	70	75	80	85	90	95
6	72	78	84	90	96	102	108	114
7	84	91	98	105	112	119	126	133
8	96	104	112	120	128	136	144	152
9	108	117	126	135	144	153	162	171
10	120	130	140	150	160	170	180	190

SQUARES

1^2 = **1**	6^2 = **36**	11^2 = **121**	16^2 = **256**	21^2 = **441**
2^2 = **4**	7^2 = **49**	12^2 = **144**	17^2 = **289**	22^2 = **484**
3^2 = **9**	8^2 = **64**	13^2 = **169**	18^2 = **324**	23^2 = **529**
4^2 = **16**	9^2 = **81**	14^2 = **196**	19^2 = **361**	24^2 = **576**
5^2 = **25**	10^2 = **100**	15^2 = **225**	20^2 = **400**	25^2 = **625**

CPSIA information can be obtained at www.ICGtesting.com
Printed in the USA
LVOW09s1444190214

374387LV00002B/5/P